The Paradox of Preservation

The publisher gratefully acknowledges the generous contribution to this book provided by the Stephen Bechtel Fund.

The publisher gratefully acknowledges the generous contribution to this book provided by the International Community Foundation/JiJi Foundation Fund.

The Paradox of Preservation

Wilderness and Working Landscapes at Point Reyes National Seashore

Laura Alice Watt

Foreword by David Lowenthal

UNIVERSITY OF CALIFORNIA PRESS

University of California Press, one of the most distinguished university presses in the United States, enriches lives around the world by advancing scholarship in the humanities, social sciences, and natural sciences. Its activities are supported by the UC Press Foundation and by philanthropic contributions from individuals and institutions. For more information, visit www.ucpress.edu.

University of California Press
Oakland, California

Portions of the wilderness material in chapter 4 were published previously in two articles: "The Trouble with Preservation, or, Getting Back to the Wrong Term for Wilderness Protection: A Case Study at Point Reyes National Seashore," *Yearbook of the Association of Pacific Coast Geographers* 64 (2002): 55–72; and "Losing Wildness for the Sake of Wilderness: The Removal of Drakes Bay Oyster Company," in *Wildness: Relations of People and Place,* eds. John Hausdoerffer and Gavin Van Horn (Chicago: University of Chicago Press, forthcoming Spring 2017). Portions of the elk material in chapter 6 were previously published in "The Continually Managed Wild: Tule Elk at Point Reyes National Seashore," *Journal of International Wildlife Law and Policy* 18, no. 4 (2015): 289–308.

All photographs are by the author unless otherwise stated. The chapter frontispieces were all taken using Polaroid peel-apart film, which represents a once-thriving photographic technology that has been abandoned not because it no longer worked, but because of a deliberate decision by the film's makers to satisfy a different, more modern, market demand.

Library of Congress Cataloging-in-Publication Data

Names: Watt, Laura Alice, 1966– author. | Lowenthal, David, writer of foreword.
Title: The paradox of preservation : wilderness and working landscapes at Point Reyes National Seashore / Laura Alice Watt ; foreword by David Lowenthal.
Description: Oakland, California : University of California Press, [2017] | Includes bibliographical references and index. | Description based on print version record and CIP data provided by publisher; resource not viewed.
Identifiers: LCCN 2016019888 (print) | LCCN 2016018943 (ebook) | ISBN 9780520966420 (ebook) | ISBN 9780520277076 (cloth : alk. paper) | ISBN 9780520277083 (pbk. : alk. paper)
Subjects: LCSH: Point Reyes National Seashore (Calif.) | United States. National Park Service. | Natural resources conservation areas—California—Point Reyes Peninsula—Management.
Classification: LCC F868.P9 (print) | LCC F868.P9 W38 2017 (ebook) | DDC 979.4/62—dc23
LC record available at https://lccn.loc.gov/2016019888

Manufactured in the United States of America

25 24 23 22 21 20 19 18 17
10 9 8 7 6 5 4 3 2

Dedicated to the ranching families of West Marin—past, present, and future—for this is their story, not mine, just as it is most truly their landscape

Contents

Illustrations

Foreword

Seldom does a book's title so aptly fit its topic as does Laura Alice Watt's vividly recounted and cogently crafted *Paradox of Preservation*. The checkered history of Point Reyes National Seashore exemplifies the manifold ironies of America's love affair with wilderness, especially as deified by its preservationist patron champion, the U.S. National Park Service.

Culminating in the late nineteenth century, the cult of wilderness reversed long-standing American antipathy toward wild nature. Centuries of settlers and frontiersmen had viewed the wilderness with hostility, as a loathsome obstacle to be vanquished and domesticated into civilized productivity. Only as the far westward march of reconnaissance revealed unparalleled scenic splendors did concern for protecting the remaining remnants of apparently untrammeled nature begin to counter pressures for development. The ecstatic depictions of western explorers and the impassioned advocacy of John Muir persuaded millions that the awesome wonders of the national domain were an inspiration, not an impediment, to America's Manifest Destiny. These divine scenic cathedrals—nobler than any Old World merely human monuments—promised spiritual and physical renewal to care-worn entrepreneurs, a refreshing antidote to the hurly-burly of entrepreneurial striving. So was born the sentiment that led to the protection of Yosemite and Yellowstone, those early forerunners of the national parks.

But this wilderness could be properly enjoyed, as Watt stresses, only if seen to be perfectly wild—that is, devoid of economic activity and untenanted save by the occasional visitor. Hence her paradox: no part of the American landscape, however remote and seemingly inhospitable, had been devoid of Indian occupance for centuries if not for millennia. Indeed, it was in fact their presence that had shaped the landscapes now celebrated as pristine. But the mystique of solitude demanded their removal, along with that of other settlers in other parks treasured for wild solitude. Just as Indians were expelled as antithetic to, and desecrators of, the early western parks, so too were miners, loggers, ranchers, and farmers later evicted from protected areas. Only by banishing those who lived in and sought to make a living from the parks could their mundane working landscapes revert to the unsullied wilderness thought natural to them.

The trouble was that by expelling these inhabitants and banning their traditional modes of subsistence, wilderness stewards were not conserving but actually destroying the very landscapes and flora and fauna that visitors came to admire as untrammeled wilderness. Far from thriving in untouched virginity, much of what the public touted as wild required continual fructifying husbandry. And as even apparent wilderness became more and more scarce, demands for wilderness experience increased, worsening the dilemma. Hence the 1964 congressional Wilderness Act mandated not only protecting existing wilderness but making *new* wilderness—fashioning a semblance of untouched nature where wilderness was wanted. This was best done by condemning existing human occupation and activity as ecologically and aesthetically deleterious, by acquiring public ownership of private properties, and by rendering ongoing "nonconforming" enterprise unsustainably hazardous. But the essential faith that drove such rewilding—the fantasy that locales stripped of human impress would revert to their previous supposed stable and unchanging "natural" character—was profoundly ecologically misguided. In fact, it takes many human hands to make nature look as if no human hands had been at work. And the ensuing rewilding never restores what was ever previously there. Nor can human stewardship, once invoked either to keep or to invent a semblance of wilderness, ever be relaxed.

At the heart of Watt's wonderful book is the saga, poignant, parlous, often perverse, of Point Reyes as a peopled landscape, and of its residents' ongoing fractious engagement with the policies, threats, and promises of federal agencies and agents. Underlying all negotiations for public management is the drumbeat of progressive rewilding. Portions of the seascape, designated as "potential wilderness" in the 1970s to protect the tidelands from overdevelopment, are being stripped of all remnants of a nearly century-old oyster farm which was closed down for conversion to use only for recreation and wildlife. This transformation is authorized by a definition of wilderness so all-embracing, observed a congressional cynic, that New York's Wall Street might in time become not merely a financial but a physical wilderness. At Point Reyes, traces of prior use—power lines, paved roads, walls and fences, garden plants, exotic trees—are to be expunged so that these long-farmed areas will appear primordially untouched. That many visitors commonly mistake the lonely grass-swept pastoral lands for "wilderness" cuts no ice with some wilderness purists who anathematize the ranches as excrescences of human make that do not "belong" in a "natural" national park.

Among the counterproductive ironies of National Park Service rewilding recounted by Watt, two in particular stand out for boneheaded adherence to a misguided ethos of wilderness purity. One is the introduction of "wild" elk into lands adjacent to fenced-off ranchland pastures. Not to locals' surprise, the elk soon surmounted the fences, fed off cattle fodder, and in the absence of predators, cyclically multiplied and died off, distressing visitors, some of whom blame ranchers for this violation of "natural" homeostasis. Another was the fate of a famed 1880s ranch building, whose disrepair initially mandated renovation. After several years' dilatory delay, park officials ruled out repair as too expensive. As demolition loomed, however, a colony of rare bats was found to be nesting in the roof, requiring the building to be maintained after all, but inhabited only by bats, not humans. The sorry consequences of such batty policy decisions make this book a riveting cautionary tract for anyone concerned with land management in America's best-loved landscapes.

Watt convincingly details the federal policies that have deranged ecology, economy, and community values at Point Reyes, depriving both visitors and residents of an example of how past and present, nature and culture, wilderness and economic enterprise can and should coexist. Her aim is not to demonize federal agencies for the politicized follies imposed on them by oxymoronic wilderness mandates. Nor is it to sanctify the steadfast hardihood of the ranchers on their ancestrally stewarded pastures. It is rather to show that in managing a special locale—special in both natural and human terms—for national *and* local *and* private interests, collaborative well-being requires cooperation that harmonizes rather than estranges various stakeholders' complementary needs. Local voices with local *nous* lend earthy reality to national diktats. Dairy farms and oyster reefs enhance surrounding sea and shore. No set-apart segment could so embellish Point Reyes as does their conjoint ever-changing history.

The untouched wilderness ethos of the National Park Service and its preservationist allies was, to be sure, a tonic rebuttal to the ongoing extractive rapine of nature by timber, mining, and power interests. For all their antipathy toward Native American and later inhabitants, wilderness crusaders successfully protected large tracts of the spectacular national domain for public contemplation, instruction, and recreation as a precious permanent legacy. Absent such preservation, private enterprise would have left these incomparable locales irreversibly scarred and degraded. Yet wilderness advocates' fixation on nature devoid of human impress increasingly encumbered the National Park Service with a mission that was not only ecologically absurd and socially regressive, but aesthetically crippling. For it helped to instill among visitors, actual and potential, the view that *only* untouched landscapes are worthy of care and concern, that traces of past occupance tarnish the purity of such scenes, and that ongoing human occupation and enterprise (except the paraphernalia of conservation stewards and visitors themselves) are incompatible with and degrading to wilderness ideals. Americans have consequently absorbed an ethos that assigns all virtue and beauty to scenes by definition only rarely if ever visited, and that abandons the everyday scenes in which most people live and work to squalid neglect.

Attachment to wilderness refreshes body and soul; it reminds us of a living plenitude whose loss we regret; it offers lessons in fortitude and self-reliance. But these benefits are necessarily scarce. Wilderness visits cannot be many or frequent; were they common the wilderness would be loved to death. Nor does the rare wilderness experience compensate for our neglect of, if not contempt for, the pervasive landscapes we fashion for lucre and shelter, traffic and transport. To privilege wilderness is to prize only the rare and the remote at the expense of the workaday realms that ought to enrich and enliven our quotidian lives. As social beings we should reclaim our inherited landscape from humdrum neglect. There is no better place to make a start than at Point Reyes, whose workaday ranches and oyster farm have been long admired and enjoyed as valued adjuncts to the wild cliffs and shores that frame them. This book points the way toward a vitally needed amalgam of the natural and cultural merits of our national landscape.

David Lowenthal

Acknowledgments

This book began life as my doctoral dissertation at the University of California, Berkeley, with research starting in 1997 and completed in 2001, but it has come a long, long way since then. When the oyster controversy documented in these pages first blew up in 2007, it took me completely by surprise, as in my earlier years researching the history of park management at Point Reyes, no one ever mentioned the oyster farm. It had been a nonissue, and in fact does not appear in the pages of my original dissertation except in a single footnote. Once I began revising that document into a book, the unfolding oyster story kept getting stranger—and more complex—yet I could see clear roots of the controversy in my earlier material on cultural landscapes and national parks ideology.

As a research project this book has been in progress, in one way or another, for so many years, it would be impossible to list everyone to whom I am indebted in some way. With a profound but necessarily general sense of *enormous* gratitude to all those myriad persons, I would like to take this opportunity to specifically thank the people (and one particular place) I consider my intellectual and inspirational parents, as their ideas and influences have left fingerprints all over this book.

First, a place: I have always believed that for anyone to do something as odd as to write a doctoral dissertation, the subject of study must almost always be rooted somehow in their early life experiences. My starting point is clear: my parents met at a biological field station high in Colorado in 1962, where my father still carries out his research, and we spent almost every summer of my childhood in this amazing, captivating place. The Rocky Mountain Biological Laboratory, located on the site of a former 1880s silver mining town called Gothic, is surrounded by high altitude wilderness that still contains traces of its oh-so-human past: a tumbled-down cabin here, the mouth of a mineshaft there, an old iron cookstove rusting unexpectedly in a field of lupine and columbine. The persistence of stories in this wild landscape, and the depth they gave to my sense of this place, have unquestionably shaped all my academic work in some essential, sometimes unexpected ways.

Next, three writers whose works imprinted on me early and often, starting most fundamentally with Laura Ingalls Wilder: as a little girl I firmly believed for a time that I might be her most recent reincarnation, not only because of the similarity of our names and status as younger sisters, but also because our births are about six weeks off from being exactly one hundred years apart. I learned far more western history from her stories than I realized until years later, but also her deep love and respect for the places in which her family eked out a living, and their intimacy with the natural world upon which they relied, made a lasting impression. Next to her works on the bookshelf would be William Least Heat-Moon, whose amazing masterpiece *Blue Highways* is one of the most dogeared in my collection, a tale of seeking out what was distinctive in 1970s small-town America on the smallest roads he could find. His celebration of the unique, the quirky, and the local, all while lamenting the increasing tendency toward homogenization of U.S. landscapes and lifeways, remains inspiring every time I reread this beloved book. And Austin Tappan Wright, whose novel *Islandia*, though flawed in some fundamental ways, is so evocative of place that reading it feels intensely familiar, even though it is a wholly invented, imaginary land. All three authors have taught me not only to "see" through writing, but also to "see" what a landscape can contain—not only ecosystems but history, and not only history but thick layers of personal and community meaning.

In a perhaps similar way, I have been enormously inspired by the photographs of Carleton Watkins and Sally Mann—working over a century apart, but with similar camera gear, and both going far beyond the more modern Sierra Club-calendar type of "nature photography," toward an intense expression of meaning in the landscape. Through their lenses, the natural world is not just pretty or even breathtaking, but full of power and mystery. Their images show landscapes in relationship with their inhabitants, shaped by and shaping those who live and work in those places—not just natural scenes to admire, but places to be drawn into, to spend time in, and to perhaps gradually come to understand.

Now a series of direct and indirect mentors, starting with my undergraduate advisor at Cal, Harry Greene: his guidance and encouragement on my never-quite-completed senior project in vertebrate zoology, focused on viper evolution, first suggested to me that new information and new insight were things I might actually be able to contribute, to *create,* not just read and absorb. Later, returning to Berkeley in the first semester of my doctoral program, I was lucky enough to stumble serendipitously into an elective offered in the law school by Joe Sax. Coming from my mostly biology background, I unexpectedly dove into his material about preservation law, focused both on natural resources preservation, which I thought I already understood—endangered species, parks, and wilderness—and on cultural preservation, everything from repatriation of museum artifacts to communities' attempts to protect their own sense of self-definition and identity. The two endeavors turn out to have so much in common, and yet so rarely speak across those disciplinary boundaries to each other; I immediately sensed a challenge that I wanted to help tackle, one that has shaped pretty much everything I have done since.

Graduate school can be a strange and murky place, and if there's one person who deserves the primary credit for ensuring that I actually found my way through to the finish line, it is unquestionably Paul Groth: always inspiring, always challenging, and somehow always in possession of a confidence in my project that I often lacked in myself. Without his occasional cheerleading, I would not have stuck with this story, which I now feel is so deeply important to tell. And through Paul's guidance, in classes and in our office-hours conversations, I also gained my acquaintance with David Lowenthal—first through his classic *The Past Is a Foreign Country,* and much later through a much-treasured friendship—and it is *far* beyond an honor that he has written a foreword for this volume. The work and influence of these two geographers, Groth and Lowenthal, combined with the legal perspective of Sax to provide me with a new vocabulary, of landscape, property, and preservation and all the intricate ramifications of each, with which to try to make sense of my place of doctoral study, Point Reyes.

I am immensely fortunate to have worked with Carolyn Merchant as my "gateway" into the field of environmental history, both learning from her and teaching with her. Her influence led directly to Bill Cronon, Nancy Langston, and Richard White as core sources of inspiration; to my lasting involvement with the American Society for Environmental History (ASEH), which despite all my crazy interdisciplinarity is my true intellectual home; and to the ASEH wilderness gang, especially Jim Feldman, Jay Turner, Michael Lewis, and John Hausdoerffer, with whom I have served on many conference panels and had many long-into-the-evening conversations, and who continue to inspire me with their insight and friendship.

I will always be hugely indebted to my thesis advisor, Sally Fairfax, for teaching me to teach, and to my many, many incredible students, for teaching me every time I attempt to teach them. I hope they will recognize elements of this book as emerging directly from our classroom conversations over the years, and so they share authorship as well.

. . .

I also wish to extend specific thanks to those whose assistance helped directly to get this research project actually finished and done. First, to many National Park Service (NPS) staff, particularly Gordon White and Carla DeRooy at Point Reyes, for their assistance and cooperation with my research process over the years, even during some trying times. The influence of Gary Machlis is almost immeasurable, for his championing of the Canon National Park Science Scholars program, as is the role played by Hal Rothman in my being granted that fellowship, without which this research never would have gotten off the ground. Similarly I am grateful for Rolf Diamant and Nora Mitchell for their steadfast work within the NPS to promote working landscapes in parks, which has blazed a trail that I hope this book continues to expand and encourage.

I have been incredibly fortunate to work with three editors at UC Press: many thanks to Jenny Wapner for bringing me on board, Blake Edgar for his persistence, and Merrik Bush-Pirkle for guiding me through getting the final thing wrapped up. Peter Alagona and several anonymous reviewers provided essential suggestions and insight into strengthening the manuscript, and Audra Wolf had enormous patience and temerity to work with me through a lengthy revision process—the result is far, far more than I ever could have produced on my own. This work is also their work, in large part.

And of course, I would have lost my mind long ago if it were not for my dear colleagues at Sonoma State, especially "the band" with whom I love to teach: Margie Purser, Michelle Jolly, and Melinda Milligan. In addition, Jeff Baldwin, Don Romesburg, Paula Hammett, and David

McCuan always delight me with their conversation, as well as their willingness to be sounding boards at times. Academic comrades across the country include Leigh Raymond, my dissertation-writing partner; Steve Corey, who makes sure ASEH meetings are always wildly entertaining; and Matthew Morse Booker, who shares my love for the Bay Area and a desire to chronicle its many environmental pasts, presents, and futures. I also have such gratitude for all the wonderful people in the San Francisco Bay sailing community who have helped me set the writing work aside for a needed day or weekend's racing, especially Henry King; and similarly for my fellow film-and-polaroid-geek photographers, especially Cate Rachford and Nick Grossman. And huge, huge thanks to my beloved longest-time friends, who have stuck with me through decades: Cissie Bonini, Becky Lekven, Craig Meyer, and Susannah Clark.

Most especially, gratitude and appreciation are due to the amazing people of West Marin—particularly Nancy and Kevin Lunny, Judy Teichman, and all of the folks from the Alliance for Local Sustainable Agriculture (ALSA), but a huge list of others as well. Phyllis Faber is truly a hero of mine, for her decades of tireless work to protect working lands and community in West Marin, and I am honored by her friendship. Dewey Livingston's extensive exploration of local ranching history forms a key foundation for this book. Tess Elliott and her staff at the *Point Reyes Light* have been incredibly gracious as I've pawed through their archives. In recent years, long discussions with Peter Prows helped to extend my insight into the wilderness designation process at Point Reyes, and friendships with Nichola Spaletta and Kathy and Gino Lucchesi grounded my academic research in the place that really matters: the landscape itself, and their families' long engagements with it.

Finally, thanks do not come close to expressing my feelings for my wonderful family: Jean, India, Wiley, Mom and Powell, Dad and Carol. I am beyond lucky to have ended up with you all, and I cannot imagine life without you.

FIGURE I. Drakes Beach, Point Reyes National Seashore, 2015. Photograph by author.

A Management Controversy at Point Reyes

On November 29, 2012, Secretary of the Interior Ken Salazar announced his decision to close an eight-decades-old oyster farm at Point Reyes National Seashore (PRNS), just north of San Francisco on the California coast. His memo cited National Park Service (NPS) law and policy concerning commercial operations and wilderness, and represented the culmination of five years of fierce, polarizing battle between advocates for wilderness and for sustainable agriculture.[1] The controversy drew national attention, as it was not the usual industry-versus-nature debate. On the one hand, national environmental organizations sought official designation of a marine wilderness, a maritime sanctuary to be enjoyed by visiting hikers and kayakers. On the other, the oyster farm operators and local foods advocates insisted that the historic operation was doing no harm and should be allowed to continue harvesting from the dark waters of the estuary, routinely cited as the most pristine on the West Coast despite the long presence of the oyster business. How did these two factions, which ordinarily are on the same "side" of environmentalism, come to be such vitriolic opponents? And why *was* there an oyster farm in a national park, anyway, and what did its closure represent?

The controversy burst into public view at a Marin County Board of Supervisors meeting in May 2007. The then-superintendent of Point Reyes, Don Neubacher, made an alarming announcement: the Seashore's harbor seal colony was being seriously threatened by the Drakes Bay Oyster Company (DBOC), which cultivated oysters on state-controlled leases over the tidelands in Drakes Estero, the glove-shaped body of water at the heart of the Seashore. This assertion was based on a NPS report first written in 2006, which argued that DBOC's boats and harvesters were negatively impacting harbor seal populations by driving seals away from their breeding sites, increasing water turbidity and affecting seal pup survival.[2] At the supervisors' meeting, Neubacher and his staff cited chronic disturbance of the seals as the reason that so many pups had been abandoned by their mothers; they described an 80 percent reduction in the number of seals observed at the site that spring.[3] NPS scientists Sarah Allen and Natalie Gates also described increases in invasive species and harm caused to eel grass in the estuary, which had been shaded by oyster racks and cut by boat propellers; beds of slender eel grass provide food and habitat for shorebirds.

One by one, members of the public stepped forward to offer alternate viewpoints. Kevin Lunny, whose family only two years earlier had purchased the oyster farm from its long-time owner Charles Johnson, expressed surprise at the NPS's presentation; he said this was the first he had heard of these alleged harms. Paul Olin, Sea Grant director from the University of California Cooperative Extension, attested that DBOC was in full compliance with state wildlife regulations, and contradicted the NPS staff's warnings by citing recent *increases* in harbor seal populations at Drakes Estero. He also referred to aerial photographs showing a doubling of eelgrass coverage since the Lunnys had taken over operations from Johnson in 2005. In contrast, representatives from environmental groups—national organizations like the National Parks Conservation Association (NPCA) and Sierra Club, and local groups like the West Marin-based Environmental Action Committee (EAC)—questioned the overall appropriateness of aquaculture in the estero, designated by Congress in 1976 as "potential wilderness." The lines of a dispute that would embroil the local community for years to come, based in claims of both environmental harms and the legal requirements of wilderness law, had been drawn.

At first glance, this battle over oysters may seem to be an outlier in parks management. Point Reyes was the only national park unit in the United States with commercial oyster harvesting within its boundaries, and remains one of the few with active agriculture: despite closure of the oyster farm, eleven dairy and beef ranches continue to operate within the Seashore's borders. While Point Reyes' story may be unique, it nevertheless reveals a great deal about parks management in the United States—not only about how such an agricultural landscape came

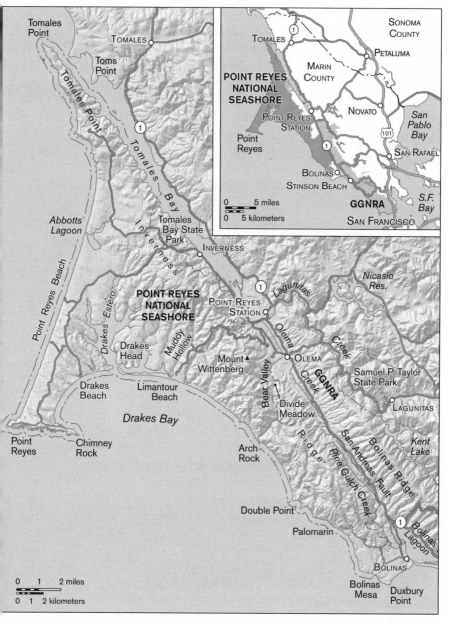

MAP 1. Map of Point Reyes National Seashore, with Marin County inset. Courtesy of Ben Pease.

to be designated as a national seashore in the first place, but also how park management has been reshaping that landscape ever since. This is not only a local controversy for the Seashore and surrounding communities of West Marin; it highlights much larger questions about what parks are *for,* what they are meant to protect and provide to the public, and how to make choices between competing uses or management priorities for park resources. Point Reyes also exemplifies the effects of preservation on a landscape, how it changes what was originally preserved in ways that can be hard to see without a sense of the place's history.

The oyster controversy at Point Reyes represents an ongoing collision of an older, wilderness-based conception of resource preservation with a newer view that aims to rethink resource management in the era of the Anthropocene, with its increasing recognition that all environments around the globe are now being deeply influenced by human activity in one form or another—that no natural landscapes are truly "pristine." National parks have long been associated with the vision of John Muir, the nineteenth-century writer who championed the earliest parks as "the people's cathedrals and churches"—but particularly since the 1960s, an increasingly diverse array of park types has demanded new approaches to management.[4] The continuing presence of cattle ranches on Point Reyes' rolling grasslands offers a vision of how working landscapes—places characterized by "an intricate combination of cultivation and natural habitat," maintaining a balance of human uses and natural forces—should be recognized as part of both natural *and* cultural heritages worth protecting.[5] The U.S. national park system contains areas that primarily aim to preserve natural scenery as well as those that primarily preserve history and cultural heritage; Point Reyes offers the suggestive possibility of protecting all types of heritage resources *together,* as a landscape whole, rather than separately, and of including the resident users' input in management decisions. The continued presence of the ranches at PRNS alludes to the strength of such a broader approach, one based in community collaboration, with implications for how we humans might better understand nature's role in a human-built world.

National parks are considered national treasures, enjoy far more public support than many other federal programs, and are often touted as one of America's "finest inventions." The founding legislation that created the NPS in 1916 directs the agency to preserve the parks' resources for use by the public, and to manage them "unimpaired for the enjoyment of future generations."[6] But preservation is not a neutral act, and the NPS is not a neutral actor. In much the same way that a beetle stuck in amber is "preserved" but no longer living, preserving landscapes aims to prevent change, to maintain places "unimpaired" indefinitely into the future. And yet the process of protecting and managing land as parks exerts its own influence on the area's landscape, gradually changing its form and appearance to reflect the preservation agency's own values and preferences. These changes often remain invisible to the public. They also frequently remain invisible to the managers, who present them to the public as part of what was originally preserved. And if the object of preservation is a working landscape, this process can gradually disconnect the residents from their own landscape, bringing their presence there into question and sacrificing their needs to the illusion of pristine nature.

While landscapes may seem timeless and unchanging to the casual visitor's eye, they are actually dynamic, continually shaped by social forces that occur within and around them, and similarly affecting the forms those social forces take. People and the places in which they live and work change each other: the people modify the landscape through their actions, and conversely their actions are limited or affected by the landscape's physical and ecological characteristics. Over time, landscapes reflect differences in social power, as groups or individuals try to dominate or influence them according to their own needs or interests; for instance, an estate controlled by a single wealthy owner will look quite different from a landscape of small family farms. And because of this interactive relationship, landscapes are continually in flux, responding to shifts in people's activities and ideas, and likewise steering social change in particular directions.

This shaping and reshaping process tends to "naturalize" or normalize a landscape's appearance, including its inherent power relations, causing landscapes to appear to be biologically and politically neutral, without having been made or produced in any way. A classic example is New York City's Central Park, which consists of former farmlands, settlements, and swamps—home to as many as sixteen hundred people at one time—engineered and designed in the mid-1800s to appear to be a wild park of thick forests and lush meadows, connected by winding trails.[7] Many people today are surprised to learn that the apparently "natural" features of the landscape—granite boulders, lakes, and entire hillsides—were intentionally placed and remain artificially maintained, because the overall effect is one of unchanging, unquestioned natural scenery. While the park was intended to be democratic, available to all people instead of just the wealthy few, its establishment displaced African American, Irish, and German enclaves whose residents both lived and worked on the very lands that became this plot of seemingly untouched yet designed nature.[8]

The rules for Central Park's use, such as restrictions on organized sports or loud gatherings, are similarly based on the ideals of its original nineteenth-century designer, Frederick Law Olmsted, who believed nature should be a place of quiet contemplation rather than rowdy crowds. Today, these ideals are still maintained and reinforced in many sections of Central Park by park management, signs, and law enforcement, creating the impression that this place *ought* to be one of quiet natural scenery and contemplation. The social ideas and values that shape landscapes often appear as unquestionable parts of reality that are taken for granted, rather than as human-created dynamics that could be altered or improved upon. This quality not only obscures the socially constructed origins of the landscape, but also maintains the power relationships that control it.

This book aims to make the effects of preservation on the land— particularly on working landscapes—more visible through a case study of Point Reyes. In the process, it also documents more broadly how our national ideas about what a park "ought to be" have developed and changed over time, and what happens once the NPS and its particular approaches to management and national heritage become involved with preserving a lived-in, still-working landscape. The particular historical development of the agency has resulted in national standards and policies—what counts as heritage, for whom parks are protected, and how best to manage park resources—that do not necessarily match how local residents live and work on the landscape, nor what they value about that place. The resulting landscapes produced by the NPS represent a series of compromises between use and protection—the terms of which are constantly being renegotiated and reimagined. In many ways, the oyster controversy is just the latest installment in that series of negotiations.

PLAN OF THE BOOK

This research recounts the processes of proposing, establishing, and managing Point Reyes as a national seashore, tracing its history as a park created from a historic working landscape as a template for understanding a number of issues that figured prominently in the oyster farm debate. First, chapter 1 steps back to address the larger question of how our ideas of what parks are for have changed over time at the national level. Why does it seem surprising to find farms in parks, and why are certain uses considered "appropriate" or not? These expectations make more sense in the context of understanding the impulse to preserve, its influence on landscapes, and the particular ways that landscape preservation developed in national parks starting in the mid-nineteenth century in places like Yosemite and Yellowstone. The NPS as a government agency was created nearly fifty years *after* these first parks, and so it inherited many of the ideals that these iconic landscapes represented.

The earliest national parks were mostly created from lands already owned by the federal government, segments of the public domain withdrawn from settlement or development and largely uninhabited once Native American communities were relocated outside their borders. In contrast, chapter 2 explores the history of making parks from *privately* owned lands, a process that at first relied on donations from states or wealthy individuals, but gradually involved the direct purchase of land. Parks are often celebrated as "belonging to the American public," but in many cases, they belonged to someone else first. What are the implications of making parks from private lands? These places come with their own history, and often with residents and *their* own uses for and meanings of the land, which do not generally fit well with the simplified "empty scenic nature" model of park management.

Chapter 3 then details the specific establishment process at Point Reyes, from its first proposal as a park in the late 1950s, through its authorization in 1962 as a national seashore intended to provide beach access and recreation opportunities to the nearby metropolitan public of the Bay Area, and up to additional legislation and funding passed in 1970. Point Reyes was one of a series of parks, mostly national seashores, lakeshores, and recreation areas, created during the 1960s and '70s by acquiring large areas of private land. As part of this experimental period of direct purchase of property, Point Reyes began as an explicit attempt to retain some private ownership within the seashore as a "pastoral zone" where agriculture could remain. As will become clear, however, the establishment of the park set in place the conditions that essentially forced the sale of the pastoral zone to the federal government within ten years.

Chapter 4 recounts the application of wilderness ideals to the Point Reyes landscape in the 1970s, which further defined the landscape and exacerbated the tension between preservation and use at the Seashore. An August 1976 resolution identifying Point Reyes as a possible location for reintroduction of tule elk, a native ungulate that had been absent from the peninsula since the 1850s, began this legal redefinition. It was followed that October by the designation of a wilderness area across roughly one-third of the peninsula. With that designation, a new, "untrammeled" version of the park's history began to replace the land's human history, with visitors and park managers increasingly envisioned as the only appropriate people within the park. The wilderness bill also defined a separate portion of Point Reyes as a "potential wilderness," a new designation that would play a key role in the later showdown over Drakes Estero. And yet these two pieces of legislation emphasizing the wild characteristics of Point Reyes were soon followed by a third congressional act, in 1978, creating a leasing mechanism for the working ranches to continue operating past the original terms of their reservations of use. Together, these three laws framed PRNS as a landscape where Congress had given deliberate sanction to both its wilder aspects and the continuity of agriculture.

The next three chapters explore specific forms of management within Point Reyes, which have tended, intentionally or not, toward steadily reducing or erasing the "cultural" human imprint on the landscape. Chapter 5 shows how in the early years of the seashore, the NPS failed to recognize, let alone maintain, many historic buildings and culturally important sites, reflecting broader national trends at the time concerning what "counts" as worth preserving. Since the Seashore's beginnings, roughly half of its built landscape has been demolished by the NPS, and even as a wider array of structures and categories of significance gradually gained importance with the preservation movement, the continuity of historic *uses* of the land is still often overlooked or downplayed. Through policy decisions, management choices, and the slow but steady attrition of ranchers, the working landscape has diminished over time, from twenty-five operating ranches on the Point at the time of park's establishment, to only eleven now—and, perhaps most importantly, with decreasing local input into management. The oyster operation's recent removal is one more loss in this long pattern.

Chapter 6 chronicles how, despite hiring a cultural resources manager and giving more attention to cultural resources, as well as making official statements about the value of the area's ranching history, PRNS has continued to steer management toward the national park ideal of scenic wild-yet-managed nature, with as little human use other than recreation as possible. This can particularly be seen playing out in the Seashore's natural resource projects and plans since 1995, when Neubacher became superintendent. These efforts to create a more wild and natural landscape have often come at the expense of the working ranches. This trend is most clearly reflected in the reintroduction of tule elk to Point Reyes, first brought to a fenced range on Tomales Point in 1978, and twenty years later relocated into the southern wilderness area of the Seashore. These free-ranging animals have spread onto the pastoral zone, and are now threatening the long-term viability of several historic ranches. The NPS's lack of action to counter the effects of free-ranging elk on ranch operations seems based in idealizations of both wilderness and wild animals as requiring hands-off management.

Finally, chapter 7 puts the oyster controversy in the context of this larger story of PRNS, noting parallels between the present conflict and earlier park management dynamics. The use of formal planning processes (or lack thereof) has moreover been applied inconsistently in ways that seem to privilege natural resources and pressure the working landscape. The Seashore's long-awaited update of its general management plan, which is intended to provide an overall sense of management direction and goals for the park, remains stalled. Nineteen years have passed without the release of even a draft document for public review—during which time numerous other plans and projects have been completed, resulting in a piecemeal approach to landscape management. An increasingly *selective* use of planning, science, and history seems to consistently downplay and erode the working landscape, even while publicly the NPS staff profess to support the ranches. Many locals fear that removal of the oyster farm will be the first domino that results in closure of the ranches as well.

The conclusion notes that, elsewhere in the world, the notion of "what a park is for" is indeed changing—from the inclusion of residents in many European parks and of indigenous peoples in parks in developing countries (that is, where they hadn't already been kicked out), to world heritage sites that recognize the importance of community connections and local landscape meanings, and even to the NPS, at particular park units where the management staff is supportive of change. It recommends specific policy modifications at Point Reyes, as well as more broadly within the NPS, that could improve the long-term prospects of collaborative management of working landscapes and better reconcile their coexistence with wilderness.

• • •

John Muir famously described his beloved Sierra wildernesses as distant cathedrals where visiting humans should experience awe and wonder.

For many years, Americans have celebrated Muir's vision of natural temples as the guiding vision for their national parks. But have we listened solely to Muir for too long? Protected areas in the United States are too varied for a one-size-fits-all approach. Another voice that deserves more attention in park management is that of Aldo Leopold, who in his pioneering advocacy for wilderness protection also wrote of the importance of reestablishing a personal and cooperative relationship with the natural world through working the land. For Leopold, visiting and admiring is not enough; we need to recognize our reliance on and coexistence with the wild through living and working with it: "Conservation means harmony between men and land."[9]

Point Reyes has long been ideally suited to be managed as a Leopoldian park, a place where the wild and the pastoral are not in competition but are complementary, thriving side by side. Despite the final closure of Drakes Bay Oyster Company in December 2014 after a long legal battle, these questions about the balance of management, and the need to find a Leopoldian middle ground, remain with regard to the Seashore's working landscape. They will only become more relevant for other parks as landscapes across the country face further changes and management challenges in the future.

FIGURE 2. Allée of cypress trees at the entrance to former RCA Point Reyes Receiving Station, Point Reyes National Seashore, 2011. Photograph by author.

CHAPTER I

Landscapes, Preservation, and the National Park Ideal

An understanding of today's controversies at Point Reyes National Seashore begins with a broader look at the transformation that takes place when a landscape is preserved as a park—specifically, what exactly the public expects from parks and why. National parks are among the most popular destinations in the United States, and many of us carry fond memories of our camping trips with family, or backpacking with friends, usually augmented by loads of photographs. Yet most of us don't really think much about what parks *are*, or how they got to be that way—their role is anticipated but unspoken: there will be scenic views, interesting wildlife, trails, historic markers or interpretive signs, and rangers in funny green hats. But where do these expectations come from? How do we choose which places "should" be parks, and how does that designation and management affect what we find there? In many ways these expectations have been built or "written" into the landscape itself through the process of park preservation and management.

And yet these expectations for parks often do not easily accommodate working landscapes, places that have been shaped by the work and lives of many individuals over generations, maintaining a distinct character yet responding to the changing needs of their residents.[1] Early parks, established in the nineteenth century, did not celebrate the working landscape, but rather overwrote it; Native inhabitants were usually forcibly removed, and new settlers prevented from claiming homesteads, so that the park's magnificent natural scenery could be preserved unchanging into the future. In the twentieth century, however, Congress established more and more parks in inhabited places. Understanding what happens when parks are carved out of lived-in landscapes, such as Point Reyes, first requires us to understand the complexity that any given landscape represents, to explore the preservationist impulse, and to see how preservation began to shape the earliest parks into an ideal, an image of what a park *ought* to be, that continues to influence park management today.

LANDSCAPES AS INTERACTION OF PEOPLE AND PLACE

To start, just what *is* a landscape? If asked to imagine a landscape, many of us would envision a view, perhaps of rolling hills or a mountain in the distance, or even a city skyline. Defining that landscape is more difficult, as it is more than the physical ground itself. Even the most natural-looking of landscapes almost invariably include some degree of human influence—trails, campsites, or any other human-made structure or modification to the land—as well as elements of personal and/or cultural meaning. Two visitors standing side by side at a Civil War battlefield site may "see" landscapes with different meanings, if one visitor's ancestors fought for the Union while the other comes from a formerly Confederate family, or if one visitor is white and the other black. Similarly, a small town looks very different to a tourist stopping to buy gas than to a person who grew up there. Our experience of a landscape is a combination of what physically lies before us and what is in our heads; how we think and feel about what we see matters, and these issues in turn influence how we interpret, use, and change what we see.

Through constant reinterpretations and changes over time, landscapes gradually come to reflect the ideas and values of the people who live within their areas. Thus a landscape can be thought of as the "unwitting autobiography" of those people, filled with cultural meanings that can be read, if you know what to look for.[2] In a recent survey of the field, Paul Groth describes landscape as "the interaction of people and place: a social group and its spaces, particularly the spaces to which the group belongs and from which its members derive some part of their shared identity and meaning."[3] This idea of landscape draws attention to human actions that result in the constant, day-to-day manipulation, negotiation, and contestation of landscape meanings. As a tangible combination of the natural environment and its social, political, and historical context, landscape is "not so much artifact as *in process of* construction and reconstruction."⁴ Even an area designated as a national park and protected in perpetuity continues to shift, not only with such physical variables as changing management regimes or tourist densities, but also with variables of meaning, such as whether the nation is at war or peace, whether neighboring communities feel enriched or limited by the park's presence, and so on. As an element of study, landscape "provides a door to understanding how individuals and societies perceive their environs and how they behave toward them."⁵

The term *landscape* can refer to the physical earth itself, with the combination of natural and human elements upon it, or a view of the same—or it can mean a representation of a landscape, such as a photograph, painting, or description in a novel. The physical landscape itself may also be symbolic of cultural ideas, either local or more generally held. A landscape is not just a passive stage on which people act out their lives, but a representational and symbolic space in which the dominant social order is materially inscribed and, by implication, legitimized. By way of example, French sociologist Henri Lefebvre asks whether religious ideology would be nearly as compelling "if it were not based on places and their names: church, confessional, altar, sanctuary, tabernacle?"⁶ Lefebvre also asserts that the spatial practices of everyday life contain traces of older traditions otherwise obscured, constrained, and reshaped by powerful societal influences, such as corporations or government agencies. These traces represent the possibility of recovery from the ways in which modernity and capitalism alienate us from our own lives.⁷ This desire to recover the past suggests one reason why "everyday" landscapes have recently become the focus of many preservation efforts; it also suggests why those efforts—particularly via landscape planning and design, including parks management—can have troubling consequences.

Thinking of a space as a "landscape" changes assumptions about who, and what, belongs in it. The process of creating a particular landscape framework through which to view the world puts the framer in the authoritative position of defining who or what is "in" or "out" of the picture; it also sets up the framework as something that seems "to exist apart from, and prior to, the particular individuals or actions it enframes. Such a framework would appear, in other words, as order itself."⁸ Defining the world as a series of landscapes allows those with power to define certain aspects of the world as important, while ignoring others, thus shaping and controlling which social relations may be expressed or reproduced. In terms of national parks, NPS management selects certain aspects of park landscapes as the primary focus of each place, while overlooking or downplaying others, thus shaping and controlling which meanings may be expressed or reproduced.

Official landscapes not only reflect power relations, but also function to "naturalize" those relations, to make them appear to be unquestionable, taken-for-granted parts of reality, rather than social relations that could be altered or improved upon. The word *landscape*, like *nature*, *culture*, and *nation*, historically contains unspoken or unrecognized meanings that bolster the legitimacy of those who exercise power in society. These meanings can also, of course, be manipulated to create new power relationships. All four words, according to geographer Kenneth Olwig, "tend to be used as if their meanings were unambiguous and God-given, thus 'naturalizing' the particular conception which remains hidden behind a given usage."[9] By freezing the constant shifting of social struggles into material form, landscapes "solidify social relations, making them seem natural and enduring."[10]

Marxist and Hegelian theories separate the concept of nature into two categories: "first nature," that which is original and prehuman, and "second nature," which consists of human alterations that overlay and remake first nature. When second nature is confused with, or defined as, first nature, the human activities and intentions that produced it become veiled, blending into the primordialism of first nature.[11] Yet the identification and management of universalized "natural" objects is always political; that which is natural is "'fixed' in specific ways from particular perspectives and with particular implications for how we might behave toward 'it' and each other."[12] Because they are defined as natural, those political associations and exertions in the landscape are disguised and made to appear as elemental as the rocks and trees found there.

Several landscape scholars have shown how powerful social actors obscure their actions by associating second-nature manipulations of landscapes with pristine first nature. Olwig, for example, shows how sixteenth-century courts in northern Europe redefined traditional conceptions of custom and law by creating popular presentations of landscape scenery, both in artistic works, such as paintings and theater, and in the physical landscape, with formal gardens and estates. These efforts, which emphasized geometry and spatial aesthetics according to the idealized past of imperial Rome, created "'natural' surroundings while simultaneously erasing the memory of custom's common landscape uses which stood in the way of gentry 'improvement.'"[13] Similarly, NPS management reshapes local landscapes into "parkscapes"—overwriting the older appearance and meaning of local memory, so that understanding of the place as a "park" overtakes all previous understandings, often even for the locals themselves—and yet only rarely acknowledges or interprets its own presence in the landscape, as if NPS management is somehow "outside" the land's history.[14]

Discussing the eighteenth-century development of private parks (precursors to U.S. national parks in many ways) in Britain, literary scholar Raymond Williams finds the intent was to "make Nature move to an arranged design . . . [as an] expression of control and command." The existence of the estates depended on the working agricultural land around them for income to support the landowners, yet all traces of work and labor were removed from the estate grounds themselves, even though considerable work was required to create and maintain these aesthetically controlled spaces. These two separate landscapes— pastoral lands and private park—remained connected economically, yet "in the one case the land was being organized for production, where tenants and labourers will work, while in the other case it was being organized for consumption—the view, the ordered proprietary repose, the prospect."[15] The owners and designers of these park landscapes aimed to make them appear unworked and "natural," thus obscuring their origins as the productions of actual landscape design and manipulation. Public parks like Point Reyes that aim to *protect* working, lived-in landscapes, therefore, contain an inherent tension, between NPS staff supporting agricultural operations on the one hand and aiming to produce more natural-appearing, "unworked" scenes for tourists and recreation users on the other. The NPS unconsciously disguises its own land management efforts by emphasizing natural resources and downplaying traces of local history.

THE EFFECTS OF PRESERVATION

If landscapes are created by continually changing social forces, what does it mean to preserve one? What is *preservation*? To begin to explore this process, imagine a pickle—a classic dill pickle. Now think of a fresh cucumber—are they the same? Of course not: we all know that the pickle started out life as a cucumber, but that the thing that once was a cucumber, through the presence of vinegar and salt and the passage of time, has changed in some fundamental ways in the process of becoming a pickle. In the same way fresh fruit is transformed by preservation into jam, or fresh pork is transformed by preservation into bacon or salami or ham, landscapes are transformed by preservation into parks and protected areas. The second state has a relationship with the first, but they are not the same.

Heritage is literally that which we inherit: the stories, meanings, and tangible evidence of the historical past that survive in the present. Heritage can include buildings, furniture, pieces of art, myths, cultural traditions, even language itself. The natural world is also often referred to as the heritage of mankind; advocates for the protection of biodiversity, wilderness, and other aspects of the environment have all appealed to the need to preserve our common heritage. In recent years the ranks of what is defined as heritage have changed markedly, "from the elite and grand to the vernacular and everyday; from the remote to the recent; and from the material to the intangible."[16] Because heritage can include almost anything, it is vulnerable to constant redefinition. Despite this, distinctions are usually made between natural and cultural heritage, and preservation efforts for each are almost always considered as distinctly separate concerns, although they actually share much in common.[17]

Specifically *which* resources deserve deliberate preservation, however, is an open question. For after all, if heritage is simply that which is passed down from history, why the need to preserve it? Preservation implies protecting something from harm, damage, or danger. For most of humankind's existence, people generally either rebuilt and reused old structures, continually adjusting or reinventing them as circumstances warranted, or ignored them, allowing them to fall into ruins, sometimes disappearing entirely.[18] Starting in the sixteenth century, however, a series of elites began to embrace classical antiquities as desirable links to the great Greek and Roman cultures, which they considered superior to their more recent history.[19] By the eighteenth century, this attitude had developed into a widespread upper-class aesthetic visible in architectural styles, art, and literature. Because this developing interest in heritage put such a primary emphasis on material items, deliberate preservation became crucial as time and social change caused artifacts to fade or crumble, buildings to be replaced, and old ways of life to disappear.[20] Heritage could then be visited and viewed by tourists, in museums or at official historic sites, where the visitors' sense of connection with the past would be learned and/or reinforced. For those in power, for whom change was a threat, preservation formed an important way to reassert and protect relics symbolic of their social prestige and control.

This desire to prevent change makes a kind of intuitive sense: known objects and stable spatial configurations allow us to maneuver through our daily lives more easily. Simply put, people often tend to want their surroundings to stay as they are now; we have a fundamental discomfort with change, preferring our world to be predictable and constant. Hence many forms of change, particularly those that are unexpected, make many people ill at ease.[21] As a result we tend to try to fix things in both time and space.[22] Historical durability is often interpreted as a sign of worthiness, according to a sense that if "it lasted this long, it must be good." While the future is murky and unknown, the past is usually thought of as tangible and clear, unalterable, providing a sense of stability, familiarity, and security. Thus it also appears to be one unbroken, uncomplicated chain of events, rather than a continually reworked narrative.

Preserving material objects is not the *only* way to conserve a heritage. In Ise, Japan, for example, the Shinto Grand Shrine is disassembled every twenty years and an exact replica, rebuilt of similar materials, is assembled in the same place. In this form of preservation, perpetuating the building techniques and the ritual act of re-creation matters more than the physical continuity of the structure.[23] Similarly, the ancient White Horse of Uffington in England was "re-created" for centuries by locals, who scraped the chalk figure every seven years to keep it from being obscured by growing vegetation. Cultures that rely on oral traditions retain their sense of cultural heritage without any tangible objects at all, but rather by retelling stories from the past. These and other traditional or "folk" ways of retaining heritage bring the past and present together, fused in a repeating, cyclical sense of custom through use and interaction in everyday life.

Despite these alternative approaches to preservation, the most prevalent modern conception of preserved heritage remains focused primarily on physical artifacts, set aside and ostensibly protected from change. Yet this kind of evidence only reveals the limitations of this vision of history, according to which anything that didn't take material form can be left out. The high visibility and accessibility of relics, especially old buildings, tends to cause people to overemphasize—and overestimate— the stability and homogeneity of the past.[24] For example, places where many artifacts survive from one particular epoch, as if they had been pickled, can give the impression that time has stood still, that the places are perfectly unchanged since the era the artifacts reflect, regardless of what the actual historical experience may have been.[25] Nor can material relics tell their own stories. They require interpretation and explanation, adding another layer of present-day attitudes and values to the understanding of the past.

Preservation's act of reshaping the past according to the views of the present effectively distances the past from the present, causing it to seem like a distinct, separate realm, rather than something intimately connected with today. Recognizing the past's difference promotes its preservation, and the act of preserving it makes that difference still more apparent.[26] Particularly in the United States, heritage is often not permitted to coexist with the present; instead it is fenced off, "always in quotation marks and fancy dress," and visited on special occasions, rather than seen an integrated part of everyday life.[27] Setting aspects of the past off as national parks contributes to this separation, implying that history is something to be visited and viewed, rather than lived with, day to day. Similarly, as geographer Yi-Fu Tuan writes, preserving particular ways of life associated with the past "turn[s] them into figures in glass cases, labeled and categorized as in a museum"—an approach more bluntly described as "geographic taxidermy."[28]

All forms of preservation advance a strong tendency to idealize the resource in question, whether the historical past, a present-day cultural system, or a natural resource or ecosystem—or a combination of all three in a landscape—to make it representative of some imagined era of perfection, thus all the more worthy of preservation. The ways in which currently held values can be projected back onto history underlie this desire to see the past not as it actually was but as it *should have been,* according to how we see it today.[29] These resources may be idealized for aesthetic reasons, or because they contribute to some group's sense of identity or heritage. They may also be preserved as examples of natural or cultural diversity, or in hopes of gaining knowledge or profit, particularly from tourism. These different motivations can result in different strategies or techniques of preservation, but in a core sense they all seek to prevent change, or at least to control the direction and degree of change in a resource—and in doing so, they distort our ability to see and understand the past as it was, in all of its messy complexity.

What, then, constitutes "authentic" heritage? In regards to the value of a piece of art, authenticity refers to the originality expressed in the art; forged masterpieces do not fetch the same market value as the real thing. But how does this idea extend to whole cultures, to nature, or to the past? Edward Bruner describes the "problem of authenticity" as being built into Western societies as "the notion of a privileged original, *a pure tradition,* which exists in some prior time, from which everything now is a contemporary degradation."[30] This conception was particularly prevalent in historical practices of anthropology, as researchers searched for "primitive" peoples and cultures, considered unchanged for eons and therefore more "pure."[31] Drawing the focus too tightly on purity, authenticity, or pristineness tends to disempower or eliminate any non-pure elements—such as people in nature, or recent changes to historical scenes or artifacts. Thus preservation has a catch-22: we ideally expect it to reveal the whole, pure, singular past, but we necessarily only have the fragments left over in the present—and the better adapted those fragments are to present-day life, the less authentic and thus less credible they seem.

An example of this problem is found California's Bodie State Park, an old mining town preserved in a state of "arrested decay" since the 1950s.[32] In her ethnographic research on the site, geographer Dydia DeLyser recorded visitors talking about how "authentic" Bodie seems, with its tumbling-down buildings, all weathered, faded, and lonesome. Visitors often contrasted Bodie with the gaudy displays in such nearby "restored" mining towns as Virginia City, Nevada, where the buildings were brightly painted and lit up, and entrepreneurs were always trying to sell them something. Yet ironically, when Bodie was still a functioning lived-in town, it *was* brightly painted and had loud saloons and restaurants; furthermore its existence was entirely about people trying to make money. So how can the preserved ghost town of Bodie be more "authentic" than Virginia City? It's an authentic preservation of the idealization of a *ghost town*, which is a simplified and more pure version of the West's past than an actual living town. Similarly, had the California State Parks Department just left Bodie alone, it would have fallen down completely long ago, the victim of time and harsh climate. Instead, the agency uses a deliberate policy of "arrested decay," propping things up and repairing wear to keep the buildings in a perpetual state of "falling down but not all the way down." Yet most of the visitors DeLyser interviewed were untroubled by this *actual* discrepancy in the site's authenticity—that it is actively and deliberately managed to keep it in a particular state—as long as the *appearance* of authenticity was maintained.

While preservation may appear to freeze things in time, in actuality preserved resources increasingly reflect the values and ideals of their preservers, through the choices they make in terms of what to protect and how to manage and display them—in this sense, the authentic past is that which the authorities have chosen to preserve. Because preservation is an exertion of power, that power is reflected in and reinforced by the preserved resource, whether an ancient vase in a museum, a wild

animal caged in a zoo, or a landscape preserved as a park. The particular kinds of animals kept in a zoo, for example, reflect the interests of
the institution itself; if the zoo wants to emphasize the diversity of life,
many rare or unusual kinds of animals may be represented, while a zoo
focused on entertaining the public may have more popular, familiar animals. No institution can display *all* kinds of animals, and many—common
insects, for instance, or rats—are traditionally not included, indirectly
conveying the message that these other species are less important, less
interesting, or less a part of nature. In these ways, the institution's
values are reflected in the act of preservation: the choice of creatures to
display (or not), how they are displayed, and the information provided
about them. In the case of Point Reyes, the National Park Service maintains this authority, through its ownership, management, and interpretation of the Seashore for the public viewer.

The ideals of preservation and the power that enforces them usually
become "naturalized" at protected sites, so that the methods and standards of preservation come to be seen as normal, predictable, and inevitable. This process changes our perception of the preserved resource to
include those values or ideals interjected through the process of preservation itself. To return to the example of a zoo: most people would be
surprised to see a display of domesticated dogs in a zoo, because we
have become so trained to believe that they "naturally" do not belong
there. This has nothing to do with the dogs themselves, or the public's
like or dislike of them, but instead reflects the values of zoos as reposi-
tories of only non-domesticated animal species. As we will see, similar
expectations often lead visitors to question the role of agriculture at
Point Reyes, simply because they have been trained to believe that
ranches do not "naturally" belong in national parks.

NATURAL LANDSCAPES AS NATIONAL HERITAGE

A recent documentary series on the national parks quoted Wallace Stegner referring to them as "the best idea we ever had."[33] But national parks
are not a single idea; they emerged from ideas about preservation, nature,
national pride, and tourism (recreation and history came later). When
many people think of America's national parks today, they envision
large expanses of pristine natural areas. The public imagines parks as
sanctuaries for wilderness, or as a means to preserve and protect ecosystems. Nevertheless, the earliest national parks did not focus on the inherent value of nature for its own sake. Alfred Runte's classic work, *National*

Parks: The American Experience, argues that the primary motivation for setting aside the first parks stemmed from a national need for cultural icons, "natural wonders" that would assure Americans that they had a heritage equal to or better than that of Europe.[34] Places of spectacular natural scenery became infused with patriotic significance, representing America's first major contribution to world heritage, and also became tourist attractions, where the public could bask in symbolic grandeur and connect more deeply with their nationalistic pride.

Parks are also not "our" idea. The national park system is often imagined as a group or collective effort, democratically conjoined through collective public ownership. More than any other type of government-owned land (such as a post office or an IRS building), parks are frequently referred to as places that "belong to the American people," as if each citizen has direct ownership of Yosemite or Yellowstone. President Franklin D. Roosevelt, for example, is quoted on an NPS website as having said, "There is nothing so American as our national parks. . . . The fundamental idea behind the parks . . . is that the country belongs to the people, that it is in process of making for the enrichment of the lives of all of us."[35] In reality, however, citizens do not own these lands; as will be discussed in more depth in chapter 2, the public has the right to access and to comment on parks' management, but we are not direct owners in a legal sense—we tend to confuse a sense of shared national heritage with actual ownership and control. And while the NPS is the federal agency with direct ownership responsibility, it is the U.S. Congress that creates parks via legislation. The impetus for the earliest parks came not from ordinary citizens or a groundswell of public opinion but rather from cultural and economic elites who exerted political power to push Congress to create these spaces. They did so both to represent specific ideals of national superiority and natural purity, and to propagate those values to the visiting public. In this way, parks became self-reinforcing expressions of a very controlled message in and about the landscape.

More specifically, the early parks came to embody the view, generally associated with the West, that nature and wilderness are completely separate from human habitation and use. Despite their importance as tourist destinations, these parks were more like museum or zoo exhibits, something to stand back from and observe.[36] Once the parks took on such powerful cultural symbolism, they had to be held static, as unchanging as the national values they now reflected. Their appearance also was enhanced and manipulated by landscape designers so as to accentuate their grandeur as an unchanging view of uninhabited, pristine nature.

These monumental parks preexisted the National Park Service, the federal agency created in 1916 to administer and maintain them. In its subsequent formation, the agency adopted these symbolic values of nature as its own foundational principles. Though never explicitly written into official park policies, these assumptions subtly undergird much of the NPS's organizational culture and management approach.

At the start of the nineteenth century, most Americans thought of wilderness, or "pure" nature, as something to be avoided, or better yet, tamed and subdued. Indeed, in 1831 the French writer Alexis de Tocqueville resolved to see some of the American wilderness while touring in the United States, but "when he informed the frontiersmen of his desire to travel for *pleasure* into the primitive forest, they thought him mad."[37] Environmental historians have identified two major components to this traditional bias against wilderness: first, a very real threat to survival, and second, a dark and sinister symbolism, inherited from a long tradition of Western thought.[38] From its ancient biblical usage, "wilderness" implied the opposite of civilization, the place Adam and Eve were condemned to after being cast out of Paradise. The early Puritan colonists carried with them this idea of wilderness as a "wholly negative condition, something to be feared, loathed, and ultimately eradicated—something to be replaced by fair farms and shining cities on hills."[39]

The romantic movement of the 1820s and 1830s, which emerged out of eighteenth-century Europe and took hold among many American writers, artists, and scientists, added some complexity to this traditional view of wilderness. The romantics saw the handiwork of God in Enlightenment accounts of an apparently harmonious and orderly universe. Nature, therefore, should be considered sublime, as something that inspires exultation, awe, and eventually delight. In the romantic view, nature is specifically empty of human habitation and influence; it is the antithesis of civilization. Any human-induced artificiality reduced the direct connection to the divine and the sublime.[40] This, combined with Rousseau's idea of primitivism, became "the belief that the best antidote to the ills of an overly refined and civilized modern world was a return to simpler, more primitive living."[41] Within a few decades, writers like John Muir and landscape designers like Frederick Law Olmsted would actively espouse these romantic ideas in their work. While this perspective might seem to contradict the earlier negative notion of wilderness, in reality they are two sides of the same coin; in both, nature is necessarily empty of, and distinct from, people. The only difference is which side of the dualism is privileged.

American romanticism (in contrast to the older European tradition) took an especially nationalist turn, with natural scenery becoming emblematic of national greatness.[42] To better understand this development, it is important to note that the new nation, having twice fought free from European political control in its first few decades, still borrowed heavily from Europe for much of its "culture"—art, architecture, and literature. While wanting to distinguish themselves from European societies, Americans depended on them as their link to the richness of the great heritage of Western civilization. National leaders and intellectuals believed that the United States was destined for a glorious future, but doubts persisted as to whether the society could really survive apart from its European parentage.[43] Having no truly ancient artifacts to point to as heritage, patriots began to rely on spectacular natural monuments as proof of distinctive national greatness.[44] The physical landscape became a way to quickly acquire a sense of national superiority, and nature tourism a primary vehicle for appreciating it.[45]

Yet there was little that was genuinely unique about most of the then-settled American landscape. The East's best hope for a symbol of greatness lay in Niagara Falls, on the border between New York and Canada.[46] Observers both home and abroad considered the falls to be America's greatest natural spectacle; romantic and nationalistic views of nature merged to form an image of Niagara as sublime nature that would produce a corresponding moral sublimity in those who associated with it.[47] The Niagara landscape represented an idealized national identity, and popular artistic portrayals began to reshape expectations of what the actual physical place should look like.

But this natural splendor was compromised, and its symbolic power eroded, by increasing visitation to and development of Niagara, especially after the opening of the Erie Canal in 1825. Newly constructed bridges, paths, and staircases rendered both sides of the falls accessible, and the nearby mill town of Manchester churned with industrial activity. By 1830, numerous small-time entrepreneurs competed to offer the best view of the falls (for steep prices), as well as food, trinkets, or tours, crowding out the famous view with their tawdry stands and signs. This disorganized and haphazard commercial development, according to historian William Irwin, "spoiled the more raw, adventurous, and reverent mode of experiencing the Falls"; reality no longer matched the idealized image of Niagara as a symbol of national strength and purity.[48] The commercialization did not go unnoticed by European visitors, who wasted no time in roundly condemning the private profiteers overwhelming Niagara.[49]

FIGURE 3. View of Niagara Falls in winter, c. 1855. Whole plate daguerreotype by Platt D. Babbitt. Courtesy of the J. Paul Getty Museum.

This stream of published criticism on tourism at Niagara hit a raw nerve: Americans, already sensitive to their lack of contribution to "world culture," stood accused of having no pride in themselves or their past. Lack of control over the private development of Niagara Falls had led to its apparent ruin; by allowing it to become overrun with ugly commercialization for private profit, Americans were seen as willingly selling their cultural legacy to the highest bidder. Word of spectacular landscapes opening up in the West soon offered a chance for national redemption.

CREATION OF CONTROLLED, UNCHANGING LANDSCAPES

By the mid-1840s, white American settlers were rapidly moving west of the Mississippi River, displacing Native American tribes as they went, and encountering truly unique landscapes along the way. Western expansion represented the country's future; Manifest Destiny, the reigning territorial philosophy of the time, carried a message of progressive advancement and historic inevitability.[50] This expansion moved through some of the boldest and most magnificent landscapes ever seen on earth; surely this must be a sign of national superiority! Journalists traveled west and published widely read descriptions of the marvelous landscapes they passed through. With most of the U.S. population still living along the Atlantic coast, the West became a stage, easterners watching as the spectacle of the West unfolded in popular journals and newspapers. Representations of the West, in paintings and literature as well as the popular press, idealized the landscape within a particular nationalistic framework that emphasized majestic and eternal natural scenery awaiting "discovery" by heroic white pioneers. The massive paintings of Albert Bierstadt, for example, exaggerated the steepness and depth of western mountains and depicted glorious landscapes empty of Indians; other paintings of the era depict railroads cross-crossing the landscape and wagon trains full of settlers pointing the way west.[51] Journalist William Gilpin, in an 1846 report read in the Senate, extolled westward expansion: "*Divine task! Immortal mission!* Let us tread fast and joyfully the open trail before us! Let every American heart open wide for patriotism to glow undimmed, and confide with religious faith in the sublime and prodigious destiny of his well-loved country."[52]

Ecstatic descriptions of the West began shaping landscapes as symbols of national pride and destiny in the imaginations of the American public—particularly because, as Niagara had been, they were romanticized as pristine wilderness, a people-less natural landscape that could be understood as sublime and grand. These images did not merely illustrate particular conceptions about the western experience; they endorsed them.[53] Mark Spence notes that "the conflation of racial, political, and geographic 'destinies' with the cant of conquest effectively erased the human history of western North America and replaced it with an atemporal *natural* history that somehow prefigured the American conquest of these lands."[54]

The "discovery" by whites of Yosemite Valley and the nearby Sierra sequoias, in 1851 and 1852, respectively, provided two early examples of natural wonders through which the United States could claim cultural recognition. Nationalistic writers began drawing comparisons between western and European mountains, going as far as to belittle the Swiss Alps in favor of Yosemite. Samuel Bowles, editor of the *Springfield Republican,* exclaimed, "It is easy to imagine, in looking upon [the sides of the Yosemite Valley], that you are in the ruins of an old Gothic cathedral, to which those of Cologne and Milan are but baby-houses."[55] Surveyor Clarence King described the giant sequoias in 1864, writing that no "fragment of human work, broken pillar or sand-worn image half lifted over pathetic desert—none of these link the past as to-day with anything like the power of these monuments of living antiquity."[56] Neither of these descriptions mentioned the long-resident Indians, who maintained a community in Yosemite Valley despite several attempts by the military to push them out.[57] They were already being edited out of the landscape, so as not to taint the national symbolism.

This natural, supposedly empty splendor soon enough had to fend off would-be settlers and concessionaires. By the end of the 1850s, private entrepreneurs were already hard at work at Yosemite, trying to make a profit by capitalizing on the grandeur of the natural discoveries.[58] Individuals attempted to claim the portions of the valley with the best access to the spectacular views, in anticipation of the sightseers sure to follow in their footsteps. Similarly, while the giant sequoias were not useful for lumber—when felled, they tended to shatter upon impact with the ground—several of the largest specimens were nevertheless stripped of their bark, cut into sections, and shipped off for sale as curiosities in the East and overseas, thus destroying the spectacular trees so that they could be sold piecemeal.[59] Both Yosemite Valley and the nearby sequoia groves represented a new claim to U.S. greatness via natural splendor, but uncontrolled land use and exploitation threatened to spoil both.

Another round of nationalistic criticism sprang up in response, lamenting the apparent repeat of Niagara Falls' fate in the Sierra. In 1853 *Gleason's Pictorial,* a popular British magazine, published a letter from an irate Californian regarding the destruction of the "Discovery Tree" for public display. Had the giant been a native of Europe, he suggested, "such a natural production would have been cherished and protected, if necessary, by law; but in this money-making, go-ahead community, thirty or forty thousand dollars are paid for it and the purchaser chops it down and ships it off for a shilling show."[60] But Yosemite and the sequoias turned out not to be like Niagara after all: to protect the national interest in these places, the federal government became involved in efforts to preserve them.

Yet it did not become involved of its own accord. While it appears that private ownership of the attractions themselves, or control of access to the views, was not to be tolerated, corporate interests like railroad companies realized they could reap great profits from providing transportation to and facilities at these new tourist destinations, if the government could prevent or control small-scale settlement and clutter.

Particularly by the 1850s and 1860s, about the same time that Yosemite was first entering the public eye, the industry of providing for tourists had become increasingly sophisticated, and shrewd businessmen were on the lookout for new opportunities. By enlisting the federal government to set western scenic wonders aside for preservation, corporate tourism providers could ensure that they would have these opportunities to themselves.

By early 1864 nationalists and western promoters began urging legislation to protect Yosemite Valley and the Mariposa grove of giant sequoias to prevent private occupation and to preserve them for "public use, resort and recreation"; the idea originated not in a groundswell of public outcry, but in a letter written by Israel Ward Raymond, representative of the Central American Steamship Transit Company—at that time, ships were the primary way of getting to the West Coast, as the transcontinental railroad had not yet been constructed—and sent to John Conness, the junior U.S. senator from California.[61] Niagara Falls served as a role model of sorts, both illustrating the possibilities of drawing tourists and highlighting the dangers of overcommercialization by small entrepreneurs. The official rhetoric written into the legislation, however, relied on the cultural symbolism of the iconic valley, rather than any economic rationales (pro or con).[62] The argument found support in Congress, and on June 30, 1864, President Lincoln signed the Yosemite Act into law.[63]

The grant itself was small—only the valley itself, the encircling peaks that formed its scenic backdrop, plus the Mariposa grove, located a bit to the south, were protected—and shortly thereafter was turned over to the State of California for administration.[64] Significantly, the grant contained a clause insisting that the protection be "inalienable"; as Runte writes, "from a cultural perspective, preservation without permanence would be no real test of the nation's sincerity."[65] This language of permanence shows an early emphasis on stasis in the establishment of U.S. national parks, a desire to protect the symbolic qualities of the grant in perpetuity by preventing change. The first western park preserved by the federal government was clearly imbued with particular ideological meanings, which the NPS would later inherit: it was to be a natural landscape frozen permanently into a static image of national greatness.

Ten years later, an almost identical process of nationalistic description and appropriation took place with the "discovery" and exploration of the area that would become Yellowstone National Park. The area had not received much attention from early explorers, mostly because of its

FIGURE 4. William Henry Jackson photographing from Glacier Point in Yosemite National Park, c. 1884. Courtesy of History Colorado.

inaccessibility. The discovery of gold in Montana in the 1860s, however, brought people to the area, and soon tales of wondrous scenery began circulating in eastern society. The proliferation of geysers and other geothermal features at Yellowstone was unique; once again, many comparisons were drawn to European ruins. Charles W. Cook, on an 1869 expedition to the area, noted a limestone formation that "bore a strong resemblance to an old castle," whose "rampart and bulwark were slowly yielding to the ravages of time."[66] And once again, most accounts of the landscape omitted the presence of the local Indians so that it could be symbolically recast as "empty wilderness," despite archeological evidence and active Native management suggesting otherwise.[67]

As had happened at Yosemite, almost immediately after the "discovery" of the Upper Geyser Basin individual entrepreneurs began fencing off the most scenic areas for future tourist spots. In response, representatives from the railroad industry started a campaign for government protection of the area. The initial suggestion of creating a national park at Yellowstone came on stationary from Jay Cooke and Company, a principal financier for the Northern Pacific Railroad, who presumably hoped to spark tourism.[68] A report written by the House Committee on Public Lands stressed the promotion of the area as a scenic oasis, and criticized private individuals' attempts to lay claim to the distinctive geologic features and "to fence in these rare wonders so as to charge visitors a fee, *as is now done at Niagara Falls,* for the sight of that which ought to be as free as the air or water."[69] Once again, Niagara served as an effective reminder of the nation's failure to protect its cultural heritage; the report stressed that the leaders must match their rhetoric with a commitment to action. The bill passed Congress and was signed on March 1, 1872.[70]

As was the case with Yosemite, Yellowstone's initial protection was largely of symbolic importance. The park's large size reflected more on the lack of accurate surveys of the scenic geologic features rather than on a desire to protect the wilderness; the boundaries were drawn large enough to ensure that all of the features would all be included. Furthermore, the park was retained by the federal government, rather than given to the state government to administer, because Wyoming was still only a territory—the "first national park" only remained national because there was no local government body to give it to. It would be several years until Yellowstone would receive any significant tourist visitation; even the Northern Pacific Railroad, the original advocate for creation of the park, didn't connect to the park until 1883.[71] Yet once the tourists began coming, the railroad monopolized transportation to and from the park, as well as all hotel accommodations, with no competition from small-scale independent entrepreneurs. Despite rhetoric that the parks were necessary to preserve nature from the clutter of development, in many ways they were actually created to provide business opportunities for corporate interests.

Significantly, both of the two first national parks, later to become "crown jewels" of the National Park Service, were initially created to prevent change in natural scenery that could be identified as uniquely American and to provide tourists with new destinations where they could gain a sense of national heritage. The establishment of all the earliest national parks and monuments—including Yellowstone, Yosemite, Sequoia, Mount Rainier, Glacier, and the Grand Canyon—followed a similar process, in which legislation created static landscapes to match society's idealizations of natural and national heritage. The parks were infused with symbolic ideology, their natural scenery representing unchanging visions of national greatness, demonstrating the lasting vitality and virtue of America's republican government; all were advocated for and served by private railroad corporations rather than demanded by "the public."[72] As landscape theory suggests, these powerful business interests utilized government protection of these landscapes to push out small-time competitors and control the views so as to attract tourists, and the parks themselves became relatively static viewscapes reflecting nationalistic ideals. This ideological meaning, and connection to corporate concessionaires rather than local settlers, have stayed with the national park system, even as it has expanded to include an ever-widening array of types of landscapes, and they influence management even in places like Point Reyes, as we will see.

PARK IDEALS AS EXPRESSED THROUGH PRESERVATION

Preserved landscapes are not neutral; they require management to produce the appearance and maintenance of stasis. What ends up being preserved is not the actual landscape as it was at the time of preservation, but those aspects of it that coincide with the values that the agency assigned with its management seeks to accentuate. The cultural uses and meanings that produced the landscape in the first place are increasingly overlain or replaced by the social dynamic of preservation itself, which comes to be built into the landscape, both in physical shape and cultural meaning. Yet these landscapes tend to be seen only as places of aesthetic wonder, with little or no consideration for how they got that way.[73]

In the case of the national parks, the agency tasked with preserving the landscape is the National Park Service. As discussed above, the first generation of natural parks stressed natural beauty and unpeopled landscapes in the service of national greatness. By the time that the NPS was created in 1916, that mandate had incorporated recreation and tourism, with services provided by corporate partners rather than locals. The NPS's institutional values, inherited from the first parks, created before its establishment, are based on long-held societal assumptions about what national parks "ought" to look like, which in turn are based on nineteenth-century conceptions of nationalism, romanticism, and Manifest Destiny. While it did not create these ideals, the NPS inherited them along with the parks it was assigned to protect, and it made them its own. The agency gradually developed standardized rules for the parks' management, with an emphasis on unchanging permanence as a goal of preservation. National parks have also idealized nature as completely exclusive of human habitation or use, aside from tourists viewing the scenery. These ideals weave in and out of the agency's history, at times becoming more predominant, at other times less, but never completely fading. Despite not being written down as formal policy, they form a powerful ideological foundation for how the NPS conducts much of its business. Through land acquisition, design, management, and interpretation they become written into park landscapes.

Of all of the ideals of the national parks, the most consequential one has almost certainly been the idea that parks are devoid of people. Time and time again, the notion of "empty" landscapes has pushed out residents—most obviously Native Americans, but also Euro-American landowners in the case of more recent parks. Because of the reciprocal relationship between landscapes and the people who live in them, the removal of people from natural landscapes has observable effects on the land. In Yosemite Valley, for instance, the valley's ecosystem has shifted from primarily open meadows and oak woodlands to closed coniferous forest in the years since Native Americans and their land management practices, which included frequent burning, were reduced to living museum exhibits.[74] Few visitors know of this transition, and the thick forests, with a few designated scenic overlooks kept clear to emphasize the iconic views, are now considered "natural" and timeless.

Yet within two decades of its establishment, the NPS was also assigned the task of managing landscapes that no one could ever mistake for "unpeopled." Starting in 1919, Congress began to authorize the establishment of parks in places that had previously been in private ownership. In the 1930s the NPS also took on management of the nation's historic heritage as well as natural scenery. More recently (as chapter 2 will explore in detail), the agency has been tasked with protecting vernacular landscapes, lived-in ordinary places that represent the everyday, rather than the extraordinary and iconic, as our society's ideas of what counts as "heritage that is worth saving" have continued to shift and change.

Yet the very concept of preserving vernacular landscapes is difficult to reconcile with the ideology of national parks in America. John Brinckerhoff Jackson, cultural geographer and publisher of the influential magazine *Landscape,* simultaneously raised the appreciation of common everyday landscapes and decried their formal preservation. He regarded protected landscapes of all kinds as political products, created by official legislative acts to be stable or unchanging, rather than remaining in the true vernacular as mobile, unpredictable and fluid. He was critical of both the environmental and historic preservation movements, asserting that their efforts result in seeing the landscape "less as a phenomenon, a space or collection of spaces, than as the setting of certain human activities."[75]

The NPS is not alone in facing this quandary; it is a frequent outcome of preservation efforts around the world. For example, the National Trust in the UK, established in 1895, aims to preserve both natural and cultural heritage, including historic homes and gardens as well as industrial monuments, social history sites, and entire lived-in landscapes such as England's Lake District. Despite the Trust's explicit focus on protecting "for ever, and for everyone," British anthropologist Barbara Bender notes that ordinary, everyday people tend to be excluded by the official protection and repackaging of landscapes.[76] She writes that "its main focus has been the landmarks of those with power and wealth, inscribed in an aesthetic, which, as it has done for centuries, bypasses the labor that created the wealth."[77]

Both in Britain and the United States, the fact that preservation protects some landscapes and not others raises the question, "Preservation and presentation for whom?"[78] Which places will be singled out for protection, and in what ways does protection modify their original appearance and meaning? Will they be places of production or consumption? Through its actions—park selection criteria and techniques of maintenance, management, and interpretation—the NPS cannot help but place its imprint on the landscape, reshaping the protected area into a national park, something with its own distinct meanings and implications. Yet the NPS managers generally do not recognize this process of creating a set of official landscapes; they naturalize the process in its own right. The shift toward a designed, orderly "national park-scape" is not interpreted or questioned by NPS managers, but accepted as "normal," desirable, and inevitable. Thus preserved landscapes, as they move from the realm of the vernacular to the official, often reveal less about the history of the place being protected than about the preserver's perception of the past.

In his work on the establishment of English estates, Raymond Williams has asserted that "a working country is hardly ever a landscape. The very idea of landscape implies separation and observation."[79] Nevertheless, the NPS has increasing become involved in protecting working landscapes, including, of course, Point Reyes. At these sorts of parks, the history of the land as having been worked and inhabited is ostensibly what NPS is trying to preserve. For example, Lowell National Historical Park in Massachusetts preserves the brick buildings and water-powered mills of the early textiles industry—yet the factories are no longer working, and the complex is more akin to a ghost town, with buildings standing empty except for exhibits and visitors. Park units such as Point Reyes, including Ebey's Landing in Washington and Cuyahoga Valley in Ohio, are different in that at least some commitment has been voiced to actually maintaining the landscape as a working landscape.[80] And yet, this is difficult, because of the uneasy relationship between these kinds of lived-in places and the background ideology the NPS has inherited concerning parks.

Landscape can be both work and an erasure of work. Landscapes are created through people's labor in their daily lives, and yet the traces of their labor become increasingly invisible as management operates to make them disappear into the "natural," taken-for-granted appearance of the park. Preservation of working landscapes, in particular, builds on the social production of the residents' lives, yet creates a new landscape that tends to diminish or eliminate their contributions to it. The accompanying interpretation, with its frequent emphasis on natural heritage, reinterprets the landscape's significance according to the agency's historical ideologies. Thus it is important for this book to make this reshaping process in the national parks more visible; as geographer Peter Jackson asserts, "recognizing the ideological dimension [of landscape] robs it of much of its power."[81] By doing so, this research can help to clarify landscape management goals and direction, and to reveal and prevent unintended consequences of park management actions.

A ROLE FOR WORKING LANDSCAPES IN TWENTY-FIRST-CENTURY PARKS?

The NPS now manages over four hundred park units, including national parks, monuments, seashores, historic sites, battlefields, recreation areas, rivers, and trails. Over its hundred-year history, the agency's mandate has both changed dramatically and maintained certain key assumptions. The NPS no longer focuses only on scenic vistas; gradually it has embraced the importance of ecological health and science-based management of its parks, as well as a broader understanding of U.S. culture and history. Native peoples and their uses of the landscape are more accurately depicted in interpretive materials today than at any time in the past. Yet because the NPS adopted the early ideology of parks—preservation of scenic heritage and provision for recreational tourism—as the core of its mission, assumptions about appropriate private residents and land uses in parks remain; in many ways they are built into the policies and management strategies of the institution itself.

These assumptions are not aspects of formal policy, nor are they solely held by the NPS—they are cultural ideals of what a park "should" be that are commonly held throughout U.S. society. And while these ideals have gradually changed over time, the values that were most prevalent at the time the NPS was established have molded the agency's structure and culture, its sense of mission and purpose regarding the lands that it manages. NPS historian Richard Sellars, in his study of the slow but gradual incorporation of ecology into the agency's management objectives, concludes that, "Given the strength and persistence of ancestral attitudes within the Service, its core values are likely to outlast any one director, even one who is stubbornly determined to change them."[82] And the parks themselves continue to reinforce and recreate these ideals—they reflect the values of preservation, which in turn emphasize unchanging scenic nature.

Understanding the effects of preservation ideology on parks is significant for two important reasons. First, preservation, while often well intentioned, can become a tool of control in a landscape, redefining the place according to the idealized image of what the preservers want it to be. By exploring the role of institutional ideology in steering landscape change, this book adds to the theoretical understanding of how landscapes are used, inadvertently or not, as tools of power. Previous researchers have shown that institutions often intentionally manipulate the landscape and its meaning as a way to marginalize others' interpretations of it.[83] As an extension of this approach, this history investigates the degree to which NPS management both intentionally and unknowingly has reshaped park landscapes to reflect the institution's beliefs and priorities. Similarly, in some instances environmental advocates or certain (often wealthy) locals have steered NPS policy and management to impose their landscape preferences on others, often privileging the area's value as a scenic vista or recreation destination over its role as a working, living landscape. The ideology of park preservation is not limited to the NPS itself.

Writers from Raymond Williams to Richard White have noted the tendency in environmentalism to separate traditional resource-based work like agriculture or forestry from aesthetic or recreation spaces, equating work with environmental destruction; working a landscape causes unpredictability and change, and can be harder to control, standardize, or "tidy up" as bucolic scenery.[84] In a similar way, because the presence of residents challenges the agency's sense of control, the NPS has struggled with parks that contain them. This at least partially explains why the NPS has insisted on full-title ownership of parklands whenever possible—early on, NPS leaders believed that private ownership was incompatible with effective preservation management. Full-fee public ownership also makes the landscape more "legible" to an administrative agency, as a bounded, simplified, manageable space rather than a more complex and contested landscape.[85] Nevertheless, legislative guidance for a number of parks created in the past fifty years has suggested other ownership models; Point Reyes is interesting, in part, because it illustrates how difficult it has been to integrate those alternative models into the NPS's preferred administrative policies.

Second, a more complete view of the effects of preservation on working landscapes will contribute to further developing and refining resource protection policy in the National Park Service. Policy makers and park managers need better awareness of the historical trends and on-the-ground outcomes of NPS management so that effective protection of working landscapes can be sustained over time. This is of particular importance and urgency in park units where the human activity that created a distinct cultural landscape is still active, as the changes resulting from NPS policies may impair the residents' own sense of landscape meaning and significance, or even their ability to persist as a functioning community. At the very least, park personnel should be aware of this dilemma, so as to have greater clarity regarding the intent and goals of management, and greater recognition of the ways in which management may affect the landscape. There is nothing inherently wrong with causing change within parks, but the NPS should be cognizant of these processes and their implications for the resources, residents, visitors, and park managers themselves.

Examining this issue now is particularly timely, as the NPS is currently struggling with the question of how best to manage its ever-growing roster of populated landscapes. Increasing numbers of new parks include existing human settlements as part of the protected landscape, and many of these places have encountered controversy resulting from implementation of management policies. In some cases NPS management has overlooked the needs of the residents, resulting in what legal scholar Joseph Sax has called "communities programmed to die."[86] At the same time, interest in and concern with the preservation of cultural landscapes has increased within the NPS.[87] In a collection of articles on managing cultural landscapes, Arnold Alanen and Robert Melnick identify the NPS as the primary force in the nascent cultural landscape preservation movement in the years since the agency first recognized them as a specific resource type in 1981.[88] Hence it is crucial to understand what likely outcomes can be expected from the NPS's involvement. Yet while some excellent research has been done on how NPS management has shaped biological resources over time, few works look at historical change in cultural resources or working landscapes within national parks.[89] This volume aims to contribute both to the theoretical understanding of working landscapes and to the current challenges facing the NPS in its efforts to improve protection of these unique and rich places, by following the agency's development through the lens of one particular landscape, Point Reyes.

The evolution of the working landscape at Point Reyes under park management also informs a series of growing questions about how best to respond both to the reality of ever-more-rapidly shifting ecological conditions due to climate change, which is affecting species distributions and the make-up of ecological communities, and to shifting ideas about our relationship with nature as represented by park management. The U.S. conception of the primary purpose of national parks has already changed, substantially, over the parks' 150-year history: from static symbols of national greatness, to commemorative spaces of war and history, to tourist playgrounds for the automotive traveler, to wilderness sanctuaries of biodiversity, and most recently to places that (slowly) reflect a newer understanding of the role of humans in nature. In this newer view, work in nature is not always seen as "unnatural," and it is understood that human management and use have long been integral parts of most ecosystems, while management based on the removal of humans (except as visitors) is actually a far more modern creation.[90] In recognizing that we are now in an era where human activity has far-reaching influence on almost every aspect of the natural world, many scientists are now arguing that looking backwards at some imagined "pristine nature," and trying to recreate it, is a fool's errand.[91]

FIGURE 5. Pierce Ranch School, Point Reyes National Seashore, 2013. Photograph by author.

CHAPTER 2

Public Parks from Private Lands

The California coastline just north of San Francisco is a surprisingly rugged and rural landscape, despite being only a short distance from the densely urbanized Bay Area. The fabled Highway 1 twists along the steep hillsides of the Marin headlands north of the Golden Gate, gradually leading through small seaside towns flanked by rolling hills dotted with cattle. Since its earliest settlement by non-Native residents—first Mexican rancheros in the 1830s, followed by northeastern dairiers in the 1850s—West Marin has been a place of pastoral beauty, an unexpected meeting of the wild Pacific Ocean with wide expanses of green pastures and white Victorian ranches. Many of the families working the land have roots that go back four, five, or six generations, stemming from several groups of European immigrants who together form the region's distinctive character.

The Point Reyes peninsula, a triangle-shaped wedge of land on West Marin's western edge, forms a dramatic core to this quiet agricultural landscape. The Point is partly cut off from the rest of West Marin by the north-south divide of the San Andreas Fault, forming Bolinas Lagoon at the southern end, and the long slender Tomales Bay which reaches down the Point's northeastern side; only the higher land of the Olema Valley, running from Bolinas up to Point Reyes Station, prevents the Point from becoming an island. A dark evergreen forest of Bishop pine and Douglas fir covers the spine of Inverness Ridge, running up the eastern side of the triangle, contrasting with the pale greens, golds, and grays of the more open hillsides that tumble down its western side to the ocean's edge. A typical day might bring bright sunshine in the morning and turn to dense fog and howling ocean winds by afternoon, making for a damp, cool, and somewhat harsh climate much of the year. But the frequent coastal fog also brings moisture year-round, allowing the grasslands to stay green much longer than in inland areas.

Although Point Reyes did not officially become a national seashore until 1962, it was first studied as a potential park location in the 1930s, as part of the NPS's Seashore Recreation Survey. Unlike the first generation of national parks, Point Reyes was not remotely "empty." Like most of the rest of California, it had been inhabited for centuries by Native peoples, in this case several groups of coastal Miwok bands.[1] Recent archeological studies suggest that the landscape experienced extensive burning and other forms of indigenous vegetation management over at least the last two to three thousand years.[2] Since approximately the 1830s, when Mexican rancheros started settling this part of California—drawn to the peninsula's lush grasslands, created and maintained by Miwok land management practices—the peninsula has been used extensively for raising cattle. Nearly two centuries of ranching has had a profound impact on the landscape, resulting in distinctive patterns of land use and meaning for local inhabitants. The strength of the dairy industry kept the land open and relatively undeveloped, making it an attractive location for a national park unit.[3]

The rancheros who settled on massive land grants from the Mexican government in the 1830s raised large herds of cattle for the hide and tallow industry. After 1850, when California became part of the United States, several settlers originally from New England established dairies on the peninsula. The cool moist climate of Point Reyes provided ideal conditions for dairy cows: plenty of grass with a long growing season and abundant fresh water supplies. Litigation over the land title to the Mexican land grants resulted in ownership of nearly the entire peninsula falling into the hands of a legal firm, Shafter, Shafter, Park and Heydenfeldt. The brothers James and Oscar Shafter bought out the interests of the other two partners. In 1865, with the help of Oscar's son-in-law Charles Howard, the men developed a network of thirty-two tenant-run dairies and cattle ranches, most named for letters of the alphabet, transforming the Point into an ordered, fenced industry. They sold one ranch site, covering the tip of Tomales Point, to a friend named Solomon Pierce (who also established a dairy), but otherwise, the Shafters controlled the entire peninsula, except for the very southern end near Bolinas.

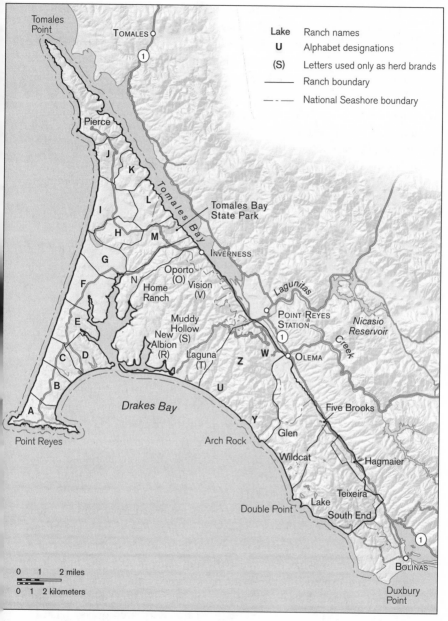

Tomales Point

TOMALES

Pierce

J

K

L

I

H

M

Tomales Bay State Park

INVERNESS

G

N

Oporto (O)

Home Ranch

Vision (V)

F

Muddy Hollow

New Albion (R)

(S)

Laguna (T)

Lagunitas

POINT REYES STATION

Nicasio Reservoir

E

C

D

W

OLEMA

Creek

B

Z

A

U

Drakes Bay

Point Reyes

Y

Five Brooks

Arch Rock

Glen

Wildcat

Hagmaier

Teixeira

Double Point

Lake

South End

BOLINAS

0   1   2 miles

0   1   2 kilometers

Duxbury Point

MAP 2. Map of the Shafter-era "Alphabet" Ranches, based on historical research and map by Dewey Livingston. Courtesy of Ben Pease.

The Shafters' ranch system, along with several other independent dairies in the area, produced record yields of the highest quality butter and cheese throughout the late 19th century. In 1872, Marin County produced more butter than any other county in California; most of the four million pounds produced came from Point Reyes, shipped on schooners to the rapidly growing city of San Francisco.[4] Because of the Point Reyes dairies' excellent reputation—in part based in the landscape itself, as the fog-watered pasture led directly to better tasting butter—their trademark star-within-a-circle stamp was often counterfeited by unscrupulous competitors. Several families who started dairying at Point Reyes later moved on to other parts of the state, contributing to the establishment of California as a leader in dairy production. Innovations in dairy technology and technique developed on the peninsula would eventually be adopted nationwide.

The Shafters developed the dairy ranches themselves, creating a distinctive pattern of architecture and ranch layout. The ranches varied in size from eight hundred to two thousand acres, and each required access to freshwater and lands open enough for grazing. Whitewashed ranch houses, dairy buildings, barns and outbuildings, and fencing systems were constructed roughly between 1860 and 1880. The Shafters selected tenants whom they believed to be steady and reliable, mostly Irish, Italian-speaking Swiss, and Azorean Portuguese immigrant families. Tenants rented the cows, buildings, and land, but provided their own furnishings, tools and equipment, and other livestock, mainly horses and pigs. Within a few decades almost all of the residents planted rows of trees, either blue gum eucalyptus (*Eucalyptus globulus*) or Monterey cypress (*Cupressus macrocarpa*), to shield the ranch sites somewhat from the ever-present coastal wind and fog. They also planted home gardens and small agricultural plots.

Almost two centuries later, these ranching practices continue to mark the landscape of Point Reyes. The physical presence of commercial agricultural activities is a large part of why Point Reyes strikes many of its visitors as an unusual national park unit. It can often be difficult for visitors to distinguish between the park-owned land on the west side of Tomales Bay and the private lands on the eastern shore; both are dotted with dairies and beef ranches, with all the associated grassy pastures, fences, farm buildings, pickup trucks, and the like. The presence of so many obviously working people sits uneasily within a national parks ideology that has traditionally emphasized empty landscapes, a wilderness aesthetic, and recreation.

FIGURE 6. Kehoe (J) Ranch buildings and road, Point Reyes National Seashore, 2000. Photograph by author.

Given this history of intensive land management and use, how did Point Reyes end up becoming a park? The answer lies in the history of the National Park Service itself, after its establishment in 1916, when its leaders activity sought to make the agency truly *national* by creating some parks along the East Coast. Yet this presented a challenge: the big western parks had been carved out of existing federal holdings, each with its own authorizing statute from Congress. In addition, in 1906 Congress passed the Antiquities Act, intended to allow protection of archeological sites on federal lands by presidential executive order, which was quickly employed by Teddy Roosevelt's administration as a direct mechanism for creating parks, setting aside vast landscapes like the Grand Canyon as national monuments.[5] Yet this still could only be used to establish parks in western states, as the East contains very little federally owned land.

The solution, as this chapter will explore, was to create new eastern parks out of *private* lands. Congress in the first half of the twentieth century was skeptical of spending money on parklands, so the fledgling NPS forged alliances with state governments and wealthy capitalists to acquire private parcels, and then donate them to the new agency to manage. In some cases, like the donation of numerous estates on Mount

Desert Island to create Acadia National Park in Maine, the private owners were willing to give up their claims to the land (although many retained private enclaves). In many other cases, however, state condemnation or private acquisition took over areas with a large number of small-scale landowners. In order to fit within the prevailing national park ideal, however, these peopled landscapes had to be emptied of their residents and reinvented as pristine wilderness.

The vision of parks as devoid of people makes an exception for two categories of people: tourists and park managers. Tourism necessarily involves human presence in the parks. The difference is between *residents,* who are seen as antithetical to the park ideal, and *visitors,* who justify the parks' existence. Even though the density of tourists in certain areas of parks may be higher on average than that of many towns, official policies still generally welcome visitors in parks, while they tend to discourage, if not completely remove, residents. Environmentalist rhetoric has encouraged this distinction, particularly in the last few decades, making nature "by definition a place where leisured humans come only to visit and not to work, stay, or live."[6] The romanticized conception of nature as uninhabited sets recreational tourists against Natives and the rural people who actually earn a living from the land, devaluing their connections to (and knowledge of) the land through work and home. It can also cause these residents to be sentimentalized so much that if they do anything "unprimitive, modern, and unnatural," they fall from grace.

One of the oddest qualities of this idealization of parks as uninhabited is the extent to which they still *are* inhabited, though by employees of the NPS rather than locals. Park managers work on a daily basis in this landscape, and quite a few of them reside within its boundaries as well, yet their work is overlooked, or treated as both inevitable and largely invisible to the public's eye. In a number of cases, long-time residents have been displaced from their homes only to be replaced with NPS employees—public employees rather than private entrepreneurs, but otherwise a different but parallel form of living and working in the landscape. Yet their presence in the landscape is not questioned in the same way.

The apparent out-of-place-ness of private residents in park landscapes stems in large part from our common understanding of the meaning of public land ownership. Some public lands have very clearly defined private uses, such as national forests that have leased out portions of the forest to timber companies, or BLM lands that contain grazing permits or mineral claims. Many environmental writers and advocates tend to contrast parks primarily used for recreation with these landscapes of extractive use, but recreation *is* itself a use—for personal benefit rather than private profit, mostly, but still a use—with its own impacts and implications. Those effects often become "naturalized" as an inherent part of the landscape itself. Public access for recreation is a presumed right that comes along with government ownership, but it brings along with it the assumption that residents "do not belong." On the flip side are private lands with public interests, for example, a family-owned ranch with a conservation or agricultural easement (representing public dollars in terms of foregone tax revenue) or a privately owned historic building designated as a national landmark. These scenarios are two poles of a continuum of public and private ownership, a concept requiring further explanation.

WHAT DOES PUBLIC OWNERSHIP REALLY MEAN?

As the previous chapter noted, popular rhetoric surrounding parks often refers to them as belonging to "the public," as if each U.S. citizen were a co-owner in some collective way. This construction, by conflating the sense of belonging to a common national heritage with actual ownership in common, conveys an inaccurate understanding of the property rights involved. The concept of "property" is best described as referring to the rights and obligations an owner has in relation to the thing owned, rather than to the thing itself. Property can refer to tangible objects, like a car or a piece of land, or it can refer to more intangible products, such as the ideas in a book or the words of a song. Property can be categorized as private, where the individual owner has the right to exclusive use, including the ability to exclude others; or it may be held in common, where the rights to an area or a thing are shared (although not necessarily equally) by a group, in which case the individual members have the right to not be excluded.[7] In either category, individuals hold the rights. With private property, however, the owner can be an "artificial individual," such as a corporation: Disneyland, for instance, is owned by the Walt Disney Company, not Walt Disney's heirs. Government-owned property is actually a case of corporate ownership in which the land or building in question is owned by a particular administrative department or agency—in the case of the national parks, the owner is the National Park Service, a division of the Department of the Interior.

In early U.S. history, private property represented a political ideal based on an understanding of rights that stemmed from the writings of

John Locke. Lockean theory argued that labor creates property rights—in Locke's view, as a man worked a piece of land, his self-ownership of his own body transferred to the land, as a product of his work—and that governments were formed primarily to protect those claims to property.[8] The founding fathers structured the U.S. government according to Lockean ideas. The Declaration of Independence's promise of "life, liberty, and the pursuit of happiness" essentially restated a phrase from Locke (except he promised property rather than happiness) and granted voting rights (liberty) only to the subset of white male citizens who owned land. Legal scholar Eric Freyfogle describes the early U.S. view of private property as "a right of opportunity, a right to gain land."[9] For the country's first century or so, the federal government held very limited authority to regulate private property, and in fact much nineteenth-century federal legislation was preoccupied with moving government property, such as the lands obtained through the Louisiana Purchase in 1803 or seized from Mexico in 1848 through the Treaty of Guadalupe Hidalgo, into private ownership. The most convincing way to claim land as private property was to work it, putting it to productive use. For example, the 1862 Homestead Act required claimants to live and work on their quarter-section claims for five years; if they managed to hang on that long, the land was granted to them essentially for free.[10]

After the Civil War, this Lockean view of property was complicated by a new metaphor for property, that of a bundle of sticks. In this metaphor, each stick represents a specific, separable right, such as right to use, alter, rent, sell, or destroy the property in question. This new legal conception made it easier for parties to exchange and transfer fragmented rights of ownership among different actors or owners, while retaining the intuitive link between the use of property and its ownership.[11] For instance, the conservation easement, a modern instrument, allows private landowners to retain all of the basic rights of private property with the critical exception of the right to develop it; the development "stick" is no longer part of the bundle, even if the property changes hands. In the same way, even as the federal government began to retain portions of its federal lands, first in the early national parks, then reserving forest lands in the 1890s, it also continued to allow some private extractive uses of those landscapes, such as grazing and timber-cutting. The control of various resources moreover became fragmented between separate federal agencies.[12]

Because recreation represents a use of the land for enjoyment of the visiting public, rather than for individual or corporate profit, public access for recreation tends to become conflated with a collective sense of ownership. While the rhetoric of parks suggests that they are a form of common property, they are not; noncitizens have identical rights of access to national parks as do citizens, even though they do not have voting rights in our governmental system, and parks can be closed to public entry, as they were during the federal shutdowns of 1995–96 and 2013. Park management is determined and conducted by NPS employees, not by citizen co-owners. Visitors to parks have the same rights that they have to any other form of government property, be it a post office or a federal office building; we can enter when the building is open for business, but we cannot determine what those hours will be, nor can we use the building ourselves for some other use—we have to abide by the rules established by whichever federal agency owns the building. In this way, visitors to Yosemite National Park have rights identical to those visiting Disneyland, a privately owned park. Anyone who pays the entry fee can walk around and enjoy the contents of the park according to particular rules that are set and enforced by each place's owner.

The general public does, however, have one additional right to the public parks. In addition to the right of access, since the 1970s the public generally has had a right to comment on federal land management plans and projects, as established by the legal requirements of the National Environmental Policy Act (NEPA) and other land management laws. But again, this role is more limited than it might initially seem. Agencies are required to consider public input, again from citizens and noncitizens alike, but they are not required to follow it. It is moreover helpful to see disputes about land management as arguments about the right to be "heard" politically by the federal agency in question. Some individuals attempt to use their political power or access to exert greater control than others over outcomes on the landscape. The public's right to comment is, in essence, an *indirect* form of property claim that seeks to control and shape the land's use without the inconvenient financial and managerial responsibilities of direct ownership.

These misunderstandings about what actually constitutes "public ownership" exacerbate the impression that permanent residents have no place in parks. When combined with notions of democracy, public ownership implies that every member of the public should somehow have *equal* access to publicly owned land—hence private residents or land uses don't seem to fit, because they provide a small group of people with different access than the rest of the public. Yet many federal properties are restricted in ways that create unequal access—for instance, military bases are only accessible by military personnel or those with special permission (or sometimes not at all!). Similarly, private individuals or companies can hold exclusive grazing or timber rights on Forest Service or BLM lands. The difference with national parks is in the *rhetoric* about ownership, and the types of use allowed, rather than the actual rights involved—parks may *represent* our democratic system, but they are not owned or managed democratically by the public at large.

INSTITUTIONALIZATION OF PARK IDEALS

Park designer Frederick Law Olmsted first suggested that the U.S. government had a *duty* to protect unusual scenery in 1865, yet by the turn of the century the national parks were more of an accumulation than any kind of an organized system or government program.[13] By 1900 there were only five national parks; by the time of the establishment of the NPS in 1916, their number had increased to fourteen, all reserved from land already owned by the federal government. These early parks were intended to be self-supporting, and were given little or no budget from Congress for either staff or maintenance; one House member, in response to Teddy Roosevelt's early conservation programs, pronounced, "not one cent for scenery!"[14] Army troops were deployed to protect Yosemite and Yellowstone early on, while other sites had little or no management, and certainly no coordination from one park to another.[15] Yet in the first two decades of the twentieth century, the government gradually took on more responsibility for preserving and managing parks, eventually bringing the existing parks under the management of a single agency, the National Park Service. As a result, the young agency was infused with the symbolic values invested in the early national parks as its own ideological approach to management.

The move toward more systematically organizing and administering parks coincides with the Progressive Era, roughly 1890–1920, a period exemplified by the belief that public policy decisions should be based primarily on scientific principles of management and efficiency.[16] This approach was best embodied by Gifford Pinchot, first chief of the Forest Service after it was established in 1905; he espoused a philosophy of conservation that aimed to integrate natural resource management so as to minimize waste and to ensure benefit to the public in both the present and the future. The Progressive conservation mentality created an additional source of pressure to formalize the parks' management, as it held that park preservation was an outdated and inefficient method of managing public lands.[17] Without a government bureau assigned to manage the parks, parks advocates worried that parks would be taken over by utilitarian interests.[18] Parks were thought of as vulnerable to congressional politics, as individual senators might resent restrictions on local economic land uses.

By 1915, Interior Secretary Franklin Lane had hired Stephen T. Mather, a Chicago businessman, to take charge of the federal parks, along with his young assistant Horace Albright.[19] Together Mather and Albright produced an extensive publicity campaign for a proposed bill to create a parks bureau. Mather opportunistically realized that the war in Europe, in preventing Americans from traveling abroad, represented a rare chance for the national parks to "expand their operations and win new friends."[20] Mather and Albright were joined in their publicity campaign not only by parks' traditional allies, the railroad companies, but also by the nascent automobile industry, both of which hoped to spark increased domestic tourism. The media blitz proved successful, and in 1916 a bill finally passed creating the National Park Service, with Mather as its first director.[21]

There is a definite mismatch between much of the rhetoric surrounding the establishment of the NPS and the actual substance of political debate leading up to the Act of 1916. The rhetorical language is most often associated with wilderness advocate John Muir, whose preservationist perspective continued the romantic view of nature as a sacred temple, everlasting and unchanging, and championed parks as the best way to protect nature from the ravages of civilization. The writings of Olmsted and his son, Frederick Law Olmsted Jr., exemplify this attitude, emphasizing the importance of the preservation of "scenery of those primeval types which are in most parts of the world rapidly vanishing for all eternity before the increased thoroughness of the economic use of the land. In the National Parks direct economic returns, if any, are properly the by-products; and even rapidity and efficiency in making them accessible to the people, although of great importance, are wholly secondary to the *one dominant purpose of preserving essential aesthetic qualities of their scenery unimpaired* as a heritage to the infinite numbers of generations to come."[22]

However, in the utilitarian conservationism of the era, parks needed some kind of practical justification for their existence. Most of the early parks' landscapes did not contain extensive resources such as timber or grazing opportunities, as they protected primarily high-altitude mountaintops of "rocks and ice." In an era focused on economic efficiency, the heritage-only focus of the Olmsteds could not gather sufficient political backing for the parks' administrative support. Historian Donald Swain writes that "the political facts of life in 1915 and 1916 simply demanded that the parks be 'used.' Unless and until the American people started flocking to the national park reservations, Congress would refuse to appropriate adequate funds for the administration and protection of the parks."[23]

Tourism, as the only "dignified" way to promote parks, provided that economic justification. The potential economic profits from tourism prompted commercial interests to use their political influence to support park protection and expansion.[24] This tension between idealistic rhetoric and economic motivation remains written into the parks statute, creating the NPS's "dual mandate" of preservation of "unimpaired" scenery together with encouragement of public use, while not ever defining the actual meaning of unimpaired.[25] The wording of this mandate is attributed to Olmsted Jr., who, as a landscape architect, saw "no inherent contradiction in preserving a place through its thoughtful development as a park."[26] In this way, the romanticized image of what a national park ought to be—nationally significant areas of unchanging nature—screens the political reality of their support and maintenance, causing it to appear that parks are not "used" at all, despite their high visitation numbers.

Progressive Era governance relied on the expertise of technically trained, scientific professionals. Such experts filled the ranks of the increasing number of newly established federal bureaucracies, each created to serve and regulate specific interest groups.[27] At the NPS, the self-appointed scientific experts arrived with experience as landscape architects and engineers, with the effect that they adopted a standardized approach to design and infrastructure development throughout the national parks. Their work shaped both the physical landscapes and the cultural expectations for the parks. The landscapes preserved and created by the NPS continue to perpetuate and reinforce this idealization of what parks "ought" to look like.

In all the discussions of preservation in the parks at the time of NPS establishment, the importance of protecting scenic beauty via careful park development remained paramount. Here, the landscape architects reigned supreme.[28] NPS advocate Horace McFarland suggested that it was the job of landscape architects to educate the public, and that "professional standards, *not politics* . . . should determine the future of the parks."[29] NPS historian Ethan Carr describes the expertise of the landscape architects as "analogous (in a more artistic vein) to the scientific expertise by Pinchot's foresters."[30] Along with engineers and park rangers, they formed the core of an emerging "leadership culture" in the NPS, all of whom considered science to be the apolitical basis for proper efficient management.[31]

The creation of the NPS also coincided with a period of growing demand for outdoor recreation in American life. Changes in labor practices and the length of the workweek made for more leisure time among the middle class, and vacations became something considered a necessity for white-collar workers. The rapid rise of automobility made people far more mobile and vastly expanded visitation to the national parks: In 1914, parks received 240,000 visitors; by 1926, the total had jumped tenfold to 2,315,000 visitors, mostly due to increased travel by automobile.[32] This new vacationing middle class now formed "the public" for the national parks, and it shared an aesthetic appreciation that stemmed from the earlier iconic association of parks with nationalism.[33] Park advocates argued that visiting their natural wonders would not only encourage the health and efficiency of the middle class, but also maintain "individual patriotism and federal solidarity" by emphasizing nationally held values.[34] These values were designed into the national park landscapes, and thus constantly recreated and reinforced the ideological association of unchanging natural scenery with national pride.

At a time when all of the major parks were located in the West yet close to 90 percent of the American population still lived in the Midwest and East, ensuring access quickly became one of the NPS's primary responsibilities. It was imperative that visitors were able to reach the parks with relative ease, and, once there, to find all the amenities they needed to make their stay comfortable.[35] Thus during the 1920s the NPS developed the parks extensively, adding roads, hotels, trails, and other amenities to allow for greater access. Mather insisted that development harmonize with, rather than impair, the scenic landscapes, so most tourist facilities were built in "rustic" architectural styles. *Not building* them was never really considered an option.[36]

By the mid-1920s, Park Service landscape architects and engineers had established a characteristic and original style of development that responded to the practical necessities of park facilities, while remaining rooted in the earlier style of American landscape park design. These efforts manipulated the landscape in very specific ways to guide tourists to particular places and views. The effect, as described by Carr, was to "choreograph visitors' movements and define the pace and sequence of much of their experience." At the same time, this choreography remained almost invisible to those being directed, so that the parks took on a "consistent appearance, character, and level of convenience that most visitors have since come to associate, almost unconsciously, with their experience of park scenery, wildlife, and wilderness."[37] Park design continued to reinforce the romanticized park ideal by creating a standardized appearance for the national parks, and naturalizing the agency's manipulation of the landscape as a necessity for proper preservation.

The NPS quickly came to regard itself as the agency in charge of objectively and apolitically institutionalizing the national park idea as a symbol of American values—values considered to be constant across the country. This early identification with objective, scientific expertise continues to shape the agency's management philosophy. The agency persists in taking a standardized, nationally consistent approach to park management, even when faced with the relatively subjective and potentially polyvocal task of preserving historic and cultural heritage—this despite the fact that parks are generally designated as such because of some unique quality, such as a distinctive geologic feature or important historical event. Tensions between national standards and local character inevitably result. Extensive design and development of the parks, with the aim of providing greater public access, further shape these protected areas into expressions of cultural idealizations of what a park "ought" to be. All of these trends have tended to continue shaping the parks into relatively static national icons, emphasizing pure and depopulated natural scenery.

RESIDENTS AS INCOMPATIBLE WITH THE PARK IDEAL

Throughout U.S. history, Native Americans have frequently been cast as being part of the past, and actual living Native communities have been ignored, have had their disappearance predicted, or have been eradicated directly.[38] Recent books by Mark Spence, Theodore Catton, and Philip Burnham explore the complex relationship between Indians and national parks, which has usually resulted in the removal of Natives from their ancestral lands. Burnham points out the irony of George Caitlin's original vision in the 1830s of "a nation's park" that included Indians as part of the protected scene, as instead Native Americans were more commonly "considered a nemesis of our national parks."[39] Spence details how the removal of Native peoples from many landscapes created a "dual 'island' system of nature preserves and Indian reservations."[40] He specifically argues that "uninhabited wilderness had to be created before it could be preserved, and this type of landscape became reified in the first national parks."[41]

John Muir, the best-known preservationist and national park advocate of the late 1800s, exemplified the general attitude toward Indians at the time. In his early essays on the national parks he erased Native people from the scene, proclaiming that the Indians were all "dead or civilized into useless innocence," leaving early parks safe for tourists.[42] In Alaska, where he encountered too many local Natives to overlook, he suggested that they were uncivilized, superstitious, and ignorant, and framed them as exotic intruders in the landscape—ignoring the fact that he himself was much more of an exotic.[43] Racism combined with idealization of the natural landscape to make the presence of Indians completely incompatible with the national parks, necessitating their removal.[44]

Part of this bias against Indians living in parks was also a prevailing sense that they were not adequately appreciative of the majestic scenery. Lafayette Bunnell, one of the first whites to explore Yosemite Valley as part of a military campaign, referred to the resident Indians' reverence of Yosemite Valley as "demonism" rather than the awe that he and his fellow soldiers felt. In his account of the experience he wrote, "In none of their objections made to the abandonment of their home, was there anything said to indicate any appreciation of the scenery."[45] Because they did not show "proper" aesthetic enjoyment of the view, but instead focused on the plentiful food and shelter of their homeland, they did not deserve to live there.[46]

Indians were not the only people considered incompatible with national parks, however; all private property holdings and residents seemed problematic. The early experience of Niagara, described in chapter 1, suggested that small-scale private development brought unwanted change, cluttering up the view and counteracting its symbolic power. Agricultural families in particular would later be criticized for being overly focused on the productivity of their land, rather than its scenic qualities, just as the Native Americans had been. In both cases, parks advocates and managers have assumed that the best solution is to remove the variable of private ownership and/or use of the land, so that the Park Service would maintain control over the appearance and meaning of these public spaces.

The early NPS continued the bias against private ownership of any parklands as part of its emphasis on efficient management of the parks. Interior Secretary Lane released a policy statement in 1918 in the form of a letter to Mather, outlining many of the agency's founding values and assumptions. Its language echoed the agency's Organic Act of 1916, and has come to represent the "basic creed" of the NPS ever since.[47]

Among other management principles, the Lane letter specified that "all of the private holdings in the national parks should be eliminated as far as practicable in the course of time either through congressional appropriation or by donations."[48] According to NPS historian Richard Sellars, inholdings were an "anathema to the Service," with potentially unsightly uses representing the "most pervasive threats of inappropriate development."[49] Private ownership was considered incompatible with the efficient management of scenic wonders and contrary to the national interest in which the parks were created. Without complete ownership, the NPS would be less able to implement its own designs and management strategies, and feared the repeat of another Niagara-like destruction of the nation's heritage.

The creation of the NPS institutionalized the bias against including privately owned parcels within park boundaries, but—at least at first—the agency lacked a means to carry it out. Congress initially refused to appropriate money to actually *create* parks from private lands; new parks had to be carved out of the public domain or other federal lands (often transferred from national forests), or donated by private individuals or states.[50] In 1935, Congress first granted the NPS funds to purchase private lands held within a park boundary, but the NPS did not have the power to condemn private property and compel its sale to the government. At the time, only state governments clearly held condemnation authority, and parks were not yet considered a valid "public purpose" to warrant this kind of forced acquisition of private land. Without congressional support, the NPS relied heavily in its first decade on private donations, several originating with Mather himself. In Yosemite, for example, Mather financed the purchase of a privately held road in the park's back country.[51] Through this and similar actions, the NPS, despite being a federal agency, remained more of a government partnership with private interests, with a majority of its funding coming from wealthy benefactors who could then wield extended influence over the direction its preservation efforts took.[52] This institutional arrangement allowed an elite aesthetic of landscape preservation to continue dominating the parks, one that specifically did not include local residents.

At the same time that Congress began establishing additional parks for the newly created NPS to manage, it also turned to a new source for the lands. Instead of public domain lands, the new parks were increasingly created mostly from privately owned lands. If the park system were to become truly "national," it would need parks in the East as well as the West. The problem, of course, was that most of the land in the

East was inhabited, and not by Indians, and dealing with large numbers of non-Native residents posed a new problem. The agency's response remained consistent with its stated position that private ownership, particularly that of small holders, was incompatible with park management. As a result, the NPS's policy toward all residents, no matter how long their communities had been established, was to remove them. The parks, emptied of habitation except for a few wealthy inholders who were generally allowed to remain in their enclaves, could then be transformed into scenic natural areas devoid of history or use and then prepared (often in partnership with private concessionaires) for the expected influx of tourists.

The first park to establish this trend was Sieur de Monts National Monument on the coast of Maine, established in 1916, then incorporated into Lafayette National Park in 1919 and later renamed as Acadia National Park in 1929. As early as 1901, portions of Mount Desert Island had been purchased by "gentlemen of means," summer residents who considered the island to be threatened by uncontrolled development for tourists; later contributions from John D. Rockefeller Jr. and others added to the protected area.[53] The privately purchased lands were then donated to the federal government for the NPS to manage. Despite the fact that the original impetus for preserving the island was to prevent ruination of its natural beauty through overdevelopment, Rockefeller worked closely with the NPS to "build and maintain a road system to open up the lovely vistas of Acadia to viewers."[54] Although Rockefeller and the other wealthy donors viewed excessive private development as distasteful, they considered allowing greater public access through government ownership and development a laudable goal—even while several of the wealthy families retained private enclaves for their own summer homes.

A similar situation unfolded in Wyoming when Grand Teton National Park was first established in 1929, carved out of existing Forest Service holdings. At first it was relatively small, covering only the tops of the mountain range.[55] Yet efforts were already underway to augment it: three years before the park's establishment by Congress, in 1926, then-Yellowstone Superintendent Horace Albright developed a plan with Rockefeller to purchase the entire northern portion of the privately owned Jackson Hole valley with the express purpose of donating the lands to the NPS. Both men were concerned about uncontrolled commercialism encroaching on the view. Albright, moreover, believed that it was the destiny of the Tetons to become part of the National Park

System; he considered their inclusion to be one of the highest priorities of his job.[56] For his part, Rockefeller was particularly offended by the apparent "immorality" of some local developments, especially dance-halls, in front of such a sublime natural landscape.[57] While their concerns for the Tetons were not identical, they were complementary. Standing in their way, however, were many of the local residents, mostly ranchers, who were interested in having their "pioneer" way of life protected, but specifically *not* by the NPS, which they did not trust.

This distrust explains why Rockefeller embarked on a plan to buy the valley lands using a front organization, the Snake River Land Company. Both Albright and Rockefeller's connections with the company were kept secret—even from the man they enlisted as purchasing agent, Robert E. Miller, who was a local rancher and banker as well as a former supervisor of the Teton National Forest, and was "bitterly opposed to Albright and the National Park Service designs on the Teton Range-Jackson Hole area." Miller apparently was suspicious the lands he was purchasing were somehow involved with NPS plans for expansion into the area, but when he asked the Land Company about it, he "received assurance that the land purchase program was not related to the park extension effort."[58] By the time Albright and Rockefeller's involvement was finally revealed in 1930, the land company had purchased over twenty-five thousand acres, provoking local outcries of trickery and prompting a congressional investigation into possible wrongdoing. Eventually Rockefeller donated a total of thirty-four thousand acres to the NPS to create an unpopulated vista framing the majestic mountains in the distance.[59] Jackson Hole went from being a working landscape filled with human history and utility to a static monument of natural scenery.[60]

Shenandoah National Park in Virginia and Great Smoky Mountains National Park in North Carolina and Tennessee also followed this model of establishment via donated lands, but relied mostly on state condemnation of private lands for acquisition (although private donations also provided a significant portion of the funding). When authorized in 1926, the proposed boundaries for each park encircled historic mountain communities of small farms. Yet for the most part, these settlements were not seen as obstacles to creating the parks; in fact, in the early planning stages of the parks they were mostly overlooked. Promoters of Shenandoah, primarily politicians and business leaders, described it as "primeval wilderness," despite its centuries-long history of farms, plantations, and logging operations, and this pristine imagery proved too powerful to be successfully challenged.[61] The State of Virginia even passed a law allowing blanket condemnation suits to be brought against all of the landowners in the eight counties affected, rather than addressing the cases individually.[62] A similar effort was made in the Smokies, which were described as "almost untouched wilderness."[63] Private individuals challenged the legality of the use of eminent domain in both parks, but failed in each case.[64] A total of 465 families in Shenandoah and 110 in the Smokies had their property condemned, mostly in the early 1930s. Once vacated, most of the homes were razed or burned to prevent others from moving in. Early management involved an emphasis on "a quick 'return to nature' while also cleaning up the landscape and preparing to receive visitors in large numbers."[65]

Not coincidentally, the southern mountain people removed from the Smokies were stereotyped by more mainstream cultural influences in ways very similar to the Indians before them. Starting in the late 1800s, southern mountaineers were typically either portrayed as backward hillbillies "living lives of stark brutality and desperation," or idealized as the last vestige of "pure" Anglo-Saxon Americans who had more in common with early colonial communities than the modern world.[66] These characterizations allowed park promoters to justify the residents' removal from the park areas as actually being in their best interest, whether the residents wanted to leave or not. While some landowners may have welcomed the purchase of their lands at a time when there were few other buyers, there is little evidence that the majority of these people wished to leave. The residents lost their homes and livelihoods in the midst of the Great Depression, when jobs and productive farms were scarce; the resulting dispersal destroyed the fabric of communities that had existed for generations.[67] Tenant farmers, who made up a substantial portion of the residents evicted from both parks, were particularly hard-hit, as they received no compensation or assistance since they did not own the lands on which they lived and worked.

One particularly troubling aspect of the creation of these parks was the apparent deception perpetrated by the state authorities involved in the process of removing people from some of them, as had been done previously by Rockefeller at Teton. In the case of the Smokies, locals expressed concerns about being included inside the boundary soon after the park was originally proposed in the early 1920s, and in fact several communities on its fringes successfully lobbied to be excluded in the final proposal. But the community of Cades Cove remained inside the park. Numerous politicians, including the governor, a senator, and

Knoxville organizers explicitly promised that the residents would not lose their homes.[68] Despite these assurances, the condemnation bill passed by the state of Tennessee in 1927 "specifically gave the newly created Park Commission the power to seize homes within the proposed boundaries by right of eminent domain."[69] Had the Cove residents brought political pressure to bear before 1927, they might well have altered the outcome, but they had no way of knowing it was necessary. These cases and others form a larger pattern of a somewhat cavalier disregard for, and often dishonesty toward, local concerns in the process of creating parks at this time.

A different approach to residents, one slightly more tolerant of their continued presence, characterized the creation of the Blue Ridge Parkway in Virginia and North Carolina.[70] Connecting Shenandoah and the Smokies, the scenic roadway was conceived of as a "make-work" project in the mid-1930s. Forming a narrow boulevard winding nearly five hundred miles along the crest of the Appalachians, the Parkway was engineered and constructed by the NPS using local unemployed labor under the provisions of the National Recovery Act. Again, the mountain people who occupied areas designated for the park were described as isolated, poor, and leading "anachronistic" lives. The Parkway was conceived of as a way to provide jobs, better transportation and a boost to local economies "without appearing to be charity."[71] In deliberate contrast to Shenandoah, the NPS wanted to retain the traditional residents and their uses of the land as "physical evidence of a pioneer way of life" that would add to the scenic qualities of the roadway.[72] Rather than purchasing all of the land required for the road, the federal government wanted property owners to donate a two-hundred-foot strip of their property to provide the right-of-way, on the theory that doing so would enhance their own property values. In response, both Virginia and North Carolina passed laws allowing state officials to procure the rights-of-way with little consultation or agreement from the owners.[73]

Beyond the right-of-way, the land still belonged to the private owners, but the NPS requested scenic easements to ensure protection of the natural setting. Yet few of the property owners at the time understood the term *scenic easement,* particularly the degree of control it granted the federal government. Owners were expected to continue their customary uses of the land, but with extensive rules about what building and farming practices could take place, plus prohibitions against "unsightly or offensive material, such as sawdust, ashes, trash, or junk."[74] Landowners were encouraged to make their land as scenic as possible, through the practice of "scientific farming," construction of rail fences (the NPS provided free rails), or shocking grain crops into neat stacks, even if these were not their traditional practices. This orchestration of land use appeared "so 'natural' that the passerby little dreams of the great amount of effort exerted to make that scene possible and to keep it, or its equivalent, available for future generations."[75] While policies for creating the Parkway did not remove the residents outright, as was the case in most of the other parks of this period, they were still not sensitive to the local culture. Instead, the NPS demanded that the residents take on a more aesthetically pleasing lifestyle, which was then "naturalized" into the scene, eradicating the actual historical use practices of the area. In this way, the values of the agency were physically inscribed in the landscape, despite its lack of direct ownership.

With the financial support of elite individuals, the NPS developed what became some of its most popular parks during this early period. These stories show the influence and personal visions of the wealthy steering the national parks, while the much poorer rural residents were either removed or had their ways of life heavily controlled. Despite a long history of having lived and worked in these places, these people and their traditional practices were considered incompatible with the parks' purposes and management. In actuality, they were incompatible only with the underlying values of the agency. As Cronon points out, "Only people whose relation to the land was already alienated could hold up wilderness as a model for human life in nature, for the romantic ideology of wilderness leaves precisely nowhere for human beings actually to make their living on the land."[76]

A BROADER VIEW OF HERITAGE

The 1930s represented a redefining decade for the NPS. Since its inception, the agency's main goal had been to "preserve the indigenous flora and fauna and primitive feeling of those places so that they would provide glimpses of North America as it looked when the white man first explored it . . . [as] windows to America's past, keepsakes of a once-virgin land."[77] Now, primarily through the interest of its second director, Horace Albright, the NPS took on a new role, expanding its scope beyond just protecting scenic natural wonders and providing tourist opportunities to embracing the remnants of the nation's historic past as well. NPS historian William Everhart suggests that, although almost no other conservationists at the time had any interest in historic preservation, Albright believed that "American heritage was made up in equal parts of the unique grandeur of the geography and the heroic deeds of the people, it was just as important to preserve historic sites as to set aside places of natural beauty. Both were essential components of true conservation."[78]

Previously, other than protecting memorial battlefields, almost all U.S. historic preservation had been accomplished by private individuals or groups. However, by the early 1930s the sense of preservation responsibility was shifting from private philanthropy toward the federal government—as were many other social programs as part of the New Deal. Rather than being an exclusive interest of the wealthy elite, historic preservation in the U.S. was experiencing a growing *public* enthusiasm, for several reasons. Changes in the American landscape stemming from industrialization, urbanization, and the increasing popularity of automobiles suggested to some that all traces of the historical past would soon be paved over, and resulted in calls for protection of some representative pieces of "preindustrial America."[79] In addition, history served to shape a kind of patriotic nationalism, as a way of both educating new immigrants about an idealized collective "American" past, and emphasizing national loyalty during times of economic troubles and world wars. This understanding of history was deeply nostalgic, reflecting on the life of the past as an "antidote for the materialistic ills of the present."[80]

In 1933 the newly inaugurated Roosevelt administration reorganized the Interior Department by presidential order, giving the NPS jurisdiction over all of the national parks, monuments, and cemeteries formerly administered by both the War Department and the Forest Service.[81] Congress followed suit in late 1933 by establishing the Historic American Buildings Survey (HABS), employing many of out-of-work architects and draftsmen to create a huge architectural archive documenting historic buildings.[82] The primary rationale for this kind of program was that the buildings themselves were not, for the most part, high priorities for preservation; because they were not worth physically saving, HABS recorded their particulars on paper, thus clearing the way for their eventual removal. Yet the new roles of the federal government in historic preservation did not end there. In February 1935, Senator Harry F. Byrd of Virginia proposed a bill "to provide for the preservation of historic American sites, buildings, objects, and antiquities of national importance."[83] The bill passed with little discussion and was signed as the Historic Sites and Buildings Act on July 21, 1935.

While the Act did not provide any authorization to the NPS to create new historical parks, it directed the agency to survey nationally historic sites and to work cooperatively toward their preservation with other units of government and private citizens.[84] In 1935 the Branch of Historic Sites and Buildings was formally established within the NPS, and the agency organized the Historic Sites Survey to create an inventory of nationally important historic places, and to recommend sites illustrating significant historical themes in the American past for inclusion in the Park System. The Historic Sites Act also mandated the establishment of the Advisory Board on National Parks, Historic Sites Buildings and Monuments, comprised of distinguished scientists and long-time park enthusiasts to consider potential preservation projects for the NPS.

Soon they were inundated with proposals.[85] By determining whether or not proposed parks had sufficient national significance, the Advisory Board served as a buffer between the NPS and both the public and Congress, protecting the agency from political pressures that might have involved the NPS with areas representing social controversy. For example, following an Advisory Board recommendation, a list of places sent in 1937 to the NPS regional offices for study under the Historic Sites Survey intentionally omitted "all sites of contemporary or near contemporary nature which might lead to controversial questions."[86] Places undergoing cultural change or contestation were not seen as deserving attention, both because of the NPS's traditional emphasis on the static preservation of nationally agreed-upon heritage, but also because of the agency's self-image as apolitical and scientific.[87] These early criteria for determining suitability for preservation continue to form the basis of all historic preservation regulations today.

In line with the tradition of viewing parks as iconic monuments, the past represented in NPS historic sites tended to be presented as static and unchanging.[88] By categorizing sites as representational of certain historical themes, NPS staff fixed the meaning of each site within a certain historical period and context, not allowing alternate interpretations. Also, in determining which historic sites were worthy of inclusion in the system, NPS staff did not have clear criteria by which to measure. They would consider both the state of the material remnants and the events in question in their analysis, but in some cases had only one or the other, and often the "correct" interpretation was not clear.[89] And yet they felt compelled to provide the public with clear, factual information regarding the single "correct" histories of the sites; Hosmer notes that NPS historians "took their responsibilities seriously, almost to the point of looking on themselves as guardians of the national culture."[90] The combination of this attitude with the lack of clear criteria for preservation and the tendency to categorize sites all contributed to the static nature of their interpretation.

This changing vision of the scope of heritage to be protected in parks created space for the next phase of parks establishment, which involved new types of parks. As chapter 3 will explore, in the 1960s and 1970s, instead of simply encouraging recreation in order to justify the creation of parks symbolic of national greatness, Congress began establishing parks explicitly *for* recreation, and specifically aimed at providing opportunities for increasingly active rather than contemplative recreation. These new parks were often located close to urban areas. Many were not only created from private lands but also were directly acquired by the NPS, rather than through intermediates. These new park units—mostly national seashores, lakeshores, and recreation areas—further challenged the popular idea of a clear divide between public and private land. They transformed private lands into public parks presumably representing the nation's collective heritage. The concomitant presumption of public access and public ownership left little to no room to imagine a role for private residents who might wish to remain working the land. The new parks set this widespread understanding of public use and access on a collision course with later conceptions of cultural heritage that include living communities as well as historic buildings and monuments.

FIGURE 7. Abbotts Lagoon and canoes, Point Reyes National Seashore, 2013. Photograph by author.

CHAPTER 3

Acquisition and Its Alternatives

Point Reyes National Seashore evolved from an idea to a reality during a period of experimentation within the National Park Service. No longer restricted to reserving remote, relatively unpeopled lands already in government ownership, the NPS had become interested in establishing parks closer to urban population centers. Almost by definition, these areas came with complicated histories of land ownership, presenting the NPS with both challenges and opportunities. In contrast to the early part of the twentieth century, when establishment of new parks from privately owned lands relied on donations of parcels (either from wealthy individuals or from state-conducted condemnation suits that then transferred the combined properties into federal ownership), starting in the late 1950s and extending through the 1970s, Congress allowed the NPS to acquire land directly, through authorized purchases in areas designated as new parks, designed to cater to more active forms of recreation, rather than scenic contemplation.

The biggest challenge, of course, was the process of land acquisition itself. The NPS had almost no direct experience with negotiating and buying real estate. Through legislation creating each new park unit, mostly national seashores and lakeshores, Congress began to explore the possibilities represented by different property regimes. Instead of establishing a park only with land the NPS owned outright, perhaps land within the boundaries might be leased back, or restricted with easements, or some other form of shared ownership. This sort of approach seemed particularly appropriate in an area like Point Reyes, where the form of the pastoral landscape, as well as the health of the local economy, depended on the presence of active dairy ranchers.

And yet, for a variety of reasons, NPS management has tended to avoid more flexible ownership regimes. Since the beginnings of the Park Service, park administrators had premised their actions within a park's borders on full ownership—that is, full control—of the land. At Cape Cod National Seashore, the first of the seashore and lakeshore "experiments" in direct acquisition, Congress established an intermediary between the federal agency and local residents, via required municipal zoning laws; the NPS could only condemn private land if local zoning ordinances were violated. Yet at many of the other seashores and lakeshores, Congress granted the NPS the ability to buy or condemn land with little or no legal restraint—and, even when given a range of acquisition options, the agency gravitated toward full-fee purchase. This tendency to ignore other acquisition options caused problems and controversy in almost every park unit where it was done, with private owners unhappy at feeling bullied into selling their homes; after nearly two decades of experimentation, Congress ended this period of allowing extensive direct acquisition in 1978.

The experience at Point Reyes exemplifies both the urge to embrace a more flexible ownership regime, and the difficulties in implementing it. The original proposal for the park included a pastoral zone intended to remain in private ownership, which was a political necessity to get the Seashore bill passed. Within ten years, however, the NPS either owned all of the property within PRNS's borders, or was in the process of purchasing it. This chapter explores why.

THE CHALLENGE OF CREATING PARKS CLOSE TO URBAN AREAS

As the idea of parks in the 1930s broadened from scenic icons to include areas of historic and more active recreational interest, particularly close to urban areas, Congress began to experiment with allowing the NPS to purchase land directly from individual landowners.[1] Congress began granting the NPS authority to purchase property directly as early as the 1930s, albeit at first only for the purpose of "completing" parks where state or private funds for acquisition had run out.[2] Over the next two decades Congress gradually allowed the NPS to acquire land to create entirely new park units in a few relatively small-scale settings. This eventually led to the experimentation with large-scale purchase of private lands that created a series of national seashores, lakeshores, and recreation areas through the 1960s and 1970s—including, most famously, Cape Cod National Seashore, but also Point Reyes National Seashore.

The NPS's earliest authority to purchase land came in the context of adding to existing historic parks in urban settings. In 1935, an executive order interpreting the Historic Sites Act, passed earlier that same year, granted the NPS authority to acquire land at Jefferson National Expansion Memorial in St. Louis. A number of private owners challenged this authority in court, but the NPS won the suit. As a result, in 1939 the federal government (not the state, the mechanism that had been used in establishing Shenandoah and Great Smoky Mountains National Parks) condemned forty city blocks in St. Louis, removing the residents and demolishing almost all the historic waterfront buildings to eventually make way for the design and construction of the Gateway Arch.[3] Just over a decade later, Congress allowed the NPS to purchase land directly at the Independence National Historic Park in Philadelphia, where the agency had been working with the city to create such a historic site since the late 1930s.[4]

In both of these cases, Congress granted funds for the NPS to complete land acquisitions for an existing park. The earliest instance of Congress granting the NPS authority to purchase private land to establish a *new* park appears to have been for George Washington Carver National Monument, authorized in 1943. The enabling legislation granted the agency the power to purchase or condemn Carver's birthplace, located in Diamond, Missouri and known as the Moses Carver farm.[5] Acquisition stalled when the owner of the property insisted on a higher purchase price than Congress had authorized. A series of negotiations over the price led to its eventual acquisition through condemnation in 1951.[6] This was a small-scale acquisition; the transaction involved only a single parcel, and the only person dislocated was the seller.

At Minute Man National Historical Park in Massachusetts, authorized in 1959, the process was more complicated. While the acreage involved was still relatively small (capped at 750 acres), the land that made up the potential park consisted of hundreds of small tracts in a heavily developed suburban area.[7] In this case, Congress amended the enabling legislation to "remove any implication that the Secretary of the Interior may acquire properties for the Park only by donation or through voluntary sales."[8] The NPS, in other words, had for the first time broad authority to acquire land directly, including via the process of eminent domain—in other words, private homes could potentially be condemned by the NPS and owners forced to sell, so as to create a public space for recreation.[9] And while the legislation for Minute Man suggested that the NPS *could* acquire only partial interests in land, such as through easements, the agency only made fee-simple purchases—in other words, it consistently opted for the full control of ownership.

Starting with the acquisition of the lands that would become Cape Cod National Seashore in 1961, Congress began to experiment with different strategies for large-scale land acquisition by the NPS, and for the first time in places that were not memorializing a specific moment in history. Cape Cod was the first of these experiments, and the second national seashore to be brought into the NPS fold. The NPS conducted its initial Seashore Recreation Survey back in the 1930s, primarily in response to mounting development pressure on the Atlantic and Gulf Coasts. The survey identified twelve major sections of eastern and southern coastal lands that warranted protection. By the mid-1950s, however, only one of these areas—Cape Hatteras—had been acquired, and that through the traditional mechanisms of state transfer and donations from wealthy individuals.[10] Ten of the remaining eleven sites from the Survey were already in private hands.[11] A second set of surveys, conducted in the late 1950s as part of the NPS's "Mission 66" program, recommended fifteen specific sites deemed to have national significance that could become future parks, and made a general recommendation that 15 percent of the U.S. coastline be acquired for public recreation by federal, state, and local agencies.[12] In response, members of Congress proposed a flurry of bills appropriating between sixteen and fifty million dollars for the acquisition of shoreline parks.[13] None of these bills passed, but the ensuing debate recognized that providing public access to beaches and preventing private development would require land acquisition at a significantly larger scale than anything the NPS had previously undertaken.

Legislation to establish Cape Cod National Seashore was the first to result from this new survey effort. Written by staff from the offices of Massachusetts senators Leverett Saltonstall and John F. Kennedy (before his election to the presidency) and signed in 1961, the legislation took an unusual and creative approach, attempting to address both the NPS's interests and local concerns.[14] The proposed boundaries of the Seashore included 660 private homes, 19 commercial operations, and 3,600 separate tracts.[15] Many lands had been in particular families for generations, and property lines were often vague or conveyed in generalities.

According to the legislation, the NPS would acquire key tracts of undeveloped land within the park's boundaries, but would continue to permit private ownership of improved property as long as the owners complied with locally written zoning ordinances, which were in turn required to conform to standards set by the Secretary of the Interior.[16] Owners who failed to comply could be subject to eminent domain by the NPS; legal scholar Joe Sax referred to this approach as the "sword of Damocles" provision.[17] Those who sold their properties to NPS, whether voluntarily or under threat of condemnation, had the option of retaining rights of use for either twenty-five years or a life estate. Beyond the zoning restrictions, commercial and industrial properties could be condemned if the Secretary of the Interior found their uses to be inconsistent with the park.

In retrospect, park historians often cite Cape Cod as the NPS's new model for acquiring land in highly populated areas. Yet with the exception of Fire Island National Seashore, Cape Cod's specific formula of relying on local zoning regulations was not followed at the other seashores, including Point Reyes. Everywhere else, zoning rules were implemented directly by the NPS rather than through an intermediary, like town or county officials.[18] Local observers at Cape Cod have suggested that NPS staff were not involved with developing the idea of suspending condemnation via zoning and remained wary of the innovation.[19] Moreover, NPS Director Conrad Wirth opposed the role of a local advisory commission created to guide the establishment and management of the park—an innovation widely hailed as a key element of Cape Cod's success.[20] Nevertheless, as of 1961, the NPS had access to alternative models for land acquisition beyond donation from states or individuals.

A NATIONAL SEASHORE FOR NORTHERN CALIFORNIA?

During this same period, the NPS began to seriously explore the idea of developing Point Reyes into a national seashore. The peninsula had been among the areas considered in the 1930s NPS survey of potential seashore locations, but Wirth—who authored the report—had considered the presence of so many private land holdings an obstacle to a potential park. As he put it, "One of the main difficulties to encounter in the acquisition of this area would be that of obliging large ranch holders and interested members of gun clubs to surrender their property for public recreation use."[21] Though explicitly addressed to the issue of property acquisition, Wirth's comments could have equally applied to the appropriateness of the landscape itself. Point Reyes was no "pristine" seashore, but rather a working landscape shaped by agriculture and other commercial activities.

Between 1919 and 1939, the Shafter brothers' heirs sold off their ranches, a process accelerated by financial troubles after the 1929 stock market crash. These land transfers contributed to the development of two distinctive landscapes at Point Reyes. The northern section—extending from Tomales Point out to the lighthouse and down to the Home Ranch and the lands around present-day Limantour—remained primarily in dairy production, with the ranches mostly purchased by the tenants themselves, many of whose heirs continue to live on the land today. The southern section, as well as the higher lands along the Inverness Ridge, were too steep and brushy to support cattle grazing well; ranches in this southern zone were in many cases converted to beef, which is both less labor intensive than dairy, and can be sustained on rougher lands. Here, most of the land was sold to newcomers, including real estate specialist Leland Murphy, who bought roughly ten thousand acres in 1929, and sold pieces off in the 1930s and 1940s.[22] (See map 2 for the Shafter-era ranching boundaries.)

During this period the dairy industry itself was also changing, with greater state regulation of sanitation and production standards. With better transportation, most of the ranches switched from producing butter and cheese to primarily producing fresh milk, hauled by truck each day to a local creamery. Ranchers modified their dairies accordingly, usually by adding a Grade A-certified barn with concrete floors and specific drainage and sanitary conditions. The ranches also gradually became better connected to the outside world by telephone and electricity, and by widened and paved county roads. With their increasing prosperity, many owners remodeled their homes and made other modifications to their ranch buildings.

A visitor to the northern portion of Point Reyes in the 1930s would likely have noted the variety of other activities taking place in the area; this was not a ranching monoculture, but a mosaic landscape of different uses. A federally constructed lighthouse had been guiding ships past the dangerous cliffs of the Point since 1870, staffed by a head keeper and three assistants (often contracted from nearby A Ranch). In 1890, the U.S. Life Saving Service added a lifeboat station on the Point Reyes Beach, which was later moved in 1927 to more protected waters near Chimney Rock. Nearby fish docks on the edge of Drakes Bay supported a local ocean-fishing fleet that brought in crab, salmon, and some bottom fish. A commercial oyster farm, established in 1932 at Drakes Estero at the heart of the peninsula, expanded during the 1940s and 1950s, furthering an industry already present at Tomales Bay since the 1870s. During this same period, several ranches rented portions of their land to Italian or Japanese farmers, who grew artichokes and peas to be trucked to San Francisco.[23] The peninsula's commercial enterprises extended even to modern communications: in 1929 RCA developed a massive radio installation on the peninsula for short-wave trans-Pacific communications, consisting of large fields of antennae; AT&T erected a similar installation a year later in 1930. Both companies were drawn by the relative isolation of the Point, where there was little human-made noise to interfere with their signals.

The southern portion of the peninsula supported an even greater diversity of uses. Several private hunting lodges had been established in the late 1800s in the forested parts of Inverness Ridge, including the Country Club, the oldest hunting lodge in the state, in Bear Valley at present-day Divide Meadow. Used primarily for sport hunting by wealthy San Franciscans through the 1930s, the club held hunting rights on seventy-six thousand acres of Shafter ranches, and built a clubhouse, lodgings, stables for horses and cows, and several stocked fishing ponds.[24] Hunting was also a popular recreation activity for the local ranchers, on their own or neighbors' lands. The Army installed some concrete military bunkers in the southern end of the peninsula during World War II, and used Abbotts Lagoon, located between H and I Ranches, as a dive-bombing target. Between the 1920s and 1950s several ranches in the southern portion also diversified, growing artichokes, peas, or cut flowers.

Nonhunting tourists could enjoy the area's landscapes and vistas in low-key, and largely informal, recreational areas. The nearby town of Inverness had primarily been a resort town since the late 1800s, with many summer homes built by city people from the Bay Area, particularly professors from the University of California, Berkeley and their families.[25] The 1930s Marin County master plan suggested setting aside the beaches and wooded areas of the peninsula for public use, and ranchers had already donated small pieces of the shoreline to the county to establish two public beach parks (McClures and Drakes Beaches). On the eastern side of the peninsula, Tomales Bay State Park got its start in 1945, when the Marin Conservation League raised private funds for the first land purchase. Ranchers often continued the Shafters' tradition of allowing visitors to walk across their lands, as long as they were careful to close the gates and avoid bothering the livestock.[26]

Not much had changed by the 1950s. Point Reyes remained primarily agricultural and rural, despite the dramatic push of suburbanization throughout the country—especially in California, as a nationwide postwar housing boom drove rapid development throughout much of the state. Particularly through the 1950s, many state and local planners presumed a wave of impending development, as suburbanization spread across the country, threatening to swallow up open spaces.[27] Much of the Bay Area was already experiencing rapid growth, and many assumed the same would happen in West Marin, where transportation planners were penciling in the possibility of new highways from the highly developed Highway 101 corridor. With Congress debating the establishment of national seashores as a way to provide beach and recreation access close to metropolitan areas, Point Reyes once again came to the Park Service's attention.

A PRESUMED PROBLEM: DEVELOPMENT

In his first legislative proposal for a national seashore at Point Reyes in 1958, then-Representative Clair Engle presumed that the area would "not long remain undeveloped unless it is acquired for public use."[28] This narrative of impending development overwhelmingly drove Washington-based policymakers' attempts to establish a park. A few hints of what might be in store for Point Reyes had already emerged: a new state tax on standing timber had encouraged several ranchers along the Inverness Ridge to sell their timber rights to an Oregon logger, Bill Sweet, in 1955 and 1956; the following year, local developers David Adams and Benjamin Bonelli purchased the former S Ranch, also known as Muddy Hollow, with an eye to developing the area near Limantour Beach into a recreational community.[29] A couple of additional landowners on the southern portion of the peninsula had inquired with the Marin County planning department about developing portions of their land.[30] At 1961 hearings related to the establishment of the Seashore, Secretary of the Interior Stewart Udall characterized urban sprawl in the Bay Area as a "relentless force," and cited on-going subdivisions within the park boundary as evidence of the increasing urgency for establishment of the Seashore.[31] Given the speed of development elsewhere in California, Point Reyes' future seemed obvious to those who were tasked with protecting the nation's seashores.

These assumptions, which seemed so logical from a distance, overlooked not only the reality of the local economy, but also the traditional patterns of land sales in West Marin. Ranches rarely changed hands, and when they did it was usually between friends or relatives, without the use of real estate agents.[32] Most of the ranches in the northern portion of the park had been continuously operating at Point Reyes for multiple generations of the same families, first as tenants, then as owners. The wave of suburbanization washing across the metropolitan Bay Area had not yet reached this far into surrounding rural areas; Point Reyes was still relatively isolated and could only be accessed via narrow, twisty roads.[33]

Over time, development pressure certainly would have increased, particularly if the roads from eastern to western Marin had been improved and enlarged. Yet the Park Service's insistence that the only way to prevent development on Point Reyes was to purchase the ranch lands was off the mark. Most ranchers were not enthusiastic for residential subdivisions, and had expressed no interest in selling off their land. Threats of subdivision development were not coming from the ranching families—a point they made vehemently in 1961 testimony before the U.S. Senate and House. Joseph Mendoza, owner of B Ranch, clearly stated to the Senate committee, "this word subdivision, this scare you heard of subdivision this morning is certainly not the case in here. These ranches have been in the same ownership for years and years and they are all family operations."[34] The ranchers' attorney, Bryan McCarthy, pointed out that the only subdivision was located adjacent to the established town of Inverness.[35] Nor did it seem likely that RCA or AT&T would sell the ground underneath their towers—so conveniently located for trans-Pacific transmissions—at any time in the near future.

McCarthy's 1961 testimony highlighted the fact that only one subdivider on the peninsula had actually sold any lots, and that this developer's lots began selling much faster *after* the Seashore was proposed; the proposal had drawn attention to the area.[36] Later that same year, Representative Clem Miller noted that the increase in subdivision had all taken place since his initial legislation in 1959, "stimulated by the fact that the area is being considered for protection and development as a national seashore."[37] A chart included in a 1962 appraisal of Wildcat Ranch shows that by the following August, the three would-be developers had planned a total of 665 lots in ten separate subdivisions. Of these, only 104 had been sold, and far fewer had begun building anything. (See figure 8, a graphic made by the NPS to illustrate the proposed parcelization.) Significantly, all of these subdivisions had been presented to the Marin County Planning Committee for approval *after* the national seashore was proposed, not before.[38] Lawmakers' consideration of the area as a

FIGURE 8. Photograph of Limantour Bay and surrounding hills, with approved or projected subdivision boundaries superimposed. The image was created by the National Park Service in 1962 as part of the campaign to get the Seashore established. None of these subdivisions had been approved before the Seashore legislation was proposed. Courtesy of Point Reyes National Seashore Museum, Photo #HPRC 32110.

potential national park unit made it *more* desirable, and therefore *more* ripe for speculation. Developers assumed that, if necessary, they would simply be bought out by the Park Service at an inflated fair market value. Of course we can never know how land development patterns in Point Reyes might have looked without the creation of the park, but it is clear that the park proposal itself contributed to and aggravated the problem it professed that its intervention was needed to solve: development.[39]

By early 1962, NPS documents were citing a total of 396 landowners within the seashore's proposed boundary, a more than six-fold increase since the Seashore had first been proposed in 1958.[40] Understandably the NPS did not trust the Marin County Board of Supervisors to rein in subdivisions, as it vacillated through the 1960s over whether to encourage or slow down development in West Marin.[41] But at the time the

Seashore was first proposed and discussed, there was substantially less threat of development than the rhetoric suggested. The enabling legislation for Cape Cod, passed in the fall of 1961, of course, offered an alternative path to limiting development through zoning restrictions, but it depended on the cooperation of local advisors and municipalities. The NPS's refusal to consider such a model in Point Reyes makes more sense in the context of local opposition to the park.

WHOSE PARK?

From the beginning, there was almost no enthusiasm for the national seashore proposal from residents of West Marin. The impetus for protecting the area primarily stemmed from San Francisco- and Washington, DC- based NPS officials, particularly Regional Chief of Recreation and Planning George Collins, rather than from the residents or local governments of West Marin.[42] It was Collins who first suggested the seashore idea to his friend Congressman Clem Miller, who, after Clair Engle's election to the Senate in late 1958, took over Engle's House seat.[43] The primary goals for their early proposals, developed at a point when Congress was just beginning to experiment with broad-scale acquisition authority for the NPS, were to work out arrangements to "transfer private property to federal ownership," and find the money to pay for it.[44]

In 1959, Miller and Engle both submitted identical bills that suggested a park size of twenty-eight to thirty-five thousand acres, but included no specific boundaries, nor budget. In early 1960, Engle reintroduced Senate bill 2428, which proposed the Point Reyes National Seashore as a single thirty-five thousand-acre unit, excluding most of the active dairies from the protected area, and gave the NPS full condemnation authority to acquire the lands within the boundary. Most of these lands were at the southern end of the peninsula, where there were fewer working ranches, less intensive agricultural use, and where the few potential subdivisions were located.[45] A document prepared for Miller's office, most likely by NPS staff, suggested that the "pastoral quality of some of the interior landscape is one of the great charms of the region and should be preserved insofar as that would be possible in a national seashore project."[46]

Yet by the time of the first congressional hearing in April 1960, held locally in Marin County, the NPS had expanded the initial proposal to a fifty-three thousand-acre park, encompassing almost the entire peninsula, and mirroring Wirth's original recommendation from 1935. Many, although not all, of the ranches were included in a twenty-one thousand-acre "pastoral zone" where ranching could continue—either through leasebacks after government purchase of the land, which was the NPS's stated preference, or possibly by remaining in private owner-ship combined with scenic easements. Ranches located in the thirty-two thousand-acre "public use zone," including four of the largest dairies, would be acquired and closed down, to be managed solely for public recreation. This plan would reduce the number of cattle on the penin-sula by roughly half.[47]

Testimony suggests that this revised proposal had only been made public ten days prior to the hearings, and that this abrupt enlargement of the proposal alarmed the ranchers and their supporters.[48] A repre-sentative of the Marin County Farm Bureau asked, "If the ranches are to be leased back, then why acquire them at all?"[49] Easements were a form of land protection not yet well understood in Marin County, and many locals assumed that, despite language in the bill that allowed for the continuation of some ranching operations, agriculture would sooner or later be pushed off publicly owned lands. Waldo Giacomini, presi-dent of the Marin County Soil Conservation District, expressed this attitude: "When it gets into the hands of the Government you can say what you want, it is gone. You may lease it back, but the only reason why you are taking it is that someday you are going to need it."[50]

Officials and residents of West Marin worried about impacts to the local economy, particularly from removing so many dairies. Any given dairy industry requires a kind of "critical mass," with enough combined milk production to keep the creamery supplied, as well as adequate busi-ness for the local feed store and other supporting industries. Many feared that, should the number of dairies fell below that critical level, all West Marin dairies could be driven out of business by increased costs. The Point Reyes proposal could potentially affect ranchers well beyond its borders.

Nor were the ranchers the only locals unhappy with the idea of such a large park. RCA, AT&T, and the Vedanta Society, all of which owned large blocks of land on the peninsula, also formally opposed the pro-posal at first, as did the local farm bureau, the chamber of commerce, and taxpayers' associations. More significantly, the Marin County Board of Supervisors voted 4–1 in September 1958 to oppose any park proposal.[51] In addition to worries about the agricultural economy, the supervisors voiced concerns about the property taxes that would be lost to the county and the increased traffic noise from park visitors interfer-ing with the radio operations at RCA and AT&T. The climate on the western side of the Inverness Ridge seemed to be too cold and inhospitable for many recreational users.[52] The ranchers also voiced the possibility of conflicts between park visitors and their ranching operations, if tourists left gates open or frightened the cows. Rancher James Kehoe summed up their sentiments: "You can't mix cattle and people."[53]

In 1960, NPS planner Collins had suggested the possibility of the pastoral zone remaining in private ownership but protected from development by easements, yet by 1961, when the Point Reyes legislation was reintroduced to the next Congress, the scenic easement option had disappeared, and only the plan for leasebacks in the pastoral zone remained, most likely due to NPS opposition to partial title. Residents of the Point Reyes peninsula were particularly upset by their own lack of inclusion in the process; in a later interview, rancher Boyd Stewart noted that early advocates of the park proposal did not contact the families within the proposed boundaries at all, which may have contributed to the ranchers' perception of the proposal as a threat.[54] This lack of inclusion was further emphasized when the NPS conducted a Land Use Survey and Economic Feasibility Report, completed in February 1961, but did not interview any of the Point's residents. The ranchers received a copy of the report only on the day of the House hearings in March, at which Joseph Mendoza testified that "we can't certainly say much in opposition to something we don't know anything about."[55] At the Senate hearing a few days later, Mendoza's aging mother, Zena Mendoza-Cabral, after recounting her family's long history on the Point, told senators in emotional testimony, "Now I am faced with the possibility of losing everything that I have worked for. The strangest thing is I was never approached. Everything was done underhanded. I am not afraid to admit it. Nobody ever came to me to ask, 'Do you want to sell your property for a park?'"[56]

The NPS's failure to engage with the people who actually occupied and worked the land produced serious errors in the planning process along the way. The economic feasibility report stated, for instance, that eighteen out of the twenty-five ranches within the proposed seashore boundary were operated by lessees, when the actual number was six, all in the southern section and not in the pastoral zone. This meant that most of the ranching families would not just be exchanging a public landlord for a private one; they would be returning to tenant status, which they had bought their way out of only a few decades earlier.[57] More importantly, over half of the twelve thousand acres first identified as the proposed pastoral zone were too steep for pasturing dairy cows.[58] Four of the largest dairies, those closest to the lighthouse on the Point (A, B, C, and D

Ranches), had been excluded from the proposed pastoral zone completely, an administrative choice that would result in their being shut down. This history helps explains why the ranchers rejected the leaseback idea as unworkable. When Representative Harold Johnson, for instance, asked why the ranchers opposed it, Mendoza replied: "We will have to be tenants to people who would not consider us and ask us as to the workability, or ask our local government, or present a report to them, and that makes us a little afraid to get into an agreement of that type."[59]

The concerns of the local ranchers apparently made an impression with their congressional representatives.[60] When, at a Senate subcommittee meeting in March 1961, Wirth tried to assert the NPS's need for legal flexibility in dealing with individual ranch cases through negotiation, one senator retorted: "You know what negotiation means. That means coercion."[61] Even writer Harold Gilliam, a member of the Point Reyes National Seashore Foundation, a new organization founded at Clem Miller's request to create an appearance of local support for the park, argued that the proposal "should scrupulously preserve the rights of individual residents who want to continue living or ranching on their property. No individual should be deprived of land that is his means of livelihood. I believe that it is possible both to protect the rights of present residents and to preserve the scenic beauty of the area for the crowded future."[62]

That August—shortly after passage of the legislation enabling the establishment of Cape Cod—the NPS amended its Point Reyes proposal to remove the troubled leaseback plan. Instead, it added a twenty-six thousand-acre privately owned pastoral zone, in which the NPS could not obtain parcels larger than five hundred acres except by consent of the owners, as long as the land stayed in a natural state or agricultural use. Wirth described this plan not as a replication of the Cape Cod model, but rather as a repeat of the process that left thirty thousand acres of agricultural lands at the heart of Everglades National Park: the "hole in the donut" lands were included within the park's boundaries but left in private ownership.[63] Unlike Cape Cod, where local governments' zoning ordinances prevented development, this "zone" at Point Reyes was written directly into the legislation, with no intermediary between the ranchers and the NPS; if the former abandoned their agricultural practices and attempted to subdivide, the NPS could condemn their land. The lands owned by RCA and AT&T, as well as by the Vedanta Society on the edge of Bear Valley, were also allowed to remain in private ownership in the final bill, based on "an understanding" with the Secretary of the Interior.[64]

Congress was receptive to the new plan, seeing it as a way to reduce land acquisition costs, maintain local traditions, and preserve the local economy.[65] By late 1961, with the pastoral zone compromise that allowed the ranches to remain in private ownership, and a newly elected slate of Marin County supervisors, local opposition eased. With confidence that the establishment of a national seashore at Point Reyes would allow those engaged in the dairy business "to carry on," the bill based through both houses of Congress. President Kennedy signed it into law on September 13, 1962.[66]

"A PATCHWORK PARK IN TROUBLE"

Point Reyes National Seashore ran into trouble almost as soon as it was authorized. Even before the bill's final passage, some congressmen remained unconvinced that the Seashore could be established with the proposed budget. One representative raised an issue that later became a problem at Cape Cod: What would the NPS do with people who were ready to sell if the government wasn't yet ready to buy?[67] A series of missteps in the purchasing process soon expended the NPS's funds, with large parts of the public use zone yet to be purchased and the ranches' fate up in the air.

The estimated cost of land acquisition for the proposed seashore had been in constant flux between 1960 and 1962; initial estimates of eight million dollars from the Department of Interior soon gave way to a ten million dollar price tag to acquire fifty-three thousand acres, representing an average price of roughly two hundred dollars per acre.[68] An economic feasibility study published by the NPS in February 1961 contained a similar estimate, valuing the proposed area at $10.3 million, but warning that acquisition costs would surely increase with any delay.[69] Only a month later, Interior proposed a budget of twenty million dollars over five years for acquisition, representing an average of $377 per acre; it is not clear why the estimates nearly doubled in so short a time.[70] The final legislation in 1962 authorized fourteen million dollars, with the presumption that the twenty-six thousand-acre pastoral zone would stay in private ownership, and that Lake Ranch, in the southern end of the public use zone, would be acquired through land exchange rather than purchase.

There is good evidence that the NPS consistently underestimated land values during the park proposal process. At the April 1960 Senate hearing, NPS planner George Collins testified that sales prices had been low in the few transactions in the area in the previous year; at this time, official NPS documents were still using a cost estimate of two hundred dollars per acre.[71] In June of that same year, however, Collins agreed to a price of seven hundred dollars per acre in a secret meeting with Bruce Kelham, owner of the massive Bear Valley Ranch (made up of the old U, W, Y, and Z Ranches of the Shafter system).[72] Testimony from numerous locals opposed to the seashore during this period suggested that the total acquisition could reach thirty to fifty million dollars. Bryan McCarthy, representing the West Marin Property Owners Association, cited data from actual sales figures for the past three years showing an average price of seven hundred dollars per acre.[73] Part of the confusion had to do with conflating prices for individual lots in small parcels, which tended to carry high average prices, with the cost per acre of ranchland. Whatever the source of the problem, it remains the case that the NPS's estimations were not realistic, and that their public estimates did not match their private promises.

When the NPS purchased its first parcel, the N Ranch, in July 1963, it paid $850,000, or $762/acre—even higher than the price promised to the Kelhams.[74] Locals later expressed shock at this price; in 1971, the Marin County assessor remarked, "We damn near fainted away when we heard what they spent for the Heims property. It had been on the market before then at about half what the government paid."[75] The Heims' ranch was located within the pastoral zone, not the public use area, and had not been planned for government acquisition. A subsequent lawsuit found that the owners had assisted the NPS in preventing a neighbor from acquiring a right-of-way that could have allowed him to subdivide his property, and that the NPS's speedy acquisition of the Heims parcel permanently blocked that possible access route. The acquisition nevertheless established certain precedents: first, that the NPS could be counted on to pay a premium for property that it wanted; and second, that the NPS wanted land in the pastoral zone.[76]

The initial effort to acquire land on the Point Reyes peninsula did not go smoothly, as the fourteen million dollars quickly ran out. This was in part because of the higher-than-estimated per-acre payments, but also because several of the early acquisitions were in the pastoral zone, not the public use zone—in other words, they were lands the congressional appropriation had not anticipated paying for. Eager to work with any willing sellers that came forward, the NPS followed their purchase of the Heims property by buying the C Ranch in 1964 and the F Ranch in 1966 (both owned by the Gallagher family) and the Murphy (Home) Ranch in 1968—all of which were within the pastoral boundary.[77]

Within the public use zone, the NPS acquired the Drakes Beach Estates properties through a negotiated condemnation, the gigantic 7,772-acre Bear Valley Ranch as prearranged with the Kelhams, and the South End or Palomarin Ranch, owned by a religious community, all in 1963.[78] These lands were not contiguous, making for a disjointed ownership pattern that could not be managed coherently. More to the point, they used up most of the authorized budget.

That the budget would run out was not unanticipated. Clem Miller's widow later recalled that Miller had known from the start that the initial allocation for the Seashore would be inadequate, and had been "very open about it," but felt it was necessary for the proposal to be accepted politically. In a 1993 oral history, she explained, "Clem said the only way to get this thing through was to get the boundaries set, ask for the fourteen million dollars, which is what everybody thinks is an acceptable amount for a seashore, then come back later and get more money."[79] The NPS exacerbated the shortfall with its initial high-value purchase of the Heims' N Ranch, which then caused fair market value for all other ranches in the area to increase, meaning the NPS would be required to offer similar amounts to other landowners. In his 1993 oral history interview, former County Supervisor Peter Behr similarly commented that he believed the NPS had "started offering prices that were too generous, and establishing bench marks which came back to haunt them later."[80] By the time the initial allocation for land acquisition ran out, only roughly fifteen thousand acres had been purchased, about a third of which were in the pastoral zone rather than the public use area. By the spring of 1969 Point Reyes was referred to in the *New York Times* as "a patchwork park in trouble."[81]

The inflated purchase prices had another consequence: increased property taxes for the remaining ranches. The tax burden placed severe financial difficulties on the ranchers; annual property taxes on the Wilkins Ranch in the Olema Valley, for example, went from approximately twelve hundred dollars in 1960 to twenty-two thousand dollars by the time it was purchased in 1971.[82] Small-scale ranches tend to operate with fairly slim profit margins, so this kind of change in a fixed cost could potentially send them out of operation.[83] California's Williamson Act, passed in 1965, allowed counties to assess ranches strictly for their value as agricultural lands, rather than for their development potential, if they were part of a designated zone and protected with easements. Yet, like ranch owners statewide, only a few Point Reyes ranch owners enrolled in this program, out of fears of becoming financially trapped by a death in the family; the Internal Revenue Service still used fair market value, not Williamson Act value, for calculating estate taxes, making the passage of ranches down generations prohibitively expensive without selling the land.[84] This escalation of property tax rates, exacerbated by NPS acquisition patterns and federal tax law, likely contributed to an increase in the property owners' willingness to consider selling, as remaining a private landowner became a more and more expensive—or impossible—proposition.

Making matters worse, the original plan to acquire the Lake Ranch by exchange (i.e., by trading it for another piece of federally owned property elsewhere) fell through, meaning that parcel would need to be purchased as well. Lake Ranch was considered a crucial piece of the seashore, as it cut across the entire southern end of the peninsula, from the Olema Valley all the way to the coast. Bill Sweet had owned the timber rights to the property since the 1950s and had begun negotiating a plan with the NPS in 1960: after the NPS condemned the timber rights in 1963, he purchased William Tevis's interest in the land itself, planning to either sell to the NPS, or exchange the property for timbered land owned elsewhere by the BLM. Yet several attempts at identifying exchange lands, first in California, then in Oregon, failed to produce results after conflicts over assessing comparable values.[85] The Lake Ranch would need to be purchased directly, another cost that Congress had not planned.

Either scenic easements or county zoning might have prevented many of the problems encountered in the pastoral zone. In 1966 NPS Director George Hartzog asked the Marin Board of Supervisors to rezone all of West Marin as agricultural lands, thus avoiding the problems of both subdivision and escalating property taxes, but the Board believed it could not legally do so, as the act might constitute an inverse condemnation.[86] Yet a few years later, in 1972, the Board changed course and successfully zoned all of West Marin as agricultural land with a minimum plot size of sixty acres, thereby reducing the property tax burden on the remaining ranchers in West Marin.[87] This suggests that locally controlled land protections, along the lines of the Cape Cod model, could have worked effectively to prevent development with the right local support, but those conditions did not materialize in a way the NPS found acceptable.

In July 1966 at a hearing for the U.S. Senate Subcommittee on Parks and Recreation, Hartzog proposed a new plan to Congress. Hartzog described a revised system of zoning for the Seashore, modeled on a land management program at Piscataway Park in Maryland. The Seashore would be divided into a preservation zone of about 13,000 acres, focused on protecting natural, historic, or scenic features; a 15,200-acre public use zone, which would include "public use facilities, such as roads, campgrounds, interpretative centers, marinas, and administrative and maintenance facilities"; and a private development zone, consisting of 25,650 acres, with such compatible private uses "as ranching and dairying in some places and *properly spaced residential sites* in others."[88] In the private development zone, the NPS planned to acquire either easements to control development, or to fully purchase property and then lease it back or sell it back with conditions. Hence the agency requested both a repeal of the original 1962 legislation's prohibition on condemnation, to avoid a situation in which hold-outs would demand sky-high prices, and an increase in authorized funds for acquisition, to $57.5 million.

Hartzog's testimony once again blamed increasing costs on the threat of development, particularly in the pastoral zone. He warned the committee of "uncontrolled subdivision," yet the only examples he produced were Drakes Beach Estates, which had already been mostly acquired; Leland Murphy's *proposed* subdivision, which had not yet been lotted, much less sold; and the possibility of trouble at the Lake Ranch, where the Sweet family had hoped to exchange, but now wanted to sell or subdivide after years of waiting. When asked by Subcommittee Chairman Alan Bible why the NPS needed to acquire the pastoral zone—"It seems to be at variance with the original intent of the act"—Hartzog asserted that "unless we get the funds to implement a land use plan as this, of course, the total environment goes down the drain at Point Reyes. This is the hazard that we face with respect to nearly all of the twenty-six thousand acres in the pastoral zone."[89] He raised the possibility that, without federal ownership and control, landowners might propose "high-rise apartments or other incompatible developments" such as tract housing.[90]

Hartzog's testimony painted a wildly distorted and inaccurate view of the pastoral zone, where *none* of the dairy operators had suggested discontinuing or selling out. Having apparently convinced himself that the ranchers, left to their own devices, planned to develop the land— despite the fact that they could not, as Section 4 of the 1962 legislation allowed ranch lands to be condemned by the NPS if they were not kept in agricultural use—Hartzog therefore proposed to take matters into his own hands. The "pastoral zone" would now become a "private development zone" that the NPS would then sell back to create "a low-density residential community with a lot of open space." As he put it, "We believe that with the development activity that is going on out there, that private developers will be interested in pursuing this kind of a development"; he suggested it might include a golf course, riding stables, and a small village center.[91] Senator Bible, at this point, broke in: "It seems to me what you are doing here is substituting yourself for Mr. Murphy's subdivision. Is the National Park Service going into the real estate business, running a real estate development?"[92] Hartzog argued that county zoning boards could not be counted on to enforce easements and restrictions on use. The only alternative, he argued, was for the NPS itself to purchase the entire property—he referred to his plan as "zoning by ownership"—this despite the fact that the only buyer on the peninsula for the preceding four years had been the NPS itself.[93]

Congress did not agree; in October 1966, it passed legislation to increase the Seashore's funding to cover the cost of acquiring the Murphy Ranch, but did not enact Hartzog's plan for a new private development zone.[94] Yet soon after, the NPS's inability to acquire additional lands caused further problems. In the public use zone, the owners of the Lake Ranch, frustrated by the long-stalled NPS land exchange plan, began to proceed with a subdivision within the park boundary to try to force the government's hand. In 1969, Representative Harold "Bizz" Johnson, a coauthor of a new bill to increase funding for Point Reyes, clarified that the Sweet family did not actually *want* to subdivide, and had been waiting for the NPS to acquire it since 1962, but no longer could afford to shoulder the burden, as their property taxes had increased to twenty-two thousand dollars on land that only brought in twenty-four hundred dollars from a grazing lease.[95] Similarly, Representative Jeffery Cohelan added that Sweet "has been most cooperative and wants very much to see the park come into full flower."[96] A year later, Johnson actually praised the family for *preventing* development on the parcel through the 1960s.[97]

The group that owned Pierce Ranch, at the northern tip of the peninsula, found themselves in a similar situation: in 1966, they had purchased the ranch as a partnership to help out the recently widowed Mary McClure, who not only faced steep estate taxes after her husband passed away, but also felt she could not run the ranch on her own.[98] The partnership had seen the land initially as an investment, but by the 1969 hearings, their tax burden had tripled, and partner Web Otis wrote to Congress advocating for the appropriations increase, stating, "we must find financial relief from this continued deficit. The only alternative apparent to us is development of the property. However, we believe that a completed National Seashore would be in the public interest."[99] They did not desire to subdivide, but could not afford to keep holding the land indefinitely.

The uncertain status of these two parcels, plus the pro-growth stance of the County Board of Supervisors at the time, added urgency to the need for Congress to increase the appropriation ceiling once again.[100] Yet at House hearings in May 1969, speaking on behalf of a bill proposing the new acquisition limit of $57.5 million, Hartzog continued to push for another private development-based plan, this one aiming to sell off 9,208 acres within PRNS boundaries to developers for residential and related commercial uses. This plan was "based on experience of some public utilities that had recovered part of reservoir projects costs by selling reservoir shoreline land for subdivision."[101] The NPS estimated the sale would bring in roughly ten million dollars, and actually suggested that Congress might therefore *reduce* the proposed appropriation ceiling to $47.4 million. Hartzog apparently thought development was an inevitable step, even as Congress considered a higher funding limit.

He also continued to blame the ranchers for increased land values, claiming that too many dairiers had sold their land to speculators: "It was simply a question that after the legislation was passed, many of these people did not want to dairy quite as badly as they did before the legislation was passed."[102] This was flatly untrue: the only ranches within the pastoral zone that had even considered subdivision were Murphy's (already acquired) and Pierce Ranch; both had been triggered by estate tax issues, not a lack of interest in ranching; and neither ever went through with the process nor sold any land.[103]

But Congress once again drew the line at Hartzog's sell-off plan. Representative Cohelan stated that, while he approved of continuing ranching on the peninsula, he strongly opposed any land being sold or leased for residential development, as it ran contrary to park objectives.[104] Representative James McClure (of Idaho; no relation to the McClures at Point Reyes) stated, "I think most of us in this country do have fears about the Federal Government acquiring private property by purchase or particularly by condemnation and then turning it over to private development."[105] Representatives from the Marin County Board of Supervisors, from environmental organizations, and from the ranchers themselves all weighed in against the sell-off idea. Representative Johnson concluded at the end of the hearing that the subdivision plan was inappropriate, "both for the National Park System and for Point Reyes."[106]

The sell-off plan was later prohibited by an amendment to the House bill. But perhaps more importantly, the specter of the NPS actively seeking suburban development in the park triggered a new advocacy campaign, called the Save Our Seashore (SOS) campaign, led by former Marin County Supervisor Peter Behr and Clem Miller's widow, Katy Miller.[107] Miller attended the 1969 hearings and looked at the NPS's presentation exhibits during lunch, which showed the extent of development they were proposing—and was "just staggered."[108] She quickly convinced Behr to chair the SOS campaign, based on the model of the Save the Bay campaign that had helped spur creation of the San Francisco Bay Conservation and Development Commission earlier that decade. The campaign was primarily one of letter-writing and petitions, aimed squarely at President Nixon, to convince him to support the additional funding for Point Reyes, as well as back-channel discussions to secure adequate support in Congress.[109]

A growing consensus outside the NPS seemed to hold that Point Reyes needed additional funding for land acquisition; the question of whether it should include condemnation authority in the pastoral zone, however, remained unresolved. In February 1970, the NPS leaders were back at the Senate, arguing not only for condemnation authority but also, once again, on behalf of the sell-off proposal. But an agreement among the ranchers ultimately brought about a solution. At a meeting held at Boyd Stewart's ranch a week before the 1970 Senate hearings, attorney Bryan McCarthy and most of the Point Reyes ranchers considered their options.[110] Stewart assured them that if they were willing to allow the NPS to have the power to condemn, the agency would probably get the appropriations to pay them a fair price. Faced with soaring property and estate taxes and a sense that the park's presence created a "cloud" on their land title, the ranchers finally agreed.[111] A decade of rising taxes and legal uncertainty, along with the constant narrative of imminent subdivision, had convinced them that dairying could not continue at Point Reyes indefinitely.

Stewart brought word of this agreement to the Senate: the ranchers would sell willingly if Congress passed the appropriation. When asked whether they wanted provisions included to allow them to continue operations, Stewart relayed the ranchers' consensus: "We will take our chances and decide whether or not we want to operate dairies after we have sold the land."[112] When faced with what seemed to be an unavoidable choice, between subdivision and park, the ranchers all agreed; Stewart explained:

> Senator, you know, we ranchers did a pretty good job of keeping that country so that it was available for a park, and one of the reasons was we thought we were the most fortunate people in agriculture. We lived in the most beau-

tiful place, and we didn't intend for anyone to come in. We would have built a fence around it if we could have, and stayed there . . . . The thing, though, that we do feel, and this is common, I wouldn't want to poll all of them, but this is common, . . . we would rather see it used by the people for a park than we would see it subdivided.[113]

The ranchers' agreement cleared the way for final action on the legislation. The NPS's sell-off proposal was dropped, and the final bill repealed Section 4 of the original 1962 Act and granted the NPS condemnation authority if needed. The bill that President Nixon signed on April 3, 1970, raised the total PRNS allocation to fifty-seven million dollars—almost exactly the amount predicted by the ranchers' lawyer in 1960.[114] An additional appropriation of $7.1 million was unanimously approved for the speedy acquisition of Lake and Pierce Ranches.[115] The NPS then rapidly acquired all the ranches within the park boundary, with most purchased in 1971. This was followed by the establishment of the nearby Golden Gate National Recreation Area in 1972, including the Olema Valley; those ranches were purchased soon after.[116] Instead of becoming a new kind of park that included private ownership, Point Reyes ended up almost completely under national ownership, rather like Shenandoah and Great Smoky Mountains before it.

STRUGGLING WITH ALTERNATIVE APPROACHES

In the two decades following the establishment of Cape Cod and Point Reyes National Seashores, Congress created ten additional national seashores and lakeshores.[117] It also began creating urban-based recreation areas, starting with Gateway (NY) and Golden Gate (CA) in 1972, which also required the NPS to acquire substantial acreage of private lands. While the Cape Cod zoning-based framework for land acquisition and regulation did not become a standard model for these new parks, it did establish a trend of Congress suggesting alternatives to the traditional NPS preference for outright fee acquisition. The late 1960s and early 1970s represent an era of transition, where writers of park legislation experimented somewhat with acquisition strategies, including private use zones or encouraging less-than-fee acquisition methods (such as easements) in the authorizing statutes for new parks. The NPS, however, does not appear to have been entirely ready for these new approaches, as it continued a pattern of acquiring land mostly in fee, often despite legislative language to the contrary. Point Reyes was one of the first in this group; legislation attempted to keep agricultural lands in private ownership, but implementation of the land acquisition program seemed to steer almost inevitably toward complete NPS purchase of the peninsula.

The story of Sleeping Bear Dunes National Lakeshore in Michigan, first proposed in 1959, is in some ways similar to what happened at Point Reyes. Creating the lakeshore involved the acquisition of hundreds of private tracts, mostly summer homes, and generated a great deal of local opposition.[118] One of the first Sleeping Bear bills introduced to Congress proposed used the Cape Cod "model" of allowing private residences to remain if local zoning was in place, and permitting hunting.[119] When eventually established in 1970, the park included three land-use categories: public use and development (to be owned by NPS); environmental conservation lands (some owned by NPS in fee but others private with scenic easements); and private use and development, in private hands and protected from condemnation. The existence of the zones, however, created a good deal of confusion for landowners and left them in limbo. Because the cost to the NPS of purchasing an easement was often close to full-fee value, NPS preferred to simply acquire in full fee, often using aggressive tactics.[120] In his administrative history for the park, Theodore Karamanski suggests that "a heavy-handed, poorly planned land acquisition program reinforced the bitterness that surfaced during the decade of struggle that preceded authorization."[121] Ultimately, over fourteen hundred private parcels were purchased, almost all in full fee (most by 1974, although acquisition continued through the 1980s), and the overall process was "unfortunately underscored with a feeling of resentment based on the perception that they [the landowners] had not been dealt with fairly."[122]

Similarly, at the Buffalo National River in Arkansas, established in 1972, early proposals included specific private use zones to protect agricultural and economic values as well as aesthetic and historical ones. The congressional hearings produced "impassioned pleas" from locals regarding the need for retaining private ownership in these areas, yet the zone concept did not get explicitly written into the 1972 legislation.[123] Instead, the legislation mentioned only provisions for reservation of rights.[124] The 1975 master plan once again included a private use zone of ninety-four hundred acres, where land use would be restricted by easements but would otherwise remain under private control. Buffalo's first two superintendents, however, pursued an aggressive acquisition program, with land agents primarily from the U.S. Army Corps of Engineers buying mostly in fee despite the "stated and written objectives" of acquiring only easements. By 1982, 75 percent of the private use zone had been purchased in fee, and many of the buildings razed.[125] The result was an "eviscerated" agricultural landscape, despite all intentions to protect traditional land uses and maintain private ownership.[126]

Two more parks authorized in 1972 caused a great deal of controversy in the NPS: Gateway National Recreation Area in New York City and Golden Gate National Recreation Area (GGNRA) in San Francisco. Both parks stemmed from the broader urge to provide more recreation opportunities to urban areas, and both converted a number of decommissioned military sites to recreation use.[127] Creating GGNRA, however, also involved acquiring a substantial amount of private land, a mix of agricultural and rural residential areas in Marin County. This effort followed immediately on the heels of the bulk of land acquisition undertaken at nearby Point Reyes (the northern portion of GGNRA is actually managed by the staff at PRNS). What made the both recreation areas controversial, though, was the expense involved in developing and maintaining them. For instance, the funding appropriated for land acquisition for GGNRA topped out at over eighty million dollars, and much of the land north of the Marin Headlands, including the Olema Valley, was included in the legislation only after a special request from then-president of the Sierra Club, Edgar Wayburn.[128]

These recreation areas involved the NPS for the first time in directly providing urban recreation facilities. In that respect, they provided a model for additional urban recreation areas in other cities. Their creation proved especially influential in establishing a park in the Cuyahoga Valley, between the Ohio cities of Akron and Cleveland. The NPS and the Interior Department, backed by the Advisory Board on National Parks, vigorously opposed this proposal, arguing that Gateway and Golden Gate had been intended as models for state and local recreation areas elsewhere, not as prototypes for future national parks.[129] But Representative John Seiberling championed the project, and despite strenuous objections from the Interior Department and the Office of Management and Budget, President Gerald Ford signed the bill establishing the Cuyahoga Valley National Recreation Area on December 27, 1974.[130]

As was increasingly the case for park proposals in the 1970s, the bill to establish Cuyahoga started off with a prohibition against acquiring land without the owner's consent, provided the land was subject to valid protective zoning law in which land use was not incompatible with character of the park. The bill directed the Interior Department to provide technical assistance in developing favorable zoning regulations, to use scenic easements where possible instead of outright purchase, and to allow sellers to retain reservations of use and occupancy for twenty-five years or a life estate.[131] After revisions in the House, the eventual bill allowed condemnation, but only if the land use threatened the purposes of the park, or if the parcels were otherwise deemed necessary to fulfill the park's purposes.[132] The Act also specifically suggested the NPS acquire only scenic easements rather than fee title to improved properties.[133]

Despite the inclusion in the Act of a strict six-year deadline for "substantially completing" the land acquisition program at Cuyahoga, small appropriations in the first few years restricted the pace until 1977, when twenty-five million dollars was allotted for acquisition. Again the NPS relied on the Corps of Engineers for land officers, and again this resulted in an aggressive program, souring the NPS's relationship with surrounding communities. The Corps' approach was "cut and dried: full fee simple, and if negotiations stalemated, the agency initiated condemnation proceedings."[134] This ethos was echoed by the park's superintendent, Bill Birdsell, who held the traditional NPS view that "if you are going to manage it, you have to own it."[135] By the deadline at the end of 1980, the park had spent fifty-nine million dollars for nearly eleven thousand acres: 10,555 acres (603 tracts, of which 334 were residential; of those, 184, or 55 percent, retained rights of occupancy) in fee and 153 acres (64 tracts) in easement.[136]

Part of the trouble stemmed from the absence of any NPS planning documents or guidelines to steer acquisition in the first decade at Cuyahoga. No land acquisition plan was written until April 1980, and none formally approved until three years later. Initial priorities for acquisition were undeveloped lands, those threatened with adverse uses, floodplain tracts, and "hardship" cases, but the NPS accepted almost all offers to sell, without any particular reasoning or systematic approach. Furthermore, the agency felt easements were not practical on large undeveloped tracts, because appraisals had to consider full potential value once developed, and the easement to prevent development was often a huge proportion of the total cost. This was coupled with a belief that, for that much money, taxpayers ought to own or at least have access to the land. Visitors were often confused by private tracts, and tended to either trespass or regard them as "intrusions" on the landscape.[137] Improved properties where the sellers did not retain rights of occupancy sat boarded up and vacant, targets for vandalism, and contributed to an overall sense of a blighted community, making some residents want to move away even more.

During the 1960s the park service also established several large "natural area" parks (as opposed to those with working landscapes included) through the purchase of private lands. The most visible and controversial of these was Redwood National Park in California, established in 1968 by acquiring over twenty-eight thousand acres, mostly from private timber companies.[138] The legislation came with an appropriation of ninety-two million dollars, far more than had been initially committed to the seashores and recreation areas.[139] In Washington State, the North Cascades Complex came from a combination of U.S. Forest Service transfers and private purchases, particularly in the community of Stehekin, where NPS land acquisition increased local property values and hence local taxes, and triggered fears of intentions to remove the residents.[140] In 1971 Voyageurs National Park was created in northern Minnesota, on lands primarily purchased from a few large paper companies.[141] It was followed by Big Cypress National Reserve in Florida in 1974 for a total cost of over $120 million in an area not in any imminent threat of development. The combination of expense and local opposition stirred up opposition in Congress; Representative Phillip Burton, acting as chair of the House Subcommittee on National Parks, requested that the General Accounting Office examine Interior and Agriculture's "policies and practices for purchasing title to land versus using less expensive protective methods."[142]

During the 1960s and early 1970s, the NPS took a large step into the business of buying private property directly for the creation of new parks, mostly as part of the its effort to "bring parks to the people." The creation of many of these parks was pushed by some subset of locals who feared overdevelopment or the spread of suburbanization or by conservation groups who wanted more open space; both saw national park status as a solution. Legislative history makes clear that Congress did not necessarily see federal ownership as the right path toward establishing a national park, but—at least through the 1970s—it continued to be the NPS's preferred mode of acquiring parklands.

Direct acquisition raises questions of equity, as certain locals or activists persuade the federal government to buy out other locals as a way of guaranteeing an open landscape and access to recreation. In several cases, as with the Huron Mountains group at Sleeping Bear, the Vedanta Retreat at Point Reyes, Audubon Canyon Ranch at GGNRA, and the Boy Scouts of America at Buffalo, wealthy or otherwise influential landowners have been exempted from purchase. These owners appear able to persuade the NPS and/or Congress that they have no intention of subdividing their land, and so federal purchase is not necessary to protect their lands from development. In each of these cases, the NPS has no management authority over these private holdings; it is not clear why these owners might manage their lands more in keeping with NPS priorities than would any others.

Overall, during this period, the NPS took on more acquisition responsibility than it was prepared to handle (and often at huge cost), so that many of the land acquisition programs fed local resentment and helped to fuel the private property rights backlash that developed in the late 1970s and early 1980s. The movement toward creating more urban parks continued, culminating in the National Parks and Recreation Act of 1978, an enormous, so-called "park barrel" bill providing for the creation of over a dozen new national park units and raising the spending ceilings for many more.[143] Most of the 1978 parks incorporated far more private land than the NPS had ever dealt with; but in contrast to previous decades, Congress directed the NPS to leave inholdings permanently in private ownership. Santa Monica Mountains National Recreation Area outside Los Angeles, for instance, encompasses over 150,000 acres, only 14 percent of which is in federal ownership; fully 54 percent remains in private hands, and the remainder is owned by the State of California.[144] These kinds of parks not only represent a new model for park establishment; they also represent a subtle but important shift in our society's ideas about what parks are fundamentally *for*. By incorporating mixed ownerships and a partnership approach to management, these new parks are less about lofty ideals of pristine nature or national pride than they are about the particular qualities, both natural and cultural, that give each a unique sense of place.

In more recent years, a number of park units have been created on even newer models. These parks tend to look less like traditional national parks and more like private land trusts. When the Presidio of San Francisco was transferred from the military to the NPS in 1996, it was set up as the Presidio Trust, with a requirement that it be self-supporting by 2014.[145] Boston Harbor Islands National Recreation Area was established in 1996 with the NPS being only one of thirteen partners in management, and owning only five acres out of a total of 1,482.[146] Tallgrass Prairie National Preserve in Kansas, authorized that same year, is mostly owned by the Nature Conservancy, a private nonprofit land trust; of the nearly eleven thousand acres, roughly thirty are held by the NPS.[147] And an increasing number of national heritage areas were established in the 1990s. The NPS retains little if any ownership interest in these cultural heritage sites; instead, the agency's role is limited to facilitating land protection efforts and partnering in management and direction. Unlike a previous generation of parks that focused on either providing recreation or preventing development, many of these new parks explicitly focus on economic revitalization via protection of cultural heritage.

ACQUISITION AND ITS CONSEQUENCES

In 2012, a new documentary titled *Rebels with a Cause* premiered at the Mill Valley Film Festival to great acclaim. The film celebrates the heroic local efforts in the 1960s to protect West Marin from urban development.[148] While *Rebels* tells a compelling story, based primarily on interviews, it tends to conflate the tale of Marincello—a 1965 proposal financed by Gulf Oil to create a massive development, big enough for thirty thousand residents, on two thousand acres at the top of the Marin Headlands overlooking the Golden Gate—with the story of PRNS' establishment. The proposed area for Marincello did end up becoming part of the GGNRA, but its protection had a different origin than that of PRNS. The documentary especially downplays the major role the NPS played at Point Reyes, both in pushing to establish the park in the first place, and in driving the threat of development, thereby creating its own justification for acquiring the ranches.

While Point Reyes National Seashore was created to provide recreation space, it did so only with an explicit commitment to retain active agricultural uses. Without the political support that came with that promise, the Seashore might never have happened. In spite of the multiple experimental attempts at land acquisition, including most notably the privately owned pastoral zone remaining as the "hole in the doughnut" surrounded by parkland, Point Reyes National Seashore ended up as a traditional park. Within a decade, the NPS owned virtually the entire peninsula. The only parcels not acquired were owned by corporations (AT&T and RCA), the U.S. Coast Guard, and the Vedanta Society, a religious retreat. The NPS trusted neither the longtime ranching families nor the county to protect the land from the specter of subdivisions. Through its disorganized acquisition plan and inflated purchase prices, the NPS created the conditions under which it became financially necessary for ranchers to sell their land, and through its constant repetition of the threat of development—including its own bizarre plan to sell off a portion of the peninsula for a residential subdivision—the agency created the justification for the additional expense.

Federal acquisition as a national park unit was not, of course, the only mechanism available to keep the Point Reyes peninsula open and undeveloped. In February 1972, for example, Marin County zoned all of West Marin as "A-60," prohibiting subdivision of parcels smaller than sixty acres.[149] Nearly a decade later, another option, the Marin Agricultural Land Trust (MALT), began acquiring easements on over forty-six thousand acres of privately owned farming and ranch lands across West Marin, preventing development and allowing families the financial freedom to continue their operations.[150] The first agricultural land trust of its kind in the United States, MALT was the result of a remarkable alliance between ranchers and environmentalists, led by dairier Ellen Straus and botanist Phyllis Faber. Although MALT came too late to serve as a model for the Park Service's desire to preserve space for recreation on Point Reyes, its three-decade record of success shows that the original plan for the Seashore, with a privately owned pastoral zone, *could* have worked with the addition of easements to prevent development and reduce land values and property taxes. Alternatively, had the NPS implemented something more like the Cape Cod model, with zoning restrictions or conservation easements across the private pastoral zone and the threat of condemnation for noncompliance, land values would have reflected agricultural use rather than development potential, the tax burden on the ranchers would not have escalated so precipitously, and the pastoral zone could have remained in private ownership.

Big Sur, located farther down the California coast, south of Monterey, serves as another example of an alternative path for Point Reyes. In 1962, the same year PRNS was approved, Monterey County established the *Big Sur Coast Master Plan*, with the support of most Big Sur residents. This plan created what historian Shelley Brooks calls "parklike zoning managed at the county level," limiting development on private lands along the scenic coast: much of the inland mountains are publicly owned, as the Santa Lucia unit of the Los Padres National Forest, and contain numerous trails and campgrounds for recreation visitors. The main draw, Highway 1, contains frequent pull-outs for drivers to stop and enjoy the majestic views. Most residents in the area were well off; "aware of the power their wealth commanded, they boldly asserted their right to steward the land without a federal landlord."[151] After the creation of the California Coastal Commission in 1976, the *Big Sur Coast Master Plan* morphed into the mandated Big Sur Local Coastal Program, which again enjoyed strong local support. When world-famous photographer Ansel Adams began to advocate for a new national park at Big Sur, locals resisted. As Brooks describes it, residents "foresaw the worst: that the federal government would 'hound the residents out and turn Big Sur from a community into a museum.'"[152] With the help of Senator Samuel Hayakawa, a Republican who knew the area well, the residents fended off the park.[153] Like the success of MALT in protecting open space and agriculture in West Marin, the Big Sur case shows that NPS acquisition is not necessary to protect a scenic landscape for recreation purposes.

The NPS preferred federal ownership with leasebacks to ranchlands remaining in private ownership with easements. Ultimately, what is the difference? Both represent a mixture of ownership—either private lands with a public interest or public land with private users—so why might it matter which side of this divide the landscape is on? The key difference is the degree of control that comes with primary ownership, rather than ownership of individual rights. As we will see in the remaining chapters, holding title to the land has allowed the NPS to change its terms of use more or less unilaterally and often with little to no justification. The difference represents control of the *relationship* people have with the land. Perhaps that is why an agency-controlled suburban development at Point Reyes actually seemed like a preferable alternative to ranches to NPS Director Hartzog in the late 1960s; a subdivision is not in a productive relationship with the land, and is a pretty stable land use once established. Ranchers, in contrast, not only depend on the land, but work it—and the landscape itself depends on their work, if it is to remain open and grassy—hence ownership matters. When the balance of rights and control shifts from the ranchers (or Native Americans before them) to a government agency, the ability to sustain that interdependent relationship with the land is cut, and the working landscape that the NPS promised to maintain is compromised.

FIGURE 9. Aging cypress windbreak at the Pierce Ranch, Point Reyes National Seashore, 2015. Photograph by author.

CHAPTER 4

Parks as (Potential) Wilderness

The southern end of the Point Reyes peninsula has steeper topography than the northern portion. It is higher and gets more rain, and thus is more prone to develop brush and forested plant communities than grass-lands. Historically, the ranches in this area were generally smaller than the northern ones, and tended to have difficulty keeping grazing lands clear of encroaching brush. A number of these dairies were abandoned during the 1930s or converted to beef ranches and used for hunting and recreation activities. Other uses of the land, such as mercury mines and military installations during World War II, mixed in with the ranching landscape. Still, by the time the peninsula was established as a national seashore in 1962, the Bear Valley (W), New Albion (R), and Laguna (T) Ranches were all still actively in use, and the Glenbrook Ranch had just recently ceased dairy operations (see map 2). Logging activity was taking place on the Lake Ranch, and the Christ Church of Golden Rule had converted the South End Ranch into a religious community. Paved roads led to the New Albion, Wildcat, and Lake Ranches. The owner of the old Muddy Hollow ranch had subdivided and platted the property for a housing development, with roughly twenty homes and a paved road along the Limantour Spit. Readers might therefore be surprised to learn that this southern portion of the Point Reyes landscape was, only four-teen year later, formally designated as a federal wilderness area.

The 1964 Wilderness Act, originally aimed at compelling the U.S. Forest Service and other federal land agencies to protect large areas of their holdings from commercial or recreation development, defined wilderness as "untrammeled," a landscape where "the nonhuman forces of nature are to be given free rein." Yet there are few places on earth that have not been trammeled at one time or another; in particular, many publicly owned lands in the United States were at one time roaded, logged, farmed, or utilized in some other way, particularly by Native Americans. In its effort to establish a national wilderness system, Congress ten years later accepted a less-pure definition of wilderness in its 1975 Eastern Wilderness Act: the Act designated that "wilderness lands were to be managed in such a way that they *would be* untrammeled and *return* to primeval conditions *in the future*."[1] Agency management of designated wilderness is the process by which these lands become increasingly untrammeled, or at least come to *seem* to be untrammeled. The appearance of natural purity remains the ultimate goal of most wilderness management, which results in erasing traces of human history.

Point Reyes is a place where a portion of the landscape has been "rewilded"; over twenty-five thousand acres are formally designated wilderness. At the Seashore, as elsewhere, the unspoken goal of the designation has been to untrammel—to create a deliberate scene of seemingly pristine nature for a particular kind of outdoor recreation, primarily hiking and backpacking. The maintenance of such a scene has produced a tension between actions that reduce or eliminate traces of human land use in the rewilded parts of the park, and actions that continue human uses and meanings in the still-working portion of the park. In the first case, buildings have been removed, native species reintroduced, and a huge portion of the peninsula was declared to be wilderness in 1976, with resulting restrictions on roads and mechanized activities. On the other, a legal mechanism was put in place in 1978 to allow the long-time ranching families to continue working much of the remainder of the Seashore in the pastoral zone, even after their lands were purchased by the federal government.

The 1976 wilderness designation is a classic case of "mission creep": the NPS started off with a vision for the Seashore's management focused on beach recreation, but within a decade shifted to another purpose, substituting wilderness for the planned developed campgrounds and oceanside restaurants—a shift pushed mostly by national environmental groups. In addition, another eight thousand acres at Point Reyes were granted the peculiar status of "potential wilderness." This new category, created by the 1976 wilderness bill especially for Point Reyes and several other parks, was intended to limit NPS's ability to develop recreational services while allowing some existing land uses to continue. The contradictions inherent in this category set the scene for continued disagreement over what wilderness means, ultimately building to a heated standoff over Drakes Estero and the continuation of the oyster farm. The continued presence of active ranchers in the park, alongside these wilderness designations, represents the uneasy, undecided future for the Seashore as it is pulled between competing visions for its core purpose: recreation, wilderness, or working landscape.

EARLY MANAGEMENT EFFORTS

During the early 1960s discussions of the Point Reyes proposal, much was made of the fact that it was being established as a national sea-shore, *not* a national park, and that its primary purpose was to provide public recreation. During the 1961 Senate hearings, sponsoring Con-gressman Clem Miller stressed this in his testimony, making the distinc-tion between national parks and Point Reyes, a national seashore close to an urban area, where a greater variety of land uses would be permit-ted. His sentiments were echoed in a written statement from the National Parks Association, which clarified that Point Reyes was "not a national park in its primeval and wilderness sense" and stating that "the combination of dairy country and wild natural shoreland is part of the charm of Point Reyes, and we think the combination ought to be preserved."[2] Yet despite this initial intention, the park has mostly been managed as if it were a national park, with protection of natural resources as its primary goal. Very little recreation development has ever been implemented, despite repeated calls for it, including in the 1980 General Management Plan. The overall trend at Point Reyes has increased emphasis on the wild and natural character of the landscape, at the expense of both cultural resources and recreation opportunities.

The cultural imprints on the Point Reyes landscape have always been relatively invisible to NPS planners and managers, who have instead focused on describing the natural qualities of the landscape. The first NPS recreation survey of the area, completed in 1935, mentioned the presence of "ranch holdings" on the peninsula, but did not specifically mention the dairy industry as a commercial development.[3] Despite the presence of previous military installations, mercury mines, and other disruptive uses on the landscape, NPS Director Wirth opened his testi-mony at the House hearings in 1960 with reference to how the area had been "left so unaltered by the hand of man."[4] An NPS informational memo described the peninsula as being "relatively unspoiled" and "relatively undeveloped"; the Sierra Club took the trend even farther, publishing its nature photography on Point Reyes in 1962 with the title *Island in Time*.[5] From the beginning, the agency's management emphasis has been on the supposed pristine quality of the natural resources, mostly ignoring the 150 years of extensive modifications of the landscape for ranching. Unlike the expensive vacation homes at Cape Cod, which the NPS was forced to work around when planning the national seashore there, the Point Reyes working landscape was all but invisible to early establishment and management considerations.

PRNS's first *Master Plan for Park Development and Use*, written in 1963 by regional NPS planners, continued to overlook the presence of the ranches while planning extensive development at Point Reyes for recreation, including a four-lane federal parkway running throughout the peninsula and developments for car camping, boating, et cetera (see figure 10, a similar proposed recreation map from the 1961 *Land Use Survey*). It anticipated a high-intensity recreation development on the Limantour Spit, where vacation homes had just recently been removed with the condemnation of the Drakes Bay Estates subdivision (see chapter 3). Because of the slow pace and high expense of land acquisition efforts during the 1960s, however, implementation of these developments stalled. A 1970 *Statement of Management Objectives* for the park reiterated the need for recreation services, calling for development of facilities "to foster recreational pursuits of swimming, boating, bicycling, horseback riding, golf, playgrounds featuring activities for both children and adults, beachcombing, skindiving, whale watching, tide pool exploration, nature education, bird-watching, and photography."[6]

Even as late as 1980, the *General Management Plan* projected that visitation to PRNS would increase, and recommended a number of new developments to serve recreation users, many of which would utilize existing ranches and their buildings. A walk-in campground was planned for the Olema Valley (technically part of the Golden Gate National Recreation Area, established in 1972, but managed by PRNS) at the Truttman Ranch, as were hostels at both the northern and southern ends of the valley, to be located in unspecified "historic structures." The 1980 plan also called for a walk-in camp and picnic area at Five Brooks, a hike-in camp on Bolinas Ridge east of Five Brooks, and an environmental education center at Rancho Baulines, if demand warranted it. Other unspecified "local ranches" were to continue operations, with one "working ranch used to interpret historical agricultural use of the valley." For the

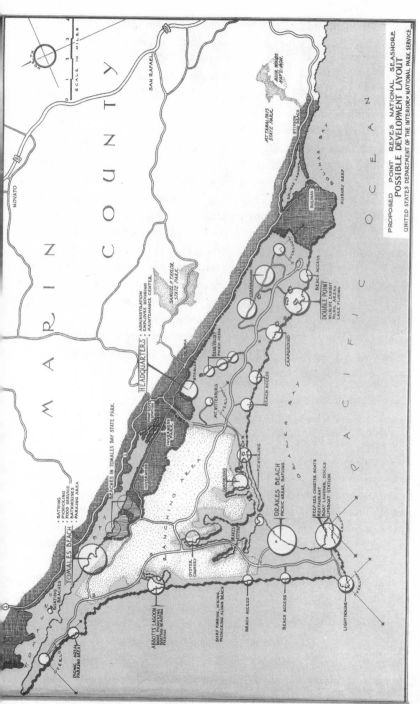

FIGURE 10. Map showing proposed recreational development, from National Park Service, *Land Use Survey: Point Reyes National Seashore: Proposed Point Reyes National Seashore* (San Francisco: Region Four Office, National Park Service, 1961).

Point Reyes peninsula, the NPS planned expanded interpretive activities and exhibits at Bear Valley and Drakes Beach, plus food service at Bear Valley. Tours and interpretive programs would be added at the Pierce Ranch and the former Coast Guard lifesaving station. Hike-in campgrounds were slated for Home Ranch and Muddy Hollow Ranch, and a boat-in camp at Marshall Beach.[7] Overall, the ideas for recreation were superimposed on the working landscape, with little consideration for how the ranches would continue to operate economically in the process.

Yet none of the recreation developments proposed in Point Reyes' first two decades as a national seashore have been realized in the thirty-six years since the last plan. The original proposals were delayed through the 1960s by the piecemeal land acquisition process and depleted budget. The NPS could not develop recreational facilities until it had a large-enough contiguous area of land for public use, and almost all of the money appropriated by Congress went to acquiring more property. By the time the NPS finally obtained full ownership of the entire peninsula in 1972, the environmental movement nationwide was increasingly embracing the ideal of wilderness, pushing Congress and the federal agencies to identify and propose areas to be designated under the 1964 Wilderness Act—which would preclude most of the planned recreation developments for the Seashore.

THE WILDERNESS ACT OF 1964

Two years after Point Reyes was first established, Congress passed the Wilderness Act, the culmination of a long effort by wilderness advocates that began in the 1930s. At first glance, the passage of the Wilderness Act would seem to be the logical endpoint of what environmental historian William Cronon has described as the "unexamined foundation on which so many of the quasi-religious values of modern environmentalism rest."[8] Yet the modern concept of wilderness, particularly as formalized in the 1964 act, must be understood as part of a broader cultural shift from a producer to consumer society. Whereas the national park ideal, as discussed in chapter 1, encouraged nature tourism as a passive, contemplative form of consumerism, the aim in creating legal protection for wilderness areas stemmed primarily from a reaction to, and concern about, those same tourists—and in particular, the vehicles that brought them to the parks.[9]

The rapid proliferation of the automobile in American society in the early twentieth century—from only eight thousand vehicles registered in

1900 to over twenty-three million by 1929—and federal support for road-building profoundly expanded the reach of the car, and the traveling public with it. While the earliest national parks in the nineteenth century had been conceived of and designed for the railroads, now parks were developed explicitly with automotive tourists in mind.[10] The car was praised as liberator and equalizer, and car touring seen as "roughing it," self-sufficient and self-contained.[11] Wartime equipment for soldiers, such as packs, tents, and portable stoves, was adapted into camping gear for these new outdoor recreationists. While many early NPS employees and supporters embraced these trends, specifically designing the choreography of parks to accommodate cars and car campers, others involved with public lands management, including a young timber surveyor for the Forest Service, Aldo Leopold, saw this welcoming stance toward vehicles as threatening something essential about the wild. Historian Paul Sutter writes, "It was to stem the growth of road-building, to control the automobile, and to temper recreational development of America's public lands that Leopold first suggested the need for wilderness preservation."[12] While the rhetoric of John Muir had critiqued the extractive industries that utilized public lands for production, Leopold instead critiqued this new consumer trend—not based on ecological concerns, but on a sense that "the opportunity for a certain kind of experience was being lost."[13]

Leopold and his fellow founders of the Wilderness Society, the organization most instrumental to the eventual passage of the 1964 Wilderness Act, wanted legislation that would protect remote areas of public lands from this car-focused burst of development. Yet they drafted the law's language so as to accommodate many existing commercial land uses, particularly for subsistence or small-scale local economies. For example, the Gila Wilderness Area, first protected by a plan written by Leopold himself in 1924, allowed existing grazing operations to remain, while prohibiting large-scale logging.[14] Leopold also specifically wanted to reconnect aesthetic appreciation to everyday land use, rather than distanced, scenic preserves: "By sequestering natural beauty from economic use, Americans were doing themselves a disservice."[15] During his work in the Southwest, Leopold helped to shape the earliest Forest Service policies in the 1920s for identifying roadless areas, protecting them from road building and other forms of development, and recognizing wilderness recreation as a dominant use. These early policies nonetheless allowed limited logging and grazing to continue, provided they did not require roads. Formalized in 1929 as Regulation L-20, this policy created the first system of protected wilderness, and referred to areas within the system as "primitive areas," emphasizing their less-developed nature while acknowledging that many of these places were not pristine, nor untouched.[16]

After World War II, though, the Forest Service managed its lands ever more intensely for timber production, using increasingly efficient technologies to build logging roads into more difficult terrain and onto steeper slopes. The economic pressure to produce lumber for a nationwide building boom led many in the Forest Service to question the strict protection of "primitive areas"; wilderness advocates believed that the agency's policy toward roadless areas needed to be strengthened by legislation from Congress, or they could be lost. At the same time, public demand for outdoor recreation skyrocketed, and in response the NPS was going through a development phase of its own, articulated most clearly by the Mission 66 campaign, which began in 1956 to upgrade facilities, expand services, and, particularly, improve and add to its road system.[17] In addition to threats of proposed dams in places such as Dinosaur National Monument, this intensification of road development across many western public lands, in service of both timber extraction and motorized tourism, drove much of the push to establish a national wilderness protection law.[18] Yet the purpose was never to "lock up" the public lands; as historian Jay Turner writes, "Even in its earliest drafts, the Wilderness Act included provisions to meet the concerns of rural Americans and the mining, timber, and grazing industries."[19] The bill was designed more to prevent new disruptions of existing wilderness rather than aggressively pushing out long-established local uses.[20]

The resulting 1964 Wilderness Act includes lyrical language, penned by Wilderness Society executive director Howard Zahniser, stating that "a wilderness, in contrast with those areas where man and his works dominate the landscape, is hereby recognized as an area where the earth and its community of life are untrammeled by man, where man himself is a visitor who does not remain."[21] The law directs agencies to identify areas with wilderness characteristics—roadless areas larger than five thousand acres—and recommend them to Congress for formal designation. These areas must remain roadless, and mechanized uses, such as the use of motor vehicles or tools like chainsaws, are severely limited. The law essentially creates congressionally designated wilderness zones that act primarily as a restraint on the actions of federal agencies, not private entities.[22] It was a pragmatic response to prevent overdevelopment of public lands by the very agencies that managed them.

Even so, the wording of the Wilderness Act makes clear that wilderness designations are not intended to displace the Park Service's other functions. Section 4(a) of the 1964 Wilderness Act clearly states that wilderness designation is *supplemental* to the purposes for which national parks, forests, and wildlife refuges are established and administered—in the case of Point Reyes, explicit protection of existing agricultural uses, including dairying, beef cattle ranching, and oyster production, while also providing for recreation use. Section 4(d) furthermore provides a list of the special provisions for established uses within wilderness areas, including grazing, motorboat usage, and other preexisting private rights and uses. Wilderness designation is intended to prevent agencies from constructing new roads and developments—exactly what was originally proposed during the 1960s by the NPS for Point Reyes—not to end ongoing activities.

PROPOSED WILDERNESS AT POINT REYES

Point Reyes was in no way untrammeled or pristine, and evidence of human use was both extensive and prominent throughout the peninsula. Still, the 1964 Wilderness Act required the NPS, along with the other public lands agencies, to review its lands for areas that might meet the wilderness criteria of roadlessness and large size, and that included Point Reyes.[23] The NPS staff at Point Reyes dutifully completed their initial wilderness study in April 1971 and proposed a wilderness area of 5,150 acres, representing the only contiguous roadless block within PRNS. The proposed area was a solid chunk of the Inverness Ridge, bounded by Mount Wittenberg at its north end and Bear Valley Gap on its south end, by the Coast Trail to the west and Bear Valley Trail to the east. At that time, the NPS did not yet own many of the ranches on the Point, but this block, made up mostly of the old U, W, Y, and Z Ranches, was in federal ownership and met the literal requirements of the Wilderness Act. The study described the proposed area as a "backyard urban wilderness," and while noting it was only twelve square miles in size, claimed that it "possesses almost as great a value as the larger High Sierra wildernesses to the east. Its close proximity to a megalopolis intensifies the inherent value—that of the restorative effect of wilderness upon the human spirit—because of the sharp and immediate contrast existing between the urban and wilderness environments."[24] This was a significant change from the rhetoric of the early 1960s, when environmental advocates and legislators argued that it was Point Reyes' pastoral qualities, *not* its wilderness values, that merited protection.

Yet the proposal immediately ran into trouble. Along with proposing the small wilderness segment, the NPS staff was also developing a new master plan for PRNS. The new plan directed traffic away from Limantour and instead would develop the Muddy Hollow corridor, an old ranch road, into a new entrance to the Seashore. This new connection would bypass the town of Inverness and bring visitors directly into the park; the head of park planning felt the traffic through town would be "more damaging than opening the Muddy Hollow Road, where cuts and fills have existed for a century."[25] But this idea triggered a backlash in the Bay Area, as many feared that the Seashore would be overtaken by vehicles. On September 15, 1971, the Marin County Board of Supervisors adopted a statement asking the NPS to reevaluate its wilderness proposal, and specifically to remove vehicle use in the seashore entirely. They also requested that Point Reyes be reclassified as a national park as a way of granting it more ecological protection.[26]

When PRNS held public hearings in San Rafael on its master plan and wilderness proposals later that month, the park staff received a high local turnout, with over two hundred attendees and eighty speakers. California's Senators Alan Cranston and John Tunney sent representatives to the first meeting, who opened it by reading statements calling for an expansion of the proposed wilderness area, specifically to include the shorelines, estuaries, and tidelands. They also pressed for phasing out private automobiles from the seashore, and developing a public transit system instead, to "prevent another Yosemite or a Yellowstone."[27] Conservation advocates insisted that wilderness status would not keep people away, but would "protect the shores from off-road vehicles and hot-dog stands."[28]

The resulting public comment period revealed some local support for the smaller wilderness area. Of all the comments received (including both testimony at the meetings or submitted separately in writing), 70 percent of those submitted by individual citizens supported the Seashore's preliminary proposal. In contrast, 74 percent of the private organizations, dominated by national environmental groups including the Sierra Club, Wilderness Society, National Parks and Conservation Association (NPCA), National Audubon Society, and the Nature Conservancy, supported a larger wilderness area.[29] Many specifically endorsed a recommendation put forward by the Sierra Club that enlarged the NPS's proposed 5,150-acre block of wilderness to 11,000 acres and added two additional sections: a 13,000 unit comprised of Drakes Estero, Limantour Estero, and the hills above them; and an 8,900-acre unit along the shoreline from Tomales Point down to the tip of the Point, for a total proposed area of 32,900 acres. The Sierra Club's plan was specifically endorsed by Senators Cranston and Tunney, as well as by other state and local legislators.[30] In their own comments on the NPS proposal, the NPCA suggested an even larger wilderness area, covering almost the entire peninsula including much of the pastoral zone, but this idea did not receive much support.[31] (See figure 11 for the boundaries of the 1974 "optimum" wilderness proposal, with boundaries very similar to the 1971 Sierra Club proposal.)

Government officials also voiced an unexpectedly strong response to the NPS's initial proposal, with a wide array of government agencies calling for a larger wilderness recommendation. One of the most strongly worded letters was from the California Resources Agency, the state's equivalent to the federal Department of the Interior. Headed by Norman Livermore Jr. (son of Caroline Livermore, who was president of the Marin Conservation League from 1941 to 1961 and honorary chair of the PRNS Foundation), the agency recommended, among other things, that "portions, if not all, of the esteros should be included, together with Limantour Spit. The presence of existing roads or the nonownership of land should not be a deterrent; unwanted roads can be obliterated, and lands can be acquired." They also demanded that at least half the total acreage of the Seashore should be made wilderness "to assure the long-term retention of its natural values and its scenic quality."[32]

Despite a fairly high level of support among local individuals for the NPS's proposal for a relatively small wilderness area, the lopsided reaction from environmental groups and government officials in favor of a larger area pushed the NPS to recommend a larger area the following year. The NPS's 1972 recommendation included 10,600 acres, consisting of the original proposed section plus a block further south, covering the old Glen, Wildcat, and Lake Ranches, plus the Laguna Ranch to the north. The NPS's planning document also considered the two additional areas from the Sierra Club's proposal, stating that they had been suggested by "many individuals and organizations" for designation, but in the end rejected them on the basis of their significant levels of development.[33]

The *Final Environmental Impact Statement for Proposed Wilderness*, published in 1974, reiterated the NPS's recommendation of only 10,600 acres, representing 16 percent of the peninsula. The document also considered three other alternatives: no wilderness at all, a smaller wilderness area (returning to the original 5,150 acre proposal), or

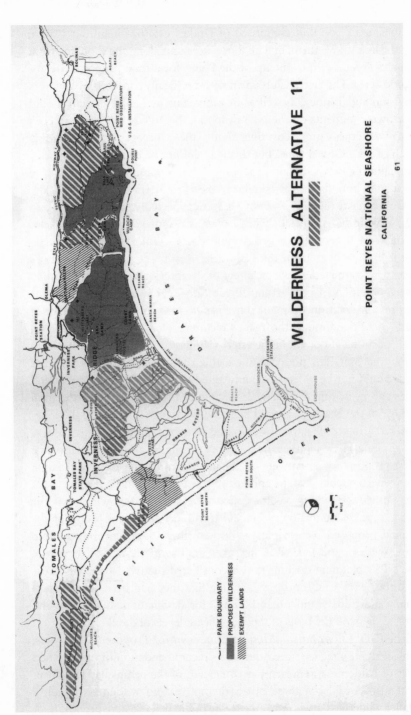

FIGURE 11. "Optimum" proposed wilderness alternative, from the *Final Environmental Statement for Proposed Wilderness* (Western Region, National Park Service, 1974). The dark gray area represents the 10,600 acres recommended by the NPS, and the diagonal lines indicate the additional 27,900 acres proposed by the Sierra Club and other environmental organizations.

a larger one. In the latter alternative, the remainder of the Point (i.e., the parts not in the 10,600 acre proposal) was divided up into ten separate "units," and each was considered for possible inclusion in the wilderness proposal. At the end of the list, an eleventh unit, labeled the "optimum unit," included the 10,600 acres plus five of the additional units listed previously, to cover an additional 14,900 acres. This optimum unit would expand the wilderness designation to include Tomales Point and basically the entire southern portion of the peninsula below the Home Ranch.

Despite the peninsula's long history of ranching, the NPS mostly ignored traces of historical uses on the Point Reyes landscape in identifying potential wilderness designations. The environmental impact statement (EIS) for the proposal included no mention of the old ranch sites within the proposed wilderness areas. The by-then unpaved ranch roads that made up the trail system were cited as "the only visible remnants of man's past activities in the area"—despite the presence of fifty miles of fence lines, some World War II-era bunkers, several old borrow pits, and a few old ranch dumps which were "being removed or obliterated."[34] However, the proposal described the remaining operating ranches outside the wilderness zone with emphasis on the environmental damage they caused, with references to "scars" from development and use.[35] Characterizing the nonwilderness area ranches as destructive to the natural environment forms a stark contrast to the total invisibility of all ranches when the Seashore itself was actually proposed, when the entire peninsula was described as virtually untouched and pristine, an "island in time."[36]

ARGUING OVER WHICH AREAS TO INCLUDE IN WILDERNESS

A major question addressed in the 1974 *Final Environmental Impact Statement (EIS)* was whether or not to include tidal lands in the proposed wilderness area. In 1965, just after the Seashore was established, the State of California transferred ownership of the tidal and submerged lands at Point Reyes to the federal government, yet the State reserved mineral and fishing rights, and NPS argued vigorously that this incomplete title precluded any wilderness designation.[37] For the Drakes Estero unit, the NPS also noted that the presence of the oyster farm, combined with its lease of the bottomlands from the State (valid at the time through 2015), made this particular estuary unsuitable for wilderness.[38] Yet the incomplete title alone caused the NPS to insist that other tideland areas, including Limantour Estero, Abbotts Lagoon, and a quarter-mile wide strip of shoreline—from the tip of Tomales Point all the way down to and around Chimney Rock, and extending down the coast—were also disqualified from wilderness status, despite no commercial uses of any of these areas.

For clarity, the NPS looked to the so-called Reed memo, written in 1972 by Assistant Secretary of the Interior Nathaniel Reed to the directors of the National Park Service and the Bureau of Sport Fisheries and Wildlife, which established guidelines for the agencies as they began developing wilderness proposals. The memo stated, "Lands need not be excluded from wilderness designation solely because of prior rights or privileges such as grazing or stock driveways or certain limited commercial services that are proper for realizing the recreational or other wilderness purpose of the areas."[39] Reed, in a memo to Wilderness Society director Stewart Brandborg, predicted that, "under these [new] guidelines, many of the areas previously excluded will be recommended for wilderness designation."[40] Just as Aldo Leopold had originally envisioned, the official policy was to accommodate some private uses of lands within wilderness, so as to designate a larger total area.

The question of whether the oyster farm's use of the estuary posed a problem played out in comment letters to the draft EIS. Numerous organizations argued back against the NPS's conservative interpretation, instead insisting that Tomales Point plus the shoreline and submerged tidelands must be included to protect the wild character of the beaches. The California Resources Agency reiterated its earlier recommendations from 1971, insisting that more of the tidal lands should be included, and specifically pointed out that, "while we agree that some activities of an oyster farm may have some detrimental effect to the estuary, we do not believe that oyster farming, *per se,* is a detrimental use . . . . Oyster farming there provides more than the 'opportunity to buy fresh grown oysters.'"[41] The Sierra Club's letter similarly stated that the draft EIS "implies that none of the Drakes Estero can be classified as wilderness because of Johnson Oyster Farm. This is misleading. The company's buildings and the access road must be excluded but the estero need not be. The water area can be put under the Wilderness Act even while the oyster culture is continued—it will be a prior existing, non-conforming use. The Reed memo cited previously seems to be speaking to such uses as this."[42] Both organizations insisted that the oyster farm was *compatible* with wilderness designation, as a prior existing use, and so should not be seen as an impediment to including the estuary in wilderness.

By June 1975, Congressman John Burton consolidated the responses to the NPS's limited wilderness proposal into a new bill, proposing 38,700 acres to be included in the designation.[43] This push to expand Point Reyes' proposed wilderness area was not unusual; across the country, the NPS was a reluctant participant in the ten-year review process that followed passage of the 1964 Wilderness Act. Congress, in many instances, "expanded wilderness areas beyond the boundaries proposed by the agencies and president."[44] With the official NPS recommendation still at just over ten thousand acres, but legislators and environmental organizations wanting more, another entity was needed to bring these two "sides" together. That entity turned out to be the Citizens Advisory Commission (CAC), established in 1972 with the creation of GGNRA. Modeled on a similar commission established with Cape Cod National Seashore, the CAC was composed of locals as well as statewide appointees, and served as a forum for airing park policies and proposals, as well as "to give advice to the Secretary of the Interior, and be the eyes and ears of Congress."[45]

The CAC's chair, Colonel Frank Boerger, appointed a wilderness subcommittee in 1975 to study the issues connected to Representative Burton's proposed wilderness bill for Point Reyes. The subcommittee held three meetings in July and August, with various guests attending and putting forth comments and concerns, both favorable to the wilderness proposal and also raising potentially negative aspects.[46] For instance, rancher Boyd Stewart, who attended the July 22 meeting, voiced concerns about the increased fire hazard that wilderness might represent, because of increased fuels, fewer maintained fire roads, and inexperienced hikers. Others argued that specific fire trails and maintenance corridors could be designated to alleviate these fears.[47]

At the third meeting, on August 5, NPS Field Solicitor Ralph Mihan emphasized the flexibility that Congress had in drafting wilderness legislation. In his legal opinion, Congress could allow the continuation of management activities that wilderness opponents were advocating. While a formal vote was not taken, those present "seemed to generally agree" to a series of specific recommendations, mostly focused on the designation of fire trails and maintenance corridors. They also stipulated that specific provision should be made in the legislation to allow two particular land uses to continue unrestrained by wilderness designation: the operation of a small portion of the Murphy ranch that had been included within the wilderness boundary, and of Johnson's Oyster Farm (which would later be renamed Drakes Bay Oyster Company after the Lunny family's purchase in 2005), "including the use of motor-boats and the repair and construction of oyster racks and other activities in conformance with the terms of the existing 1,000 acre lease from the State of California."[48]

By the time congressional hearings were held in March 1976, environmental groups and the legislators they supported had pushed the NPS to recommend the full "optimum unit," minus a twenty-acre private parcel, bringing its recommended total to 25,480 acres.[49] The bills proposed to Congress cited wilderness designation as the highest value for public use, education, and enjoyment of Point Reyes, and suggested that these could be "lost or degraded by management policies designed for *other, less natural* areas of our National Park System."[50] Point Reyes was contrasted with neighboring GGNRA as being more wild and natural, thus deserving of the wilderness designation as well as a reclassification from a recreation area to a natural area—despite the fact that public recreation and beach access had been the primary driver for the Seashore's establishment only a decade earlier. The concept of wilderness had become so popular with the general public that it was almost impossible to resist politically. The late 1970s saw enormous increases in designated wilderness areas nationally, expanding the system from its initial start of nine million acres in 1964 to nearly eighty million acres by the end of 1980.

None of those testifying at the Point Reyes wilderness hearings mentioned anything about the proposed wilderness area's former history as a heavily ranched landscape. One senator questioned the appropriateness of calling areas with so many man-made encumbrances and nonconforming uses "wilderness," suggesting that the original intent of the Wilderness Act had eroded to the point where any place, even downtown Manhattan, could be designated.[51] Yet his concerns were not shared by the other members of Congress. Senator Cranston, who with Senator Tunney had cosponsored the Senate's version of Burton's bill in October 1975, pointedly endorsed the "compromises" that had been worked out by the CAC's wilderness subcommittee.[52] Senator Tunney further elaborated that the NPS's original proposal had been too small and fragmented by segments of nonwilderness; in contrast, their legislation would protect nearly 60 percent of the Seashore from "invasion by motor vehicles or highways," apparently referring back to the NPS's original plans for recreation development.[53] Like Cranston, Tunney highlighted the CAC wilderness subcommittee's recommendations, and emphasized that "established private rights of landowners and lease-holders will continue to be respected and protected. The existing agricultural and aquacultural uses can continue."[54] Representative Burton, who had introduced the original bill on the House side, similarly supported the work of the CAC, and specified that the oyster farm and a portion of Murphy's ranch were to be included as prior, nonconforming uses within the wilderness designation.[55]

Despite the legislators' testimony, the inclusion of tidelands in wilderness remained an issue for the NPS, which still only recommended 25,480 acres of wilderness as well as a twenty-acre parcel of privately-owned (but in process of being acquired) potential wilderness. They continued to leave the tidelands outside the recommended wilderness boundary because, as NPS Director Gary Everhardt explained, of issues relating to the title. "The state," he explained, "has retained fishing and mineral rights on these tidelands, and these areas are also open to navigation, so therefore we have not included these areas in our recommendations because we consider those uses as incompatible with wilderness."[56] The agency also disagreed with designating PRNS as a "natural area," arguing that all three resource types (natural, historic, and recreation) could be found at the Seashore.[57]

The Park Service's arguments were countered by testimony from environmental groups. The Wilderness Society's legal counsel wrote that California's ownership of mineral and fishing rights in the tidal areas was not inconsistent with designation as wilderness.[58] James Eaton, testifying on behalf of the Sierra Club, argued that there was little commercial fishing in the quarter-mile tidal strip, since the surf was so rough and dangerous, and that, "even if there were, the Congress could allow that to continue, since it was a prior existing use."[59] Eaton additionally mentioned the proposed "Estero Unit," pointing out that, in the past, the NPS had proposed building a highway through the corridor.[60] This testimony suggests, as is consistent with the original intention of the 1964 Wilderness Act, that these activists saw wilderness designation primarily as blocking additional development of the Seashore, particularly for motorized recreation, but not as cutting off existing uses.

The hearings wrapped up with letters from California Assemblyman Michael Wornum (Ninth District) and local spokesperson Jerry Friedman—then serving as chair of the Marin County Planning Commission and representing numerous local organizations—that specifically referenced oyster farming at Drakes Estero as a "nonconforming use" that should continue under the designation.[61]

The environmental groups' focus on preventing further development makes sense, given the NPS's early plans for adding roads and recreational facilities, as well as its proposal during the late 1960s to sell a portion of the peninsula off to private developers. The 1964 Wilderness Act is best understood as a restraint on agency action, and that appears to have been the primary rationale for expanding the designated area well beyond the Park Service's early proposals, and particularly to include the tidal lands. In his testimony, Friedman quoted Senator Frank Church, who oversaw the 1964 Act, as stating that wilderness designation was intended to "strengthen the protective hand of the NPS."[62] And as was the original intention of the Wilderness Act, it appears that almost all participants in the wilderness proposals had a common understanding that wilderness designation would not hinder or eliminate any of the existing private uses at Point Reyes. This was most clearly expressed by the Sierra Club in a 1975 letter to the California Resources Agency: "Wilderness status does not mean an end to the harvesting of oysters in the Estero, or a prohibition on the use of motorboats by the company in carrying out its operations (it *would* prohibit any other use of motorboats in the Estero). The Wilderness Act permits prior non-conforming commercial uses to continue, and the Secretary of the Interior can authorize the continued use of motorboats in support of the enterprise. Departmental memoranda express this quite clearly and the regional solicitor has interpreted the act to permit specifically this commercial operation."[63]

But the tidelands remained a sticking point for the NPS. Ever since the first wilderness proposals in the early 1970s, the agency had insisted that none of the tidelands should be designated as wilderness. In a letter from John Kyl, Assistant Secretary of the Interior, dated September 8, 1976, the executive branch continued to recommend that Drakes Estero (among other tidal areas) not be included in wilderness designation, because the reserved rights by the State, as well as the commercial oyster farm, made it inconsistent with wilderness. Representative Burton broke through the impasse in the final hearings a week later by suggesting a new compromise: the revised proposal designated 8,003 acres, mostly the submerged tidelands, as *potential* wilderness.[64] The NPS's representative, Richard Curry, agreed with this approach, stating that the NPS had no objection to the areas where the State had retained rights being designated as potential wilderness. In his written statement, he specified that the NPS would now recommend tidelands as potential wilderness, "to become wilderness when all property rights are federal, and the areas are subject to National Park Service control."[65] This compromise was hailed in the local newspaper the following week, which quoting Burton saying, "This will allow the ownership question to be resolved in the future."[66]

In October, Congress passed Burton's compromise, with a wilderness designation for 25,370 acres and "potential wilderness" additions of 8,003 acres—and oddly, Congress passed it twice.[67] There are two pieces of legislation establishing the PRNS wilderness area; the first is Public Law 94–544, dated October 18, 1976, designating 25,370 acres at PRNS as wilderness, and an additional 8,003 acres as potential wilderness. It enacts H.R. 8002, but contains no definition of potential wilderness within the law itself. This was followed by legislation passed on October 20, 1976, as Public Law 94–567, designating wilderness areas within thirteen units in the National Park System, including PRNS.[68] It enacted H.R. 13160, and included a vague description of potential wilderness in Section 3: "All lands which represent potential wilderness additions [eight of the thirteen units in the Act had potential wilderness designations], upon publication in the Federal Register of a notice by the Secretary of the Interior that all uses thereon prohibited by the Wilderness Act have ceased, shall thereby be designated wilderness."

These two laws were the *first* congressional use of the potential wilderness designation, but their text did not clearly define the term. Shortly after the final wilderness hearings for Point Reyes in September 1976, House Report 94–1680 was written up to accompany H.R. 8002; in its "Section-by-section Analysis," it declared, "As is well established, it is the intention that those lands and waters designated as potential wilderness additions will be essentially managed as wilderness, to the extent possible, with efforts to steadily continue to remove all obstacles to the eventual conversion of these lands and waters to wilderness status. The committee specifically noted that the utility lines, easements and rights-of-way through the Muddy Hollow Corridor should be eliminated as promptly as possible."[69] The first sentence in this paragraph is odd, as the concept of potential wilderness had not yet been used formally by Congress—so it was not in any way "well established"—yet coming only a week after the last House hearings, in which potential wilderness was the primary topic of discussion, perhaps the language of the report makes more sense. It also echoed the 1971 concerns about the NPS road proposal for Muddy Hollow.

Similarly, there is no clear consistent policy within the NPS today on conversion of potential wilderness to "full" wilderness status; for example, the 1980 Yosemite General Management Plan states that "the Ostrander ski hut and the High Sierra camps will be reclassified as potential additions to wilderness. They will continue to be available for public use."[70] In 2007, the same year that the oyster controversy exploded, an NPS official testified before Congress that Southern California Edison could continue its use of a check dam for hydroelectric power within proposed potential wilderness in Sequoia-Kings Canyon "as long as it wants," with no hints toward its steady or eventual removal on any time frame.[71] "Potential wilderness," when one considers the historical record, was not conceived of as a way to gradually phase out nonconforming uses; it was a way to *accommodate* them, allow the rest of the area to be managed as wilderness, and simplify the administrative process of declaring those acres to be part of the "full" wilderness if/when the nonconforming uses cease. It is, in essence, a placeholder, protecting areas that nearly meet the requirements of the 1964 Wilderness Act from any further development, and avoiding the need to go back to Congress to fully designate those areas in the future if conditions change.

In contrast, the discussions in the 1976 hearing make it clear that the primary obstacle to the wilderness designation of Point Reyes' tidal areas came from questions concerning title—issues of control—rather than the presence of any particular enterprise—issues of use. A second legislative report that accompanied the bill, this one from the Senate, revealed that many legislators were in fact uncomfortable with the vagueness of the "potential wilderness" designation, noting that the Senate Committee was not "advocating" for the designation and that it reserved "the right to question this procedure at future wilderness hearings."[72] The idea of including the tidelands as potential wilderness, in other words, was a compromise measure to allow the legislation to move forward, in hopes that the legal issues between the federal and state governments could be resolved in the future. Friedman described the designation to the local newspaper as a "compromise wilderness plan that had the support of numerous factions," and specifically explained the "potential" status of the tidelands as being based on the State's reserved rights. He clarified that Congress had directed the NPS to resolve the title issue with the state as soon as possible, "but in the meantime to manage the land as though it was wilderness."[73]

THE "FEELING" OF WILDERNESS

In his excellent book on the NPS's rewilding of the Apostle Islands National Lakeshore, Jim Feldman discusses the importance of legibility of the landscape for public land managers. Permanent names, distinct boundaries, and standardized management approaches all contribute to simplifying the landscape and making intentions clearer.[74] Potential wilderness, with its lack of clear legislative definition, confuses legibility; it is perhaps too much of a hybrid, and doesn't fit with our tendency to see wilderness as an all-or-nothing designation. Yet following the arguments presented in 1976 at the various hearings reveals the intention of this confusing designation. The phrase was invoked to solve the problem not of an existing use—the oyster farm—but rather that of the federal government's incomplete title. This lack of legibility has fueled much of the controversy over Drakes Bay Oyster Company in recent years.

The case of Point Reyes also highlights the importance of visual impressions of the landscape in determining wilderness status. In many cases, these visual aspects seem to matter more than actual previous uses, roads, or ecological functioning of the area in wilderness policy. This conclusion is echoed in work by geographers David Nemeth and Deborah Keirsey, who found that urban riparian restoration programs are often based more on aesthetics than ecology.[75] The emphasis on the perceptual nature of wilderness is stressed by the 1964 Wilderness Act itself: Section 2(c)(I) says wilderness "generally *appears* to have been affected primarily by the forces of nature, with the imprint of man's work *substantially unnoticeable*" (emphasis added). Mark Woods's analysis of the legal history of wilderness preservation suggests that the relative naturalness of an area is based almost exclusively on this kind of perception.[76] What matters is not the actual degree of naturalness, solitude, or isolation; it is the experience that determines whether an area can be considered wilderness or not.

The Point Reyes wilderness area is not a single contiguous unit, but is scattered in pieces across the peninsula, including approximately twenty-six hundred acres isolated at the end of Tomales Point, small narrow strips of the Point Reyes Beach, and most of Drakes Estero but *not* the lands that surround it. The bulk of the wilderness is in the southern section of the Seashore, but is bisected by the Limantour road, and several access corridors, which have been "cherry-stemmed" out of the designated wilderness boundary, cut extensively across it.[77] The entire wilderness area was at one time or another used for dairy ranching, and a portion of it was also logged. Many buildings and paved roads were only removed a few years prior to wilderness designation; natural processes had not yet had much time to exert their force toward reducing the imprint of human history. Yet none of these facts detracted from the asserted "wilderness character" of the area.

As suggested by language of the Wilderness Act, the feeling of wilderness seems to stem primarily from the visual experience of the landscape. At the 1976 PRNS wilderness hearings, the question arose of whether to include Point Reyes Beach in the designated area. NPS staff argued that it should not be included because of its accessibility to cars; two parking lots, at North and South Beaches, are located immediately behind the sand dunes. In contrast, the Sierra Club representative testified that the beach could count as a wilderness experience as long as people walked a short distance away from the parking lots, as they would not be able to *see* any developments or grazing lands. He argued, "In my own personal opinion many parts of this wild beach have some of the wildest views I've ever seen, since you're faced with the wild, very violent surf and beachlands behind you. You really do not have the feeling of being close to civilization along most stretches of the Point Reyes beach."[78] The actual proximity of the beach to parking lots and several large dairies did not matter; the restricted view was enough to warrant wilderness protection.

In 1999 the Muddy Hollow corridor, originally designated as "potential" wilderness, was deemed to have received full wilderness status, based on NPS actions such as the removal of utility poles. The electrical lines are still there, but were rerouted underground following the 1995 Mount Vision fire. The physical presence of the lines is less important than their appearance. Without the lines visible across the landscape, visitors are not reminded of the actual overlay of civilization on the landscape; they can pretend they are in a pristine, isolated wilderness untroubled by such human imprints.

Nor did the presence of such a large wilderness area seem to have much of an effect on the working landscape, at least at first. In 1978, tacked onto the gigantic omnibus bill creating numerous new parks and wilderness areas elsewhere, Congress added language that created a pathway for agriculture to continue at PRNS. Section 5 of the original 1962 legislation establishing Point Reyes was amended to create a specific mechanism for establishing reservations of use and occupancy (RUOs) on agricultural properties to be converted to leases or special use permits, and giving the historic ranching families "first right of refusal" for those leases.[79] This legislative addition to Point Reyes' management suggests that the move toward re-creating the Seashore into a natural, wild landscape was not as all-encompassing as it might appear—the legislators, at least, still saw room for both wilderness *and* the working landscape.

Yet the same 1999 posting in the *Federal Register* that converted the Muddy Hollow corridor to full wilderness status also included Abbotts Lagoon and the Limantour Estero, stating that all three areas were "entirely in Federal ownership" and any activities prohibited by the 1964 Wilderness Act had ceased—despite the State of California still retaining rights in the tidelands at Abbotts and Limantour.[80] It is not clear why the NPS changed course on the legality of designating lands with incomplete federal title as wilderness, after insisting so vigorously in the 1970s that such a designation was impossible. In the process, the true origins of the "potential" wilderness designation category disappeared into the overlooked legislative history.

FIGURE 12. Horick (D) Ranch, Point Reyes National Seashore, 2015. Photograph by author.

CHAPTER 5

Remaking the Landscape

During the early discussions about establishing Point Reyes National Sea-shore, the NPS generally overlooked the role of ranching in making the place so desirable. At the 1960 hearings, for example, Director Wirth described the area as "left so unaltered by the hand of man."[1] Yet the ranchers had a different perspective, one that emphasized their families' role over the past century or more of working the land. As described by their representative, Bryan McCarthy, at the 1961 House hearings: "We have some beautiful ranches out there today, and delightful and beautiful green hills you will hear described here, with dairy cattle scattered all over them. But these are not accidents. Those green hills were made by man, tractors, and cattle. If you took man off that area and took the cattle off and took the tractors off of it that plowed those hills and reseeded those grasses and fertilized them, you would have nothing more than just an area covered with scrub brush."[2] Similarly, the Nunes family wrote in their letter to Congress, "It has been stated by the proponents of the bill that the prime purpose of the Point Reyes seashore plan is to preserve the wild and natural state of the area. To us, this seems completely paradoxi-cal. It is we who have lived in the region who have preserved the natural beauty of the land."[3] In other words, this landscape not only represented the peninsula's cultural heritage since the time of Anglo settlement, but also the area's *natural* heritage, as the rolling grasslands were kept open by grazing (and likely by Native Americans' burning before that). With-out the cattle, a very different ecosystem would likely develop.

But at the time, the NPS did not make the connection between peo-ple's use of the land and the actual heritage they then began to attempt to protect. Since then, although PRNS was established in such a way as to perpetuate agriculture within the Seashore boundaries, NPS policies have most often worked in the opposite direction, neglecting or disman-tling the built landscape, playing favorites among the ranching families to encourage divisions between them, and replacing ranches with other uses of the buildings or pastures. Thus management practices at Point Reyes, whether intentional or not, and despite some movement at the national level of the NPS to recognize the value and importance of cul-tural or working landscapes, have tended to devalue the area's history and use in favor of creating a new landscape and a more depopulated peninsula. Both goals are in keeping with early national park ideals and the more recent overlay of wilderness preservation.

While this book does not detail the history of Miwok land tenure before the ranching era, nor does it address cultural resource manage-ment approaches to reaffirming connections between present-day Miwok and their tribal history on the landscape—in part because of the very dif-ferent legal context for Native American heritage management than for non-Native-associated landscapes—the Miwoks' legacy in this landscape remains an important precursor to European settlement and warrants further research and attention. Archeological evidence suggests that the Coast Miwok used fire extensively to promote and maintain open grass-lands at Point Reyes for several thousand years, creating the very condi-tions that made the peninsula so attractive first to Mexican settlers raising cattle for the hide-and-tallow industry, and later to Anglo dairiers.[4] While the new settlers brought with them both domesticated livestock and non-native plant species, such as European annual grasses, that have altered the area's ecosystems, the "precontact" landscape of Point Reyes was unquestionably not one that was untouched by human influence.

PRNS has had management authority over the northern portion of the adjacent Golden Gate National Recreation Area (GGNRA) as well as the Point Reyes peninsula itself since 1975. These lands consist of the Olema Valley, formed by the San Andreas Fault, running between the Bolinas-Fairfax Road in the south to Tomales Bay in the north, and the Tocaloma area, east of the towns of Olema and Point Reyes Station. The entire northern portion of the GGNRA has historically been used for ranching, forming an agricultural district along the eastern foot of the heavily forested Inverness Ridge. Because this area is more similar to the Point Reyes peninsula than most of the rest of the GGNRA, which primarily consists of old military sites in and around the city of San Francisco, PRNS has administrative authority over these lands. Thus the Olema and Tocaloma Ranches are included in this discussion of PRNS management of the area's working landscapes. Together, the agricultural lands of this westernmost portion of West Marin stand in stark contrast to those privately owned east of Tomales Bay, where almost all of the landscape remains working, while active use for agriculture has dwindled on NPS-managed lands.

This chapter will detail much of the transformation of the Point Reyes and Olema Valley landscapes from the 1970s through the early 1990s, as many historic buildings and culturally significant places were gradually downplayed or removed—sometimes through neglect, but also often through direct intention. Over this same time period, at the national level, the NPS developed a stronger commitment to managing historic resources and cultural landscapes, as a new approach to heritage as a living part of the landscape took root. Yet the NPS is a highly decentralized agency, as each park unit has its own legislated priorities, and the new national interest in cultural landscapes was slow to have much influence at Point Reyes.

A NEW NATIONAL APPROACH TO CULTURAL LANDSCAPE MANAGEMENT

The new "greenline"-type national parks established after 1978, featuring shared ownership and management of the landscape, forced the NPS to deal more directly than ever before with residents and their land uses inside park boundaries. Yet the federal government had already been struggling for several decades to better define and protect cultural as well as natural heritage. These efforts stem from the late 1950s and early 1960s, when the United States was in the midst of a building boom; much of this activity, such as highway construction and urban renewal projects, was publicly funded, making protection of historic sites an increasingly federal concern.[5] In addition, the historic preservation movement was changing, seeking to preserve architectural and aesthetic values as well as static historical elements. Increasingly, preservationists were endorsing the adaptation of historic structures to "living," contemporary uses as the best means of preservation. Activists were redefining historic preservation in terms of environmental conservation.

In 1964 a presidential task force recommended the establishment of a program, drafted by NPS staff, to provide federal loans and matching grants to state and local governments for historic preservation. The program would additionally provide tax deductions for private property owners who would participate in the effort. A year later, the Special Committee on Historic Preservation (also known as the Rains Committee), sponsored by the U.S. Conference of Mayors, released a report titled *With Heritage So Rich* to publicize the issue.[6] The report called for a "new preservation" that would be integrated with, rather than isolated from, contemporary life: "we must be concerned with the total heritage of the nation and all that is worth preserving from our past as a *living part* of the present."[7] In October 1966, Johnson signed the result of these efforts, the National Historic Preservation Act (NHPA), utilizing the language of "new preservation" from *With Heritage So Rich*.[8]

The NHPA specifically called for the creation and maintenance of what would become known as the National Register of Historic Places, as well as for a process of evaluating sites for possible inclusion on the Register.[9] The National Register already existed in an embryonic form as the Historic Sites Survey and the National Historic Landmarks program, both established by the 1935 Historic Sites Act and housed within the NPS. The NHPA formalized and expanded the Register, and importantly, it added coverage of "properties of less-than-national significance."[10] Yet the National Register maintained the traditional NPS approach to historic preservation, which focused primarily on architectural history and protection of buildings.

The Register's rules and guidelines continue to shape cultural and historic resource management in the United States. In order to be protected by the NPS, landscapes must be determined eligible for listing according to the National Register's criteria of landscape components, use patterns, and structures. The first guideline published by the NPS for applying these criteria was Robert Melnick's 1984 manual for identifying, evaluating, and managing cultural landscapes, defined as rural historic districts.[11] Primarily intended to provide tools for park managers to deal with sites already within the NPS, Melnick's guidelines could be applied to sites on non-NPS lands as well. By setting out these detailed guidelines and management options, the NPS hoped "to avoid a problem that has occurred in the past: the nondecision," that is, allowing deterioration of landscapes, either through neglect, or through alteration with the intent of returning the area to its "natural" character.[12]

Melnick's manual was updated in 1987 by the publication of *National Register Bulletin 30*.[13] The manual's definition of a "historic rural landscape district," carried over to *Bulletin 30*, is "a geographically definable area, possessing a significant concentration, linkage, or continuity of landscape components which are united by human use and past events or aesthetically by plan or physical development."[14] The manual specifically does not apply to "historic sites, scenes, or landscapes," because other policies and guidelines address these, nor does it apply to what Melnick referred to as "socio-cultural landscapes," characterized by "the use of contemporary peoples."[15] The definitions of these various types of landscapes (cultural, historic, rural, sociocultural) are often vague and overlapping, leaving the determination of which kind of landscape one is dealing with somewhat open to interpretation.[16] Clearly, however, Melnick was referring only to a certain subset of the landscapes possible within parks; for example, areas containing tourist shops or campgrounds would not be included in his criteria. The emphasis on continuity in the landscape, and on human use, in defining rural districts seemed like a significant step away from the architecture-based approach that had dominated most historic preservation efforts. Yet park visitors, commercial or industrial interests, and current-day residents were distinctly left out of the definitions or designations of meaningful landscapes.

The goal of protecting these rural landscapes is, according to these guidelines, "to maintain, through management and, perhaps, preservation of individual landscape components, the essential historic character of the district."[17] Each district is required to have one or more designated periods of historic significance, which become the benchmark for measuring whether subsequent changes contribute to or alter its historic integrity. Similarly, one or more areas of significance, or "historic themes," must be identified for the district according to the National Register's requirements. In this way, the interpretation of that particular landscape becomes frozen within a particular historic framework; if it changes too radically from that period, it may lose its eligibility under the National Register criteria. However, neither Melnick's manual nor Bulletin 30 has specific instructions for how to select among the myriad possible periods of historic significance. Nor do they suggest that historical significance could be considered from different points of view. It appears that this crucial decision, pegging the landscape with a single, fixed meaning from "our" undifferentiated history, is left up to the individual park manager to decide with little more than vague guidance.

As discussed in chapter 1, all landscapes are created and re-created by human uses and values, and these cultural landscape management guidelines attempt to recognize the importance of continuity of use over time. Yet the National Register's requirements about historic *integrity* bias protection toward places that are "pure" or "well preserved," which are usually not areas that have been the sites of either continued use or cultural struggle. As with the concept of wilderness, the more adapted places are to present-day life, the less authentic and thus less credible they seem as "historic." Any changes or modifications in the architecture or structures from the designated historic period simply render the property ineligible for listing. Landscape historian Catherine Howett has suggested that determination of satisfactory integrity may be based more on the degree of documentation for the historic period in question than on historical value per se.[18] Similarly, preservation efforts can compromise the long-term viability of working landscapes, as guidelines focus far more on the artifacts of cultural landscapes (specific buildings or landscape elements) than on the *processes* of landscape formation and change. "The real challenge," as geographer Arnold Alanen has put it, "is to develop both the will and the means to maintain those landscapes that continue to exist and evolve as places where people reside, work, and pursue their everyday activities."[19]

The National Register guidelines for protecting rural historic districts represent a tentative embrace of the concept of cultural landscapes at the national level. At a few individual parks, like Cuyahoga Valley National Park in Ohio and Ebey's Landing National Historic Reserve in Washington's San Juan Islands, the shift has explicitly incorporated traditional agriculture into park management. Yet the existence of national guidelines has not necessarily translated into new approaches to management throughout the system. At Point Reyes, as the rest of this chapter explores, park policies and practices have continued to de-emphasize the residents, their uses of the landscape, and the built traces of their long history on the peninsula.

TENURE ARRANGEMENTS AT POINT REYES

In the earliest days of considering the Seashore proposal, local residents worried what the park's establishment would mean for agriculture. At a November 1959 meeting in Marin County of the Citizens Advisory Commission on Development, Bryan McCarthy argued that, even if the Seashore's creation removed only six or seven ranches, "The proposal will surround the dairy ranches and without any doubt, according to those who know, will kill all the ranches, probably sooner . . . most certainly later."[20] Many Point Reyes ranchers testified at the congressional hearings that the proposed acquisition and leasebacks simply would not work for the dairies, arguing that limited-term leases would not provide enough security for such a capital-intensive industry. For example, Albert Bagshaw, the attorney for the Mendozas, Nunes, and Kehoes, as well as for the Point Reyes Milk Producers Association, expressed concerns that ranchers might lose milk contracts if forced to move their dairy operations to conform with the NPS pastoral zone boundaries.

In addition, most of the families had only recently become owners; to have worked so hard only to once again become tenant-operators was a disheartening prospect. In Zena Mendoza-Cabral's rather emotional testimony at the Senate in 1961, she described immigrating to the United States as a young girl and her family's difficult life out on the Point early on: "We worked awfully hard and I am not ashamed to tell you that every bit of land was acquired with the sweat of our brow .... If my ranches would be taken for [military] defense, well, you have to sacrifice . . . but for recreation, what kind of recreation did I have when I was a youngster? Work and save so my children would have a sense of security and a heritage that I felt belonged to them. Now every inch of my land is supposed to disappear."[21]

The fact that there are still operating ranches surrounding the Seashore throughout the rest of West Marin today suggests that the park's ranches might have survived under private ownership in the pastoral zone with the right kind of support or protection from the NPS and/or Marin County. Yet during the 1960s the ranching families became so financially constrained by inflated land values and property taxes, as discussed in chapter 3, that buyout by the NPS seemed the only viable option. But even after the ranchers agreed to acquisition of their land, Congress's explicit intention was that ranching activity remain part of the seashore, as was made clear in 1970 by Senator Alan Bible, who stated that, despite their amendment to the 1962 Act, "the federal government *in effect made a promise to the ranchers* in the pastoral zone that as long as they wanted to stay there, to make that use of it, they could do it. We must keep our word to these people."[22] In case his meaning was not clear, Bible continued, "it is the firm intent of the committee [on Interior and Insular Affairs] that the amendment shall in no way operate to impair the integrity of the dairyman who wants to continue dairy farming."

As most properties were acquired by the NPS, residents were given the option to retain a reservation of use and occupancy (RUO), allowing them to remain on the property for a set number of years, as if they still owned it privately.[23] The reservations were "paid for" by reducing the government's purchase price by 1 percent per year of the reservation.[24]

Most of the ranches received identical twenty-year reservations, although one opted for a longer term of thirty years, and Johnson's Oyster Company took a forty-year reservation. Nonranch private residences were usually offered fifty-year reservations at Point Reyes; the GGNRA's legislation in 1972, by contrast, allowed a maximum of twenty-five years within its boundaries. Most tenant operators were allowed to stay on special use permits, but a few were required by the NPS to relocate for unspecified reasons.[25]

As the RUOs expired, mostly in the early 1990s, they have been replaced with "leases" or special use permits, as directed in the 1978 legislation, which specified that the original landowners or leaseholders be given the first right of refusal.[26] The term "lease" was originally used, according to former Assistant Superintendent Frank Dean, to "convey a longer term and more secure arrangement than the five-year special use permit document provided."[27] Early permits varied in their terms and language, but after 1995, when Don Neubacher replaced John Sansing as superintendent, they became more standardized and contained more precise legal language. Since that time, all new or updated documents are officially called "permits," after an NPS-wide legislative change allowed all parks to retain rental funds at the park level (rather than directing lease revenues to the U.S. Treasury) as a cost recovery mechanism for permits.[28]

Through the late 1990s, the park used two different permit documents. The first, a shorter permit, dealt with unimproved lands and was generally written for only five years but was routinely renewed. The other governed dairies and other improved properties, and contained more detailed language and a renewal clause beyond the first five years. Some of these latter documents still have the word "lease" on the cover along with the permit title. Older "leases" were generally written for twenty years with five-year renewal increments, and were considered by PRNS management a way to give ranchers "more security, as they've got some serious investments in the buildings, equipment, etc."[29] Yet the legal language in both "leases" and special use permits is nearly identical, and the NPS has authority to cancel either for any reason. A 1978 letter from the regional Interior field solicitor described them as "revocable at will."[30] Should the NPS revoke a permit, the permittee has little recourse; they can take the case to the Federal Court of Claims, which can be very expensive, and even if they prevail there, the only remedy is damages. Permittees cannot obtain an injunction against the breach of contract. In practice, these terms limit ranchers' willingness and ability to criticize park management, out of concern that their specific permit terms could be changed.

Being in a permittee relationship with the federal government also reduces ranchers' ability to manage their agricultural operations. Under private ownership, even after selling or donating development rights or other easement controls, ranchers can still decide to sell to another operator, lease the ranch out, or change management methods. Similarly, operators leasing private land have a fair amount of autonomy in making their management decisions; both parties must agree to changes in the lease agreement. With government-controlled leasebacks, in contrast, ranchers are much more constrained in their choices: each ranch is limited to a specific number of cattle or animal unit months (AUMs), the amount of forage a cow and calf will eat in an average month, and park management has almost unlimited leeway in changing permit terms, revoking permits entirely, or otherwise steering land use decisions. Everything from choosing what color to paint the ranch house to deciding to switch the dairy operation over to organic methods must be cleared with NPS staff first. Researcher Vernita Ediger has described this relationship as "a constant negotiation between disparate goals of ranchers and those of the NPS."[31]

Short permit terms not only prevent long-range planning and investment, they also render ranchers ineligible for grants or cost-share programs for doing landscape improvement work, such as erosion control and other types of projects supported by local resource conservation districts or other national agencies like the Natural Resources Conservation Service.[32] At a more fundamental level, however, the ranchers' status as permittees diminishes the clarity of their long-term connection with, and influence on, the Seashore's landscape. From her numerous interviews with ranchers, park personnel, and local environmentalists, Ediger concluded that ranching at Point Reyes is merely "tolerated" rather than actively supported, and specifically that some locals outside the agricultural community, including some NPS employees, felt that "the ranchers operating on PRNS needed to realize that the land wasn't theirs anymore. Ranchers no longer owned the land since they'd been paid for it, so the ranchers shouldn't expect have rights of use after signing the bill of sale."[33] Particularly given the tendency for turnover in park staffing (Ediger estimates most NPS employees move on after two to four years), a lack of institutional memory about the ranchers' relationships with the land, or the founding of the Seashore itself, further contributes to disconnecting residents from the places they live and work.[34]

REMOVAL OF BUILDINGS

One of the most immediately visible changes in the landscape after Point Reyes became a national seashore was a sizable reduction in the number of structures within the park's boundaries. The NPS tends to demolish vacant buildings if they do not have an immediate use; maintaining buildings is expensive, and dilapidated ones represent potential safety and legal hazards. In cases at Point Reyes where the NPS had purchased property and the previous owners did not retain any rights of occupancy, most structures were quickly torn down, buried, or burned for fire crew training, reducing the visible remains of historical uses of the land.[35] Precise records were not kept of how many structures were demolished or when, but at least 170 buildings had been removed from PRNS-administered lands by 1995.[36] This contributed to a more "natural" appearance to the landscape, and effectively erased human history.

One example of how this process played out is the old Gallagher (F) Ranch site. Historically this ranch, probably the oldest ranch site on the Point, was a center of commerce for residents of the peninsula; it was actually marked on California maps as a town until the 1940s. It supported the post office until 1920, and the site of the schooner landing for shipping products to San Francisco was located just below on Drakes Estero. The ranch buildings themselves, abandoned in the late 1940s, have been described as "some of the most significant on the coast" by former PRNS historian Dewey Livingston.[37] Yet after the NPS purchased the property in 1966, all of the then-deteriorating buildings were demolished. The site, now marked by an aging cypress windbreak and remaining corrals, is still leased to the Gallagher family for grazing and remains culturally important to the local ranchers, often serving as a picnic site for gatherings and celebrations. Similarly deteriorated ranch buildings were removed at a number of other ranch sites throughout the Seashore, often leaving no obvious trace of previous settlement and use. All of the Seashore's four hike-in campgrounds are former ranch sites, yet other than the occasional remaining eucalyptus or cypress tree, there is no indication at any of these sites of their historic uses.

Even in the first few decades of the park's administration, some of these removals created controversy. The Lower Pierce Ranch is one such case. The Pierce Ranch encompasses all of Tomales Point at the northern end of the peninsula, and historically supported two ranch complexes, the Upper Ranch dating back to 1858, the Lower Ranch from the 1870s.[38] Prior to the Seashore's establishment, the Pierce

Ranch was owned by John and Dorothy McClure, but John passed away in 1963, and an investment group bought the ranch in 1966 to help settle the estate taxes, as described in chapter 3. The entire parcel was purchased by the NPS in 1973, and the tenants, Merv McDonald and his family, were allowed to remain; they lived at the Upper Ranch, but leased all of Tomales Point for grazing. The buildings at the Lower Ranch had been only occasionally used by the investment partners in the few years prior to NPS purchase, and a wind storm in 1972 caused some damage.[39] In 1975, NPS staff demolished all of the structures at the Lower Ranch without giving warning to the McDonalds.[40] This raised both local and national concerns among historic preservationists that a similar fate would befall the Upper Ranch as well, with a Smithsonian historian commenting, "The immensely popular recreation potential of historic structures seems to be entirely overlooked here."[41]

In the first few decades of the Seashore, NPS staff not only removed buildings in dilapidated condition that would have been expensive to rehabilitate; they also routinely removed structures in good condition if there was no immediate use for them. Consider the situation at the Heims (N) Ranch, the first property purchased by the NPS. At first, the park managers used the buildings adaptively as park housing, but when the ranger who lived there moved out, the buildings were destroyed rather than maintained. One of the buildings demolished was the old Point Reyes I.D.E.S. Hall (a Portuguese religious and social organization), which was built in the 1890s and served as a meeting hall and polling place for all Point Reyes residents for decades. Approximately sixty structures were burned at the South End/Palomarin Ranch in 1966, where the Christ Church of the Golden Rule had developed a religious community in the 1950s. Most of the houses built as part of the Drakes Beach Estates subdivision on Limantour Spit were also burned after the land was condemned, even though they were mostly no more than a year or two old.[42] Once the Lake Ranch property was condemned in 1971, the old ranch buildings were removed; similar demolitions took place at the New Albion (R) Ranch.

Some historic buildings were demolished illegally. At the Lupton Ranch, NPS staff knocked down the main ranch house in late 1994 in direct defiance of the agency's own cultural resource management policies that no historic buildings would be removed—the local paper described it as Sansing's last act as superintendent before retiring.[43] Subsequently, the water tower and milking barn were stripped by wood salvagers and effectively destroyed.[44] Similarly at the Truttman Ranch the majority of the ranch buildings (ten in all) were demolished in February 1994; one small house had been previously burned for fire training in 1993.[45] All of these buildings had been previously determined eligible by the California State Historic Preservation Officer (SHPO) for inclusion on the National Register as part of the Olema Valley Historic District; their removal directly violated both the Section 106 process of the NHPA and the park's official cultural resource management policy.[46]

Overall, roughly half of the built environment of Point Reyes and the Olema Valley has disappeared since the Seashore was established.[47] Some of these buildings were in advanced states of disrepair and might have been removed anyway, but many were destroyed by the NPS for reasons ranging from a lack of immediate adaptive reuse to disregard for historic preservation laws. The disappearance of visual evidence of the landscape's cultural past creates more and more of an impression of a natural or wild landscape.

SHIFTING LAND USE

Historically, the Point Reyes peninsula has supported as many as thirty-two operating dairy ranches. As of 1960 on the Peninsula (not including GGNRA lands in the Olema Valley and Lagunitas Loop) there were fifteen dairy ranches on nineteen thousand acres, and ten beef cattle ranches on twenty-three thousand acres.[48] They held a total of seven thousand dairy cows, representing 20 percent of Marin County's dairy stock, and thirty-five hundred beef cattle, or 90 percent of the county's beef stock.[49] As discussed in chapter 3, the original 1962 legislation was written to allow the continued functioning of the ranching industry within the park boundaries. Even when additional legislation in 1970 gave the NPS condemnation authority within the pastoral zone, Congress clearly directed that ranching remain a permanent part of the Seashore, and it reaffirmed this with its 1978 legislation. When a senator asked NPS Director Conrad Wirth in the 1961 Senate hearings whether the proposed pastoral zone would be kept in grazing in perpetuity, Wirth answered affirmatively.[50] Director Hartzog similarly stated in 1966 that the NPS still believed dairying and ranching were "indeed compatible uses and should be allowed to continue."[51] In forming a general plan for the park over fifteen years later, NPS staff wrote, "Although the establishment of the seashore and influences within the dairy industry have resulted in a reduction of agricultural activity at Point Reyes, Congress clearly intended that the ranches continue to operate."[52]

Despite these intentions, the number of working ranches within the boundaries has dwindled significantly since the Seashore was established. Today there are only six dairies still in operation on the peninsula: Nunes (A Ranch), Double M/Mendoza (B Ranch), Spaletta (C Ranch), McClure (I Ranch), Kehoe (J Ranch), and R&J McClelland Dairy (L Ranch). There are also five beef cattle ranches: Lunny (G Ranch), Grossi and Evans (H Ranch), Grossi (M Ranch), Rogers (formerly part of M Ranch), and Murphy (Home Ranch). In addition, several old ranch sites are leased for cattle grazing.[53] In total, roughly seventeen thousand acres on the peninsula remain in grazing, which represents about 40 percent of the total when the Seashore was established.[54] (See map 3, which shows the changes in operating ranches since the seashore was established.) All remaining ranches operate on special use permits, and the families leasing them have all been working these landscapes for at least four, and in several cases six, generations.

In the Olema Valley, on lands owned by the GGNRA but managed by PRNS, there are no longer any dairies. There remain instead three beef ranches (R. Giacomini, Rogers, and Genazzi), five former ranch sites now grazed (Randall, Lupton, Truttman, McFadden, and E. Gallagher), and one horse ranch (Stewart).[55] On other GGNRA lands administered and managed by PRNS, referred to as the Tocaloma area or Lagunitas Loop, there are two beef ranches (Zanardi and McIsaac) and two ranch sites used for grazing (Cheda and N. McIsaac). The total acreage utilized for grazing of both PRNS and GGNRA is approximately twenty-eight thousand acres.[56] (See map 4, which shows the changes in operating ranches in the Olema Valley and Lagunitas Loop.)

While the trend has clearly been toward fewer operating ranches within the park boundaries, is it possible to attribute this change to NPS management? Certainly other causes have contributed to this decline. The number of dairies operating in West Marin overall has decreased since the 1950s, due to increased competition from large dairies in the Central Valley and more stringent environmental regulations passed during the 1970s, particularly those concerned with water quality.[57] Many of the dairy operations in West Marin switched to beef ranching; for example, the Lunny (G) ranch switched to beef in the mid-1970s, when the family felt that compliance with state environmental regulations had became too expensive.[58] Through the 1980s and 1990s there was a widely held belief in the region that agriculture was generally dwindling, unable to compete economically with industrial-scale farms elsewhere, and might disappear of its own accord. However, the past

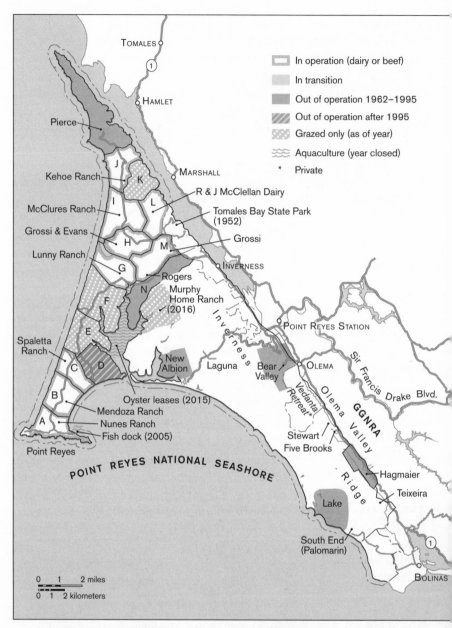

MAP 3. Changes in operating ranches in the Point Reyes National Seashore, 1962 to the present. Courtesy of Ben Pease.

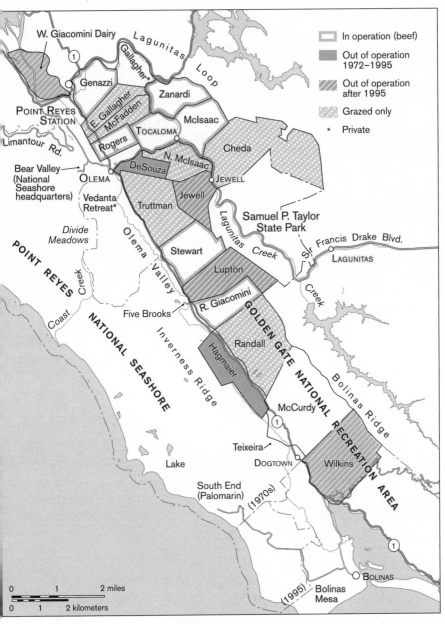

MAP 4. Changes in operating ranches in the Olema Valley, Golden Gate National Recreation Area, 1972 to the present. Courtesy of Ben Pease.

two decades have seen a resurgence of agriculture in West Marin, particularly as localism and the Slow Food movement have called increasing attention to the value of community-based and particularly organic production.[59]

In a number of situations within the Seashore boundaries, ranches, whether beef or dairy, have ceased operation as a direct result of some action taken by park management. These actions take three general forms: NPS purchasing land outside the pastoral zone, where owners were not given an option to remain; NPS not allowing new tenants to take over after owners in the pastoral zone or GGNRA left ranching entirely, either at the time of sale or upon retirement; and NPS deciding not to renew a lease or special use permit. The first issue was mostly fairly straightforward; the three ranches still in operation outside the designated pastoral zone were all closed down upon acquisition by the NPS.[60] In addition, one landowner in the pastoral zone, at the K Ranch, was not offered a right of reservation to continue his livestock operation when his parcel was acquired, for reasons that are not clear.[61] In the second instance, the founding legislation put no specific mechanism in place for continuing a ranch operation if the original owners decide to leave. There are examples of grazing lands on a discontinued ranch site being offered for lease to neighboring families, and there have been discussions about leasing former ranches to new tenants on an ad hoc basis, but no closed ranches have been re-started by another operator since the Seashore was established. And, while some families decided to leave the Seashore independently from the NPS, what exactly constitutes "voluntary" departure can be unclear. Pressure from NPS staff appears to have contributed to a gradual loss of ranches by attrition.[62]

The third case, involving evictions and/or nonrenewals, is more complicated. One example is the story of Rancho Baulines, otherwise known as the Wilkins Ranch, a historic horse ranch located at the southern end of the Olema Valley, near the turnoff to the town of Bolinas. In a three-year process that started in August 1997, the NPS declined to renew long-time tenant Mary Tiscornia's agricultural use lease so as to make way for a new tenant, the Point Reyes Bird Observatory, and a new Environmental Education Center.[63] The 1980 PRNS general management plan (GMP), which named the ranch as a likely site for such a center *if demand warranted,* served as the justification for this action.[64] How the PRNS decided that there was, in fact, demand for such a facility is not clear, as there were already numerous environmental education centers within a short driving distance.[65]

Tiscornia had been a tenant at the ranch since 1969; when it was purchased by the NPS in 1973, she was initially granted the same standard special use permit that most tenants were offered, in the standard five-year increment. In 1987, then-Superintendent Sansing turned down a proposal from the nearby educational nonprofit Slide Ranch to relocate to Rancho Baulines, citing the excellence of Tiscornia's stewardship, calling it "among the very best in the park."[66] A year later, Sansing proposed that Tiscornia switch from a special use permit to an agricultural lease, specifically to give her a greater sense of security as to her rights of occupancy.[67] She had invested substantial sums (over two hundred thousand dollars from 1979 to 1987) into the upkeep of the ranch buildings, and a memo from Sansing to the NPS Regional Director argued that a five-year lease with an option to renew for an additional five years would allow her "to recapture a portion of her investment."[68] When her lease was renewed in 1993, to run through 1998, Sansing added an additional five-year renewal, which should have lasted through 2003.[69]

The security, however, proved to be unreliable. As 1998 approached, the new superintendent decided to not honor the additional renewal clause. Tiscornia brought a lawsuit against the NPS in 1998 to try to prevent her eviction, arguing that the NPS was obligated to conduct environmental and historic preservation review for any proposed changes to the property, but lost in court.[70] The NPS argued that the lands needed to be made available to the public, yet Tiscornia had allowed a wide variety of local uses of the ranch, including hosting weddings and horse shows and providing access for school groups and artists wanting to paint the hillsides. She offered to continue to do so if she were permitted to stay. Instead, the NPS proposed to establish an exhibit hall in the old barn to give a history of Olema Valley ranching.[71]

The proposed eviction generated significant opposition in the local community, as expressed at numerous public meetings, in a flood of letters to the editor of the *Point Reyes Light*, and in a guest column by Steve Matson, chair of the Bolinas Committee on Park Planning, Bolinas Public Utility District, who argued that the current land use at the ranch was an important component of the Olema Valley historic cultural landscape.[72] Ruth Rathbun, whose grandfather first acquired the Wilkins Ranch in 1866, told the CAC that she was saddened by the NPS plan for the property, specifying "I hate to see the human aspect of the lower Olema Valley disappear . . . [and] have trouble understanding the reason for converting an existing ranch into an exhibit teaching kids

FIGURE 13. Barn at Rancho Baulines, Golden Gate National Recreation Area, 2012. Photograph by author.

about how ranching used to be."[73] These local concerns did not affect the superintendent's decision, and Tiscornia was evicted in May 2001.

Yet a few months after the eviction, at an October 2001 CAC meeting, PRNS staff suggested that it might be "inappropriate for there to be a large scale educational center [at the Wilkins Ranch]," and that instead the ranch would be used as a "field station" or research center.[74] Despite a vote by the commissioners a month later to endorse the conversion of the ranch to an education and research center, the Point Reyes Bird Observatory never moved in, and no education center was ever developed.[75] Some historic rehabilitation work has been done by NPS work crews on the residences, but the only use of the ranch has been to house PRNS staff, and the old horse barn sits marked as unsafe to enter (see figure 13). Meanwhile, land management experts have raised concerns that the lack of livestock grazing would contribute to an increase in invasive plants and fire-prone vegetation on the former ranch site.[76] The former pastures are now thick with thistles, and willows along the roadside have grown so tall as to block the view; what had been a highly visible and beloved landmark for the local community has all but vanished from sight.

The cumulative loss of dairies and beef ranches since the Seashore's establishment has seriously reduced the working character of the Point Reyes landscape. NPS management has also had the overall effect of dividing and discouraging the ranching community, thus contributing to its diminution. It is difficult for ranchers to plan for the future when the Seashore's policy toward their operation is unclear. When ranches go out of operation for any reason, there is no established mechanism for finding another tenant, so gradually the number of ranches has dwindled. Some community members fear that eventually all active ranching will disappear from Point Reyes.

INTERPRETATION OF RANCHES AS "HISTORIC" RATHER THAN CULTURAL RESOURCES

Point Reyes National Seashore offers few interpretive materials regarding the ranches, and those few tend to situate the ranches as "historic" rather than current-day interests. This in part reflects past attitudes within the NPS as a whole concerning what constitutes a "historic" resource. While these attitudes have broadened extensively in the past decade or so, not all park units have kept up. At Point Reyes, park management seems particularly reluctant to provide the public with information acknowledging the ranches as present-day cultural resources. Instead, where they are interpreted at all (such as at the Pierce Ranch, discussed below), ranches are posited as only interesting due to their historic lineage.

The earliest interpretative history materials at Point Reyes omitted agriculture entirely, focusing instead on the more "ancient" past. Between exhibits at the Bear Valley Visitor Center and the Kenneth C. Patrick Visitor Center at Drakes Beach, visitors encounter far more information about Sir Francis Drake (who, if he landed at Point Reyes at all, only stayed for a brief time) than about the more than 150 years of agricultural uses of the land.[77] Discussions of historic resources at the Seashore soon after its establishment similarly focused intensely on the possible Drake legacy, overlooking the dairies, radio and military history, and other agricultural uses almost entirely.[78] A preliminary draft of a Historical Resource Management Plan written in May 1968 identified the main historic theme for PRNS as "Early Explorations," with Native American populations and early American settlement as sub-themes.[79] This draft plan recommended that a formal historical structures report be done "to ascertain whether or not Point Reyes has any Historical Structures, beyond the Lighthouse and the Olema Lime

Kilns."[80] The numerous other buildings dotted across the landscape were simply too local or commonplace to be considered as part of authentic "history" for the purposes of this first report.

By 1973, a few ranch complexes began to be recognized for their historic value. A memo from that time indicates that an NPS historian visited the park and recommended six structures or complexes of buildings—the Point Reyes Light Station, the Coast Watchers' Stations, the Point Reyes Life Saving Station, the Home Ranch, the Pierce Point School, and Upper Pierce Point Ranch—be protected as historic resources.[81] As late as 1980, the official Historic Resource Study recognized only two PRNS ranches, Home and Pierce, as qualifying for nomination to the National Register.[82] The Olema Valley, including the seven ranches between Truttman in the north to Rancho Baulines (Wilkins) in the south, was recommended in the same study for preservation as a potential historic district.[83] None of these recommendations, in either PRNS or GGNRA, included working dairies.

Park management provides no comprehensive public information about the pastoral zone itself. One interpretive plaque describes the pastoral landscape, but, tucked away on a spur road to Drakes Beach with almost no space to pull over near it, it is very hard to find. A 1989 "Interpretive Prospectus" document mentions a possible orientation kiosk for the northern district of the park, including the pastoral zone, but it was never implemented. Nor has the park provided any interpretation of the GGNRA ranches in the Olema Valley and Lagunitas Loop, either in the Seashore's visitor centers, on the official park website, or out on the landscape. No interpretive materials mention other agricultural uses of the land, such as production of artichokes and other vegetables in the 1930s and 1940s, numerous hunting lodges, a commercial flower-growing enterprise, and so on. This lack of interpretation contributes to the invisibility of the working landscape, and may cause visitors to question the presence of agricultural operations within a national seashore.

In 1989, in response to many visitor inquiries regarding the presence of ranches inside the park, the superintendent had official "historic ranch" signs posted by some of the old Shafter ranches in the park. These brown and white signs give each ranch's alphabet designation (not the current family name) and the year it was first established, but lack explanatory text to help visitors understand why some ranches are still working and others are not. One actively operating ranch, Rogers, lacks a sign; since it was divided out from Shafter-era M Ranch in the

1940s, it apparently was not considered "historic" enough to warrant a sign.[84] Shafter ranches outside the pastoral zone were not given signs at all.[85] In addition, none of the GGNRA ranches have historic signs, even though many date back to the same era as the Shafters. Because of these discrepancies in how they are allotted, the official signs tell a muddled story, skipping over some ranches entirely and making no distinction between ranches that have disappeared, those that have been reused for other nonranch purposes, and those that remain a living, working part of the landscape.

The NPS moreover reinforces the notion of ranching as a "historic," rather than "contemporary," activity by using the old Shafter alphabet names. This emphasizes the families' status as tenants, and de-personalizes the ranch operations, linking them less to the current occupants than to the system devised by lawyers in the 1850s. In contrast, Shafter ranches outside the pastoral zone, if they are referred to as ranches at all, retain their nonalphabet names (e.g., Bear Valley rather than W, Laguna rather than T, etc.). The two hike-in campgrounds that were originally alphabet ranches were given new names, Coast Camp (originally U Ranch) and Sky Camp (originally Z Ranch). In the case of Rancho Baulines, administrative staff tend to use the old family name of Wilkins, yet most locals refer to it by its 1970s-era name, which the former tenants preferred to use.[86] All of these naming patterns decrease the connection of existing ranches to their current occupants and erase the agricultural history of former ranch sites.[87]

According to the PRNS chief of interpretation, many of the ranchers have not wanted to be "part of the exhibit," in the sense of having their ranches marked on the Seashore's official map. Still, the NPS provides no interpretive material on how long the various families have been there, where the milk from the dairies goes, or why the ranches are still an important part of the seashore. Plans to develop a "living farm" exhibit were discussed many times in the Seashore's first few decades by NPS staff, but were never implemented. This lack of interpretative attention contributes to the ranches, when they are recognized at all, being seen as static things of the past, rather than living and working entities in the present.

INFRASTRUCTURE AND MAINTENANCE ISSUES

Infrastructure designed at the national level within the NPS has the effect of standardizing the landscape, making it more immediately recognizable as a national park unit. Some of these changes are as basic as adding the familiar brown and white park signs, which are standard-issue throughout the park system. Since most of the dairies still maintain their own signs, featuring the family's name and which creamery they ship their milk to, the NPS's historic signs create a stark contrast between the vernacular and official definitions of the landscape.

Similarly, as is the case at most national parks, visitors to PRNS receive a free map of the Seashore. Initially designed locally, by the early 1980s the map had been redesigned according to the nationally uniform "uni-grid" format. The local NPS staff still provide most of the text for the map's descriptions, but the actual design and production takes place at Harper's Ferry, West Virginia, where the NPS maintains a centralized interpretive media center. This consistent design and format generates "brand recognition," signaling to visitors that they are indeed in a national park unit, and perhaps triggers expectations based on previous experiences in other national parks. These wayfinding materials take the landscape out of its local context and resituate it as part of a national system with consistent standards for appearance across the board.

The design and redesign of local roads and trails can also alter the landscape's appearance toward a more standardized format. Almost all of the roads and trails in Point Reyes are historic ranch roads, routes used by locals for generations, yet none have any indication of their origins. For example, the main road through the park, Sir Francis Drake Boulevard, dates back to the 1850s. It is actually maintained by Marin County, not by the NPS, and it remains a narrow, uneven, twisty road. In contrast, roads added by the NPS—the spurs to North and South Beaches, Chimney Rock, the lighthouse, and Drakes Beach, plus the Limantour Road on the eastern side of the ridge—were designed at the centralized NPS Denver Service Center, and are much wider and smoother, with broad shoulders and curbs, and usually correspondingly higher speed limits.[88] These specifications were not designed for the particular traffic patterns at Point Reyes; they are the standardized design of a national agency. They also create a different landscape character, more suggestive of a heavily trafficked parkway than a backcountry lane.[89]

Buildings added by the NPS also steer the landscape away from its vernacular appearance and toward a typical "national parkscape." One of the most central buildings in any national park unit is the visitor center, as it usually serves as the primary source of information for visitors, and is often the first place they encounter in the park. Originally one of the old bunkhouses at Bear Valley Ranch doubled as the visitor

center for Point Reyes, but before long staff felt it did not provide adequate space for exhibits and visitor information. During the 1970s,
park administrators discussed moving the empty Randall House (more
on its story in the next section) from the Olema Valley to Bear Valley for
use as the visitor center. This idea was eventually abandoned, however,
when private funding became available to build a new visitor center.[90]
The resulting structure, built in 1983, was designed by a San Francisco
firm specifically to look pastoral, in keeping with the NPS's "rustic"
style.[91] Yet it does not look anything like the typical 1880s whitewashed
Victorian-style barns that dominate the landscape in and around Point
Reyes; rather, it follows a more generalized idea of a barn, with
unpainted wood meant to blend into the natural surroundings. Similarly, the visitor center at Drakes Beach, built in the mid-late 1970s and
remodeled in 1993, looks nothing like any of the local buildings; it con-
forms aesthetically more to the appearance of wharf buildings found in
San Francisco or Monterey, emphasizing maritime linkages more than
the pastoral surroundings. These additions create an "official" park-
scape visually distinct from the vernacular buildings.

Even ranches that remain occupied by former landowners are affected
by NPS management rules and guidelines. Maintenance choices for the
residents are somewhat limited by the fact that they are tenants of the
Park Service, and must comply with its aesthetic standards and other
rules regarding maintenance. Ranchers own their equipment and cows,
but not the ranch buildings; they are required to get formal NPS approval
prior to making any major changes. If they need to add anything the NPS
considers "unsightly," such as a doublewide trailer for employee housing, ranchers are usually asked to put it somewhere where it can not be
easily seen from the road, or at least to fence the "eyesore" off from
public view.[92] Such requirements decrease the vernacular appearance of
ranches and homes, such that residents' particular choices and needs
become overlaid with official aesthetic preferences.

NPS ownership can also indirectly result in deteriorating conditions
of the buildings themselves. Unlike most regular tenants, park residents
are responsible for "cyclical maintenance"—painting buildings, fixing
fences, and so on—with the NPS paying for major "capital improvements," although the distinction between these two categories is not
always clear-cut or consistent. Tenants are also required to carry insur-
ance on the buildings, even though they do not own them, to help cover
repairs. However, uncertainty regarding the future of their operations,
due to questions about permit renewal or distrust of park promises, can

cause residents to invest less in maintenance of the ranches than they would if their tenure was more secure.[93] Ranchers may put off cyclical maintenance, or skip it entirely, because they do not own the structures, resulting in poorer condition; for resources that the NPS wants to maintain as historic, this can be a problem.[94] Geographer Arnold Alanen has noted that historic preservation guidelines are often imposed on those who cannot afford to meet them.[95] Ranchers may also prefer lower-cost fixes that are less compliant with historical appearance or integrity, such as metal roofing instead of wood shingles, that the NPS may not give them permission to use. While PRNS's cultural resources staff have attempted to find solutions to this problem, including splitting the costs of more historically accurate repairs with ranchers, doing so has often not been feasible, due both to lack of funds and a lack of any formal mechanism in the special use permits. Most recently, the Seashore staff have provided technical assistance for small-scale repairs that ranchers might be undertaking.[96]

Larger scale "capital improvements" require not only permission but funding from the NPS. Local staff must apply for these funds through five-year, competitive proposals. The grant pool is subject to changing funding priorities at the national level: some years the emphasis might be on disability access, other years on projects related to climate change. As a result, large projects, such as structural work on barns, can often wait for years, or even decades, for funding to come available.[97] Capital repairs are also subjected to an internal process of triage, as the most urgent needs get attended to first—balanced by an eye for equity among the ranches, a desire to "spread the work around" so that it doesn't all go to a single ranch in a given year.[98] As a result, a number of ranches are in a state of relative neglect, with a great deal of uncertainty as to when the repairs might occur.

ADAPTIVE REUSE (AND DISUSE) OF RANCHES AT POINT REYES

In the early days, Superintendent Sansing's attitude toward historic preservation was commonly referred to by NPS staff as the "D-8 policy," referring to the particular model of Caterpillar tractor used for demolition.[99] Yet not everything was torn down; some ranches that went out of operation after NPS purchase have since been used for agency purposes instead being demolished. Known as adaptive reuse, this practice maintains the former ranch buildings for such park needs as administrative offices, staff housing, and equipment storage.

Within the Seashore, these include Bear Valley (W) Ranch, which became NPS administrative offices; Laguna Ranch, transformed into a youth hostel; and a few houses scattered through the Seashore—near Limantour Beach, by the lighthouse, on the former D Ranch, near M Ranch, and at Pierce Ranch, the main ranch house—that are used for staff housing.[100] An old school building on the former South End/Palomarin Ranch was long used to house the Point Reyes Bird Observatory field station. In the Olema Valley, the Hagmaier Ranch was used for equipment storage and staff housing; in 2000 it was converted to office space for the new Pacific Coast Science and Learning Center. Rancho Baulines has been used off and on for staff housing since the former tenant was evicted. The Red Barn at Bear Valley, renovated in 2002, is now used for park archives and research center. Around the same time, rehabilitation work was done at the two RCA/Marconi wireless facilities, one of which is now used as the North District Operations Center for NPS staff. Except for the hostel and environmental education center at the former Laguna Ranch, none of these administrative-use buildings are indicated on the official park map, leaving them in the strange state of being visible on the landscape, but unexplained.

While adaptive reuse keeps historic buildings occupied and maintained, it does not necessarily maintain traditional land uses or the community "fabric." It retains the appearance of a historic landscape, but one that is arrested, not working—no longer actively the part of locals' lives; the result, as described by David Lowenthal, is that, "bereft of social meaning, landscapes become vacant, vacuous, void of context."[101] From a cultural landscape perspective, adaptive reuse of buildings is preferable to complete removal of historic structures and/or construction of new additions, but it increasingly seems to be seen by NPS staff as "as good as" continued traditional use. One of seven guidelines for the park's general management plan update process (begun in 1997, but long stalled) states: "Before any new construction is planned, historic structures will be considered for rehabilitation and adaptive reuse." This wording suggests that if no historic structures are currently available for use, management may remove current residents to make way for adaptive reuse. Recent actions suggest that the NPS is, in fact, interpreting this wording in this way. For example, when sixteen RUOs expired in 2000 in the towns of Tocaloma and Jewel, located along Sir

Francis Drake Boulevard within GGNRA lands administered by PRNS, none of those residents were permitted to stay. The community protested these removals, but the NPS had already proposed a number of adaptive uses for the buildings instead.[102]

In some cases, NPS-owned buildings were allowed to stand empty after purchase and removal of local residents, rather than being adaptively reused. For example, despite the superintendent's promise to "do everything possible not to have vacant structures," all but one of the former residences in at Tocaloma and Jewell remain empty as of this writing.[103] Unfortunately, unoccupied buildings deteriorate much faster than occupied ones. They eventually become too expensive to repair, at which time they are often torn down. At Tocaloma and Jewell, the empty buildings are slowly rotting away and are now slated for demolition once funding becomes available so that restoration of the floodplain along Lagunitas Creek can take place.[104] Similarly, on the eastern side of Tomales Bay, the oyster-farming enclave known as Hamlet, which had been the home of Jensen's Oyster Beds since the 1930s, was purchased as part of the GGNRA in 1987 and determined eligible for listing on the National Register. According to the former owner, Virginia Jensen, "They may not look like the Taj Mahal, but the buildings were all livable when I moved out."[105] Ten years later, the NPS announced plans to replace the then-dilapidated buildings with a parking lot and walking path. They were eventually taken down in 2003.

This is not a new problem. In the 1977 *Assessment of Alternatives for the General Management Plan*, "neglect of cultural features" was identified as one of three primary problems facing GGNRA/PRNS at the time: "Some of the most important cultural features of the park are suffering from lack of use."[106] Even if they are not removed, empty ranch buildings that have been boarded up contribute to a sense of an arrested landscape, one whose original inhabitants are long gone and their land use activities a thing of the past. Even if the buildings are maintained and/or restored, the result, as described by historical geographer Richard Francavigilia, is "a serene, bucolic place, a still-life rather than an active, bustling place full of strife and difficulties."[107]

One such example is the Upper Pierce Point Ranch, which went out of operation in 1979 to make way for the tule elk herd on Tomales Point (more on this in the next chapter). Between 1968 and 1976, Pierce Ranch was often proposed as an ideal site for a "living ranch" exhibit, which would allow Seashore visitors to see how a working dairy operates.[108] Even after the site was surrounded by designated wilderness in

1976, the NPS apparently continued to consider adaptive reuse options; a request for proposals (RFP) to lease the ranch was issued as late as 1986, for either private for-profit or not-for-profit undertakings by individuals, associations, or organizations.[109] No tenant was ever chosen. The ranch complex, which had been occupied for roughly 120 years, is now completely frozen in time; it was painstakingly rehabilitated in the late 1990s by the Seashore's new-at-the-time historic carpentry crew, at a budgeted cost of approximately $640,000. The outbuildings, such as the former hay barn, horse barn, bunkhouses, and schoolhouse, are now used for self-guided historic interpretation of the ranching industry circa the 1880s. Visitors read interpretative plaques along a path that runs past the buildings' exteriors, which stand empty and boarded up, except for the main house, which is not part of the exhibit and is used for NPS staff housing.[110] "Living ranch" proposals continued to be considered through the 1980s, but none was ever established.

A somewhat similar fate has befallen the Randall House in the Olema Valley. The two-story Victorian residence, built roughly in 1885, was occupied until 1973. Shortly after NPS purchase in 1974, the property was evaluated by the Regional Historic Preservation Team, and was described as being "in good condition," as was the barn to the rear of the house.[111] As the house sat empty close to the main road, it was soon vandalized and began to deteriorate. The superintendent at the time made it clear that he preferred to demolish the building and restore the site to natural conditions.[112] When locals began to protest the park's plans for removal, however, the NPS staff requested funds for basic stabilization and repair.[113] First the building was proposed for use by the Coastal Parks Association, which at the time was using space in Bear Valley Information Center; later the NPS considered moving the house to Bear Valley, where it would be used as a visitor center.

While the NPS went back and forth between possible plans, the empty house continued to suffer. By 1981, the cost estimates for repair had skyrocketed, and, despite the fact that the building had been determined eligible for listing on the National Register, the superintendent once again proposed to tear down the building.[114] Although a total of twenty-one people or organizations inquired about leasing the property between 1979 and 1981, PRNS officials did not pursue any proposals, nor did they put out a formal RFP until 1983, which also did not result in any accepted proposals.[115] The last prospective leasee, who proposed leasing the building and accepting all responsibility for restoration, wrote in 1983 that he had "only encountered discouragement" from PRNS staff,

FIGURE 14. The Randall House, Olema Valley, Golden Gate National Recreation Area, 2000. Photograph by author.

and that removal of the Randall House would only serve "an administration that wishes to eliminate unwanted responsibility."[116]

Eventually a colony of Townsend's big-eared bats *(Plectotus townsendi)*, a rare species, was discovered roosting in the attic of the house. Ironically for a house that faced demolition at the hands of the NPS for years, it is now protected as natural habitat. Because the presence of humans would disturb the roosting bats, the historic house's exterior is neatly painted and maintained in good condition, but it cannot be used for any other purpose. The boarded-up house is highly visible, located close to the road along a curve of much-traveled Highway 1, and is often cited by locals as an example of poor park management and lack of concern for neighboring communities (see figure 14).

Adaptive reuse keeps historic buildings occupied and in good shape, but removes them from their original local contexts of use as ranches and family homes, converting them instead into offices, employee housing, and public facilities. Over the years, PRNS management has seemed unperturbed by the prospect of evicting tenants to make way for adaptive reuse, even though the buildings often then sit vacant so long as to become unusable. Particularly given the large numbers of structures previously removed by the NPS because they had no immediate adaptive use, this tendency continues to raise concerns about the future of the working ranch landscape. Both adaptively reused buildings and vacant ones contribute to the movement toward an increasingly static, museum-like landscape. Adaptive reuse also recasts the landscape away from local history and more toward the official "parkscape."

NEW ATTENTION TO THE CULTURAL LANDSCAPE

In his introduction to a special thematic issue in 1993 of *Cultural Resources Management,* dedicated to cultural and historic landscapes, NPS staff member Charles Birnbaum suggested that the landscape preservation movement needed a "reality check." He quoted architectural critic Ada Louise Huxtable as saying that there had recently been a shift in perception, emphasizing that in much preservation work, history had become "bowdlerized," that is, simplified and made accessible to the masses, but not authentic.[117] Underlying this concern, however, was a crucial assumption that seemed pervasive throughout the Park Service's approach to cultural landscapes: that there exists a single authentic past, and that it can be discerned by experts, using the proper guidelines and criteria. This assumption does not allow for multiple interpretations or experiences of the past, nor does it allow for changes in the meaning of "the past" over time.

There admittedly must be some sort of criteria to decide which places warrant special protection; otherwise the landscape protection program would be dangerously ambiguous and arbitrary. However, NPS must at least recognize and acknowledge the implications of the built-in biases of the program.[118] By tying significance and integrity to a specific historic period and particular physical characteristics of the landscape, the criteria steer the definition of the "authentic" landscape away from more intangible or present-day cultural meanings. Alanen and Melnick have noted an inclination in cultural landscape preservation to "simplify rather than clarify the values inherent in cultural landscapes and, correspondingly, to simplify responses to those values."[119] In addition, they suggest that NPS's approach to preservation "holds the potential to negate the very idiosyncratic landscape qualities that set one place apart from another."[120]

The interest in cultural landscapes that began within the NPS in the early 1980s has, since 1990, grown into the small, national-level Park Historic Structures and Cultural Landscapes Program. The primary focus of this office, working with support staff in regional offices around the country, is establishing a thorough Cultural Landscape Inventory, creating a new database of all cultural landscapes within the Park System that have potential to be nominated to the National Register.[121] Official NPS policy guidelines for cultural resource management now recognize cultural landscapes as a distinct resource type.[122] At the same time, however, many parks' staff have not yet embraced the meaning of this term. One aim of the national program is to educate park staff to ensure that they understand the cultural landscape concept and how to include it in planning processes.[123]

Overall, the historic and cultural preservation side of NPS management is very top-down, based more in the centralized offices than the park units themselves. Traditional park management, in contrast, tend to be very decentralized and park-specific. Park superintendents often prefer to have extensive discretion than to follow guidance from Washington. Historic and cultural preservation also clashes with the underlying ideological themes that inform almost all aspects of park management, which idealize natural scenery, national rather than local significance, and unchanging views. The new cultural landscape program has admirable intentions, but faces an uphill battle against these historic biases of the agency and its staff, the general weakness of the historic/cultural branches within the agency, and the preservation structure and requirements imposed by the NHPA and National Register that conflict with the reality of cultural landscapes. It remains to be seen how effective this program will be in changing these attitudes.

Within PRNS, there is some movement afoot toward greater recognition of the cultural landscape. The first step in that direction came in 1991, as part of a draft Historic Resource Study. At a meeting at the Seashore in June of that year, the director of the NPS's fledgling national cultural landscape program described Point Reyes as "the most significant cultural landscape in the System," and one that warranted greater attention.[124] In 1995 the entire peninsula was found eligible for the National Register by the California SHPO as a rural historic district under three of four possible criteria.[125] The period of significance was set to match the Shafter era of ranching, 1857–1939, the latter being the year in which the last Shafter-owned ranch was sold.[126] The period of significance for the district has since been extended to 1956, based on the continuity of use from the Shafter ranch system. The decision to put the endpoint immediately prior to the first Seashore proposal, however, elides the substantial effect park management has had on the landscape, and again discounts the significance of working ranches that continue the same agricultural use patterns into the present.

While it is not yet definite, the boundary of the nominated historic district is likely to change as well. Most likely, it will include only the pastoral zone portion of the peninsula. This reduction is seen as necessary to get the nomination approved; if too much of the "parent landscape" does not contribute to the area's overall historic integrity (i.e., the wilderness area), the entire cultural landscape designation could be rejected.[127] In effect, the criteria are forcing a definition of the cultural landscape as what is left of it today, rather than the complete history of the area. Sections of the park outside the district boundary may not be recognized as having any cultural or historic value at all, continuing the erasure of history from them.

In the late 1990s PRNS hired a small cultural resource staff, including a landscape historian, an archivist, and a historic preservation carpentry team. These managers were eager to work in concert with the natural resources staff, and put forward some innovative ideas for reemphasizing the importance of the cultural landscape. In 2004 park staff completed a series of cultural landscape inventories at the Seashore, preparing to submit a formal nomination of the area for inclusion in the National Register as a rural historic district; additional inventories were completed for six Olema Valley ranches in 2011.[128] This effort represented a marked change from earlier management at the park, which tended to minimize the historic and cultural qualities of the landscape. Yet efforts to write a cultural landscape report, with management recommendations, have been hampered by the fact that the guidelines are not well designed to address large-scale, lived-in landscapes.[129] Similarly, it is not yet clear whether this documentation will translate into management actions to instill a stronger sense of the importance of maintaining an actively working landscape at the Seashore.

While this documentation effort is an important step toward greater recognition of the cultural resources at Point Reyes, it may further contribute to creating an arrested landscape, as the criteria for nomination focus only on those elements of the landscape with "integrity" and "significance." The level of historic significance has not yet been decided upon; a determination of national significance will bring greater scrutiny to ranching practices, upkeep of buildings and so on, putting more pressure on ranchers for compliance with preservation rules.[130] "Integrity" is defined as "the degree to which the landscape continues to portray its historic identity and character."[131] Some landscape elements that do not meet these criteria may still be considered as "contributing" to the overall integrity of the cultural landscape, but others may get overlooked.

Because the criteria were originally derived from architectural models and still rely heavily on material evidence, resources that have been removed (such as the many buildings demolished by the NPS) drop out of the story entirely. And, perhaps most importantly, there is no place in this process to consider what the residents think or feel about their own place, the *meaning* of the land to them. Nor does continuous use by the same family over generations merit any kind of special weight or consideration. The focus remains primarily on the buildings and other structures themselves, and not on the people and their experiences.

As Alanen and Melnick have pointed out, an overreliance on codified standards for management "holds the potential to negate the very idiosyncratic landscape qualities that set one place apart from another."[132] In addition, many of these efforts to recognize and manage the cultural landscape run counter to the institutional ideology already in place; already the more "traditional" values in park management have conflicted with some proposed changes. It remains to be seen how many of these ideas will actually see implementation, and what the overall effect will be on the Point Reyes landscape.

Emily Wakild's excellent book on the history of Mexico's national parks in the early twentieth century, which were established with the express intent of maintaining rural peoples within park boundaries, hints at what could have been possible at Point Reyes. Her research explores how the parks' radical inclusion of local residents and their uses of the land served to "confirm the connections between social stability, economic productivity, and landscape conservation."[133] In that time and place, the prospect of a revolutionary government committed to upholding the rights of laborers then pushing them off the land to make way for parks was politically untenable. In contrast, the U.S. national park ideal, developed in the late nineteenth century, allows no space for residents and their relationships with the land, even if those relationships are the very activities that shaped the land into something worthy of preserving. Attitudes about protecting "ordinary" landscapes are shifting, but slowly and unevenly through the NPS, and the park ideal is still a powerful force employed by some environmentalists to pursue a more "pure" vision of nature in parks.

FIGURE 15. Tule elk herd on the Spaletta (C) Ranch's pasture, Point Reyes National Seashore, 2013. Photograph by author.

CHAPTER 6

Reassertion of the Park Ideal

Despite the reduction in the number of working ranches and built land-scape features over the first few decades at Point Reyes, most locals regarded Superintendent John Sansing, who held that role from 1970 to 1994, as a friend of the ranching community. He made improving rela-tionships with the ranchers a major objective of his administration, holding regular informal meetings with them at one of the ranchers' homes, and throwing an annual end-of-summer picnic for the agricul-tural operators to get to know park staff.[1] Ranchers did not necessarily consider him as an easy administrator to work with—he could be brusque, and played favorites—but they certainly trusted him. Sansing tended to be fairly hands-off with management, and if he said he'd do something, he could be counted on to follow through.[2]

In January 1995 Sansing retired. The new superintendent, Don Neu-bacher, brought with him a more explicit goal of expanding the Sea-shore's resource management programs.[3] In contrast to Sansing, whose background was in accounting and land acquisition, Neubacher held a masters' degree in natural resources management and had extensive experience with formal planning processes in the NPS, including having served as the team chief on the Presidio master plan in the early 1990s. Neubacher hired a cultural resources manager, gave more attention to cultural resources, and frequently made statements as to the value of the area's ranching history. Nevertheless, PRNS under Neubacher's lead continued to steer management toward the national park ideal of scenic, wild-yet-managed nature with as little human use other than recreation (and park management) as possible. Despite substantial shifts in general understanding of the natural world and in outdoor recreation patterns since the NPS was first established in 1916, the park ideal remains surprisingly powerful.

In keeping with the park ideal, the plans and projects undertaken at PRNS since 1995 have favored a much more purist vision of the area's ecology and wilderness. Two examples include the restoration of the Giacomini Dairy, at the southern end of Tomales Bay, into a tidal wetland and the removal of nonnative deer from the Seashore despite local opposition. A third, slightly earlier example—the reintroduction of tule elk in 1978, followed by establishment of a free-ranging herd in 1998—most centrally represents the tensions between managing natural resources and maintaining the working landscape. The natural resource management perspective tends to overlook the fact that Point Reyes is not an intact ecosystem. It has very few predators, and with public hunting banned in most national parks, an overabundance of grazing animals is inevitable without some kind of population control. This landscape requires active, hands-on management, yet the NPS keeps responding to demands from certain sectors of the public for at least the *appearance* of hands-off, "wild" nature.

NATURAL RESOURCES MANAGEMENT IN THE NATIONAL PARKS

Despite a common association of national parks with natural open spaces, the NPS's move toward ecologically informed resource management occurred only fairly recently. Writing in 1997, NPS historian Richard Sellars concluded that the NPS's approach to ecology had not been particularly scientifically informed throughout most of the Park Service's history. Rather, he wrote, "nature preservation—especially that requiring a thorough scientific understanding of the resources intended for preservation—is an aspect of park operations in which the Service has advanced in a reluctant, vacillating way."[4] In the NPS's first century, the overriding focus on pleasing tourists drove resource management far more than science. This on-again, off-again relationship with the biological sciences is perhaps best exemplified in the 1963 Leopold report (written by Aldo Leopold's son Starker, a wildlife biologist), which remains one of the primary policy statements for the NPS on its natural management goals. This wildlife management study, along with a research review issued by the National Academy of Sciences in the same year, urged "a potent infusion of science into national park management."[5] Yet while recommending that "naturalness should prevail" in the parks, and despite acknowledging continuous change in nature, the Leopold report reiterated the agency's longtime goal of managing parks to "pre-European contact" appearance, deliberately re-creating or maintaining frozen examples of "primitive" America.[6]

The protected landscapes of the national park system are commonly imagined as wild sanctuaries for thriving ecosystems, even though, as science writer Emma Marris observes, "a historically faithful ecosystem is necessarily a heavily managed ecosystem."[7] In other words, the places that look the most pristine are likely the least wild in the original sense of the word. Wildlife populations in particular have been extensively manipulated and regulated in national parks. Parks have long been managed to support large populations of game animals, both to delight visiting tourists and to serve as a source of game for hunters on surrounding lands. For decades, the NPS pursued this goal with a vigorous predator control program. Public hunting is generally not allowed in parks; without any external limits on populations, herds of animals such as elk or bison frequently grew larger than local ecosystems could support. In some cases, park management has chosen to feed these animals rather than to allow them to starve to death during harsh winters or other lean times. Yet, as Sellars pointed out, "For many, spectacular scenery may create an impression of biological health . . . the public may take for granted that unimpaired natural conditions exist, especially in the larger parks. To the untrained eye, *unoccupied* lands can mean *unimpaired* lands, even where scientists might quickly recognize that human activity has caused substantial biological change."[8]

With an increasing public focus on environmental preservation through the 1970s and 1980s, the NPS did not escape the critical eye of environmental groups. Some activists regarded the agency as a bit of a turncoat; many environmentalists, having internalized the parks' mythology, expected the NPS to be primarily concerned with preserving pristine nature. They were disappointed, then, to find instead a "typical" agency behaving in its own interest, rather than in the apparent interest of the public or the environment.[9] Environmental groups increasingly pressured the NPS to take a more preservationist approach to its management policies, particularly for natural resources and ecological systems.[10]

By the 1980s and 1990s, the environmental movement had also embraced a less pragmatic and far more absolute, purity-based ideal of wilderness as perfectly pristine. Groups like Earth First! and the Wildlands Project demanded public wilderness proposals based on the new science of conservation biology and a new approach to wilderness recreation exemplified by the "leave no trace" motto. This newly idealized landscape, which activists envisioned as pristine, unpeopled, and ahistoric, excluded hunters and rural westerners, who had their own uses of and relationships with wilderness.[11] It also began to reorient mainstream environmental organizations' goals, as groups like the Wilderness Society and the Sierra Club increasingly adopted this more fundamentalist conception of idealized wilderness. Yet the ideal of wilderness obscures the power dynamics of those who assert it; as historian Richard White points out, "An appeal to nature is always an appeal to a certain *kind* of nature, but it masks the choice by making it appear that humans are not making a choice about which nature—but rather that we are letting nature itself choose and nature always knows best."[12]

During this same period, the NPS's center of authority shifted away from the director's national office, making agency subunits increasingly autonomous. Individual park superintendents began to act more and more independently, sometimes with little guidance from, or in complete opposition to, the agency leadership.[13] This shift has had profound implications for park management, as parks are often managed in particular ways that individual supervisors prefer, leading to a wide variety of approaches, often with little consistency between parks. Policy changes initiated from the top of the agency's structure are not always implemented at the ground level. Change that involves a move away from the embedded ideologies of national parks as natural, uninhabited places has been particularly difficult to implement.[14]

Overall, NPS management culture can be characterized as fairly utilitarian and pragmatic, emphasizing expediency and quick solutions. It tends to avoid information gathering through long-term research and seeks to minimize outside interference.[15] Legal scholars Joseph Sax and Robert Keiter, in their study at Glacier National Park, found that park officials' behavior tends to "resist being rule-bound"; managers prefer broad mandates to specific rules and regulations, often regard officials from both other agencies and interest groups as "outsiders," and usually seek to protect their discretion as much as possible.[16] A NPS draft report produced for the 1991 Vail Conference on National Parks articulated the agency's reaction to environmental concerns as "sporadic and inconsistent, characterized by alternating cycles of commitment and decline."[17]

Sellars's examination of NPS history also described the agency as having an attitude of "deep-seated indifference" toward national environmental and preservation legislation, often regarding compliance with laws such as the National Environmental Policy Act (NEPA) and the NHPA as merely "jumping through hoops."[18] NEPA requires agencies to complete environmental assessments before taking any major federal action that might cause substantial impact to the environment. In cases where environmental impacts are likely, agencies are required to issue a formal environmental impact statement (EIS). In practice, however, NEPA regulations are inconsistently observed and enforced. NEPA also requires public involvement in the environmental review process, something not overly welcomed by many park managers, who reportedly believe "there is already too much public involvement in NPS decision making."[19]

In 1998, Congress gave the NPS a new science mandate by passing the National Parks Omnibus Management Act, authorizing new scientific studies and encouraging an ecosystem management approach to public parklands.[20] The agency responded in 1999 with its Natural Resource Challenge program, which provided funding for improved natural resources management and research in the parks. Parks scholars have hailed this program as substantially increasing the role of science in NPS decision-making. The program has also strengthened the NPS's relationships with nonpark scientists and expanded its natural resource programs. There has been no similar systemwide effort, however, for cultural resources.[21]

A NEW EMPHASIS ON NATURAL RESOURCES AT POINT REYES

Both staff and funding for the natural resources division at PRNS expanded after Neubacher became superintendent in 1995. The Seashore has extensive science and management programs for various kinds of wildlife, including elephant seals, snowy plovers, coho salmon, and steelhead. In addition, agency staff have studied the peninsula's vegetation in depth, making efforts to identify and map plant communities, manage endangered species, and so on. Restoration efforts focus on coastal dune habitats and wetlands. To some degree, these priorities reflect the NPS's increased attention and funding at the national level over the past two decades.

Natural resource management is unquestionably a major part of the NPS's mission, but in many instances, it comes at the expense of cultural or historic resources, as well as the working landscape. For example, one major management priority at Point Reyes is the control of nonnative plant species such as iceplant, broom, and yellow star thistle. Management often relies on volunteer labor to hand-pull plants, or uses prescribed burning to keep the populations under control. Yet in some instances within the Seashore, exotic species have special cultural meaning or significance. A prominent example is a mile-long line of eucalyptus trees between the Lunny (G) and Grossi/Evans (H) Ranches, marking the old boundary between the Howard and Shafter holdings. When first planted in the 1870s, the line was a colonnade of trees, but it has since expanded outward. This grove now presents a management dilemma, as the trees are an exotic species, yet they also constitute a major component of the historic cultural landscape. The reverse problem exists for many lone eucalyptus or Monterey cypress trees that mark former ranches, building sites, or cemetery plots; in many cases the trees are beginning to die of old age, yet it is not clear whether, or how, the NPS plans to replace what are often the only remaining markers of these cultural sites. At the former Horick (D) Ranch, NPS staff have recently attempted to get a new windbreak started, using Douglas fir rather than the more traditional Monterey cypress because they are considered "more native" (the cypress is native to the Monterey peninsula, only a bit farther down the coast, but does not occur naturally at Point Reyes), but the resident population of tule elk (more on this below) keeps eating the saplings.

In a similar vein, in 2000 the NPS acquired the Waldo Giacomini Ranch, located just outside the original park boundaries at the southern end of Tomales Bay, for inclusion within the GGNRA, so as to restore the area's historic tidal floodplain. In the early 1900s, ranchers constructed levees to transform the tidal marshes into pastures. The recent restoration project, with most construction done in 2007–8, removed those levees, tidegates, and culverts, reconnected the historic floodplain to Lagunitas Creek, and aimed to improve water quality, flood protection, and wildlife habitat. The NPS ended the ranch's historic dairy operation to make way for this major wetlands restoration project, removing all of the buildings except for the old hay barn. The NPS also shut down the ranchers' small gravel quarries out in the pastoral zone, used for over a century to obtain rock for maintaining ranch roads, and filled them with the dredge material from the restoration project.[22] This forced ranchers to import rock for their ranch roads from beyond the seashore boundary, a practice that is both more expensive and intro-duces geologically foreign rocks to the peninsula. It is also harder on the cattle's feet, as the "native" rock is softer.

One of the most controversial natural resource campaigns at the Sea-shore was the elimination of two species of nonnative deer between 2007 and 2009. The axis deer (*Axis axis,* also known as the chital or spotted deer and native to Southeast Asia) and fallow deer (*Dama dama,* native to much of Eurasia) had been residents of the peninsula since the 1940s, when local eccentric Dr. Millard "Doc" Ottinger purchased thirty of them from the San Francisco Zoo and released them on his own land. The deer soon spread across much of Point Reyes and into the Olema Valley. The deer are either spotted or white, making them easier to see than the native black-tailed deer (*Odocoileus hemionus columbianus*), and some-thing of a novelty. Their populations were kept in check by locals' hunt-ing expeditions until the Seashore was established and new regulations against public hunting went into effect in 1967.[23] After public hunting ended, the deer populations boomed; between 1968 and 1994, NPS rang-ers culled hundreds of both axis and fallow deer to limit the herds to about 350 animals each.[24] The Seashore's 1976 *Natural Resources Man-agement Plan* called for reducing their numbers. While there was some support for granting "depredation permits" to the ranchers to resume their former hunting practices, public opposition to hunting kept the rangers saddled with the culling duties.[25] But culling also proved unpopu-lar with the public, and by 1992 the natural resources specialist at PRNS predicted that they would eventually eliminate all the nonnative deer.[26]

Citing resource management shortfalls and continued public criti-cism, the NPS ended culling of the axis and fallow deer in 1994.[27] Pre-dictably, their populations once again expanded; in 2004, PRNS released a draft management plan for the nonnative deer that proposed either eliminating them all, or shooting most and treating the remainder with contraceptives. Animal-rights groups and others concerned for the animals' welfare, including primatologist Jane Goodall, protested the killing, and many locals spoke out about their fondness for the unusual deer.[28] Despite the outcry, in 2006 the Seashore decided to eliminate the deer. Sharpshooters hired from White Buffalo, Inc., were brought in to do the work, mostly from helicopters.[29] When public opposition contin-ued, including letters from both California senators and from House Speaker Nancy Pelosi, the program was brought to an end in 2009. By that point, however, most if not all of the deer had been wiped out.[30]

MAKE WAY FOR THE ELK

Under Sansing's tenure as superintendent, only one ranch operator was forced to relocate. When the Pierce Point Ranch was acquired in 1973, it came with a long-time tenant rancher, Mervin McDonald and his family. McDonald's family has been in West Marin for five generations, since 1888, and he had been foreman of the ranch when one branch of the McClure family still owned it. But after John McClure died in 1963, his brother David did not feel he could run the ranch on his own, and John's wife Dorothy could not afford to buy out David's share. The McClures sold the ranch in 1966 to a land-holding group, called the Bahia Del Norte Land and Cattle Co., comprised of some friends who wanted to assist the family.[31] McDonald stayed on as a tenant rancher with beef cattle; he also held a lease for grazing rights at the former Heims (N) Ranch.

Yet the Pierce Ranch, even before it was acquired, was being considered by the NPS for other purposes. Park staff had identified Pierce as a possible candidate for a "living ranch" exhibit since 1968, but by the early 1970s, all of Tomales Point had been tentatively included in the Seashore's wilderness area proposal. Now the ranch was being considered as a prime location for establishing a transplanted tule elk herd. Tule elk *(Cervus canadensis nannodes)* are a subspecies endemic to California and particularly adapted to the seasonal wetlands of the Central Valley; their estimated population size once exceeded five hundred thousand.[32] By the late 1800s this number was reduced nearly to the point of extinction, decimated by market hunting, the early Mexican-era hide and tallow industry, and conversion of much of the Central Valley to agriculture.[33] In the late 1870s, a lone and isolated population of fewer than ten individuals was found on a private cattle ranch near Bakersfield, in Kern County, and protected by the ranch owner, Henry Miller.[34] Their numbers recovered to a herd of roughly four hundred elk by 1914, when they were reported to be causing damage to local crops and fences.[35] The U.S. Biological Survey and the California Academy of Sciences tried to transplant some of the Kern County elk to parks and private refuges around the state, eventually sending them to over twenty locations throughout California.[36] Only a few of these transplants became established populations, but at almost every location where they have resided there have been reported conflicts with local ranching or agricultural interests, chiefly damage to crops, rangeland, and fences.[37]

By 1971, the statewide population of tule elk had reached about six hundred. The population consisted chiefly of free-ranging herds in the

Owens Valley and at Cache Creek in Lake and Colusa Counties, plus the small captive population at the Tupman Refuge in Kern County. In response primarily to pressure from Los Angeles-based wildlife activist Beula Edmiston for unrestricted range for the tule elk, despite the long history of conflict with agricultural operators, the California Legislature that year adopted Senate Bill 722, sponsored by newly elected Marin County representative Peter Behr.[38] Known as "the Behr bill," the legislation encouraged expansion of the statewide population to two thousand tule elk and prohibited any hunting until the population hit that number.[39]

Later that same year, an interagency task force named Point Reyes as one of four suitable reintroduction sites for the herd, chosen from a list of twenty-three possible locations around the state.[40] Park historian Paul Sadin writes, "When discussions regarding the possibility of elk reintroduction to Point Reyes began, the biggest concern among locals and park staff was the *potential for disrupting peninsula dairy and grazing operations.* State Fish and Game officials wanted the reintroduced elk to remain inside an enclosure, because of problems that free-ranging elk had created in the agricultural sector of the Central Valley."[41] NPS proposed releasing the elk onto Tomales Point, as a place where elk could be separated from neighboring ranchlands by an unusually high and sturdy fence.[42] According to an *Audubon* article at the time, the fence was considered "necessary to keep the elk from competing with cattle for feed and knocking down ranchers' fences in the pastoral zone of the national seashore."[43]

Shortly after acquiring the Pierce Ranch, the NPS began proceedings to remove McDonald from the ranch.[44] Federal law required the NPS to assist McDonald with finding a replacement ranch.[45] However, finding a ranch with equivalent carrying capacity to Pierce, with its roughly one thousand head of cattle, that McDonald could afford proved difficult. McDonald and the NPS conducted their search across much of the West; a list generated in 1978 details real estate queries as far away as Idaho and Montana.[46] All the ranches they investigated were far out of McDonald's price range, and by March 1978 the NPS concluded that an affordable replacement was simply not available.[47]

In the meantime, McDonald continued his beef operation on two successive two-year special use permits, the latter of which expired on October 31, 1978. NPS actions after 1976, when most of the ranch's pastures were included in the designated wilderness area, made maintaining the property more difficult. Park authorities cut electricity to a water pump serving the cattle, for instance, and prohibited McDonald from repairing ranch roads after winter rains created potholes.[48] These actions were in keeping with wilderness policy, but could have been accommodated if the administration had been willing; for example, the utility lines could have been moved underground instead of cut. McDonald interpreted them as underhanded attempts to force him and his family off the land.[49] His lawyer suggested that, given all the time and expense that had gone into a fruitless search for a replacement ranch, that the elk simply be located elsewhere.[50]

As the deadline on his last permit approached and the NPS refused to extend it further, McDonald filed suit in District Court in September 1978, asking that he be allowed to remain on Tomales Point. The court ruled against him.[51] McDonald appealed, and was still waiting for a court order to be allowed to remain through April 1980, when Sansing set a new permit deadline of December 1, 1979, after which he threatened to impound the cattle herd. Fearful of losing the animals, McDonald sold them at lower prices than if he'd been allowed to wait a few more months. A few hours after the sale and shipment of the herd was completed, the court order allowing his cattle to remain six more weeks was granted, but the news came too late.[52] McDonald finally left Pierce Ranch in April 1980, after having been offered a leased ranch near Marshall, on the opposite side of Tomales Bay, by his friend, Alvin Gambonini. The new ranch was only about half the size of Pierce, meaning McDonald could only run a much smaller herd. Nor did it have space for all of McDonald's children and their families to remain on the ranch with him, as had been the case at Pierce.[53]

The first herd of ten elk (two males and eight females) arrived at PRNS in 1978. The animals were relocated from the San Luis National Wildlife Refuge in the Central Valley, where they had been moved just four years earlier from their long-time home at the San Diego Zoo.[54] Notwithstanding the presence of a ten-foot high fence separating the elk's range from neighboring ranches, NPS Western Regional Director Howard Chapman declared that the relocation realized the goal of "reestablishing a relatively wild, free roaming tule elk herd on Tomales Point."[55] The animals were initially kept and fed in a temporary enclosure at Pierce Ranch, to keep them separate from McDonald's cattle until the latter were sold and shipped. In the second year, several elk died, and several bulls developed malformed antlers, a phenomenon later attributed to copper deficiencies. A number of sub-adults also died, likely from infection by Johne's disease, or paratuberculosis, which is thought to have been contracted from cattle.[56]

A population study estimated the carrying capacity for tule elk on Tomales Point at 140 individuals. The NPS believed that "once the elk reached that level, the population would naturally stabilize."[57] After the drought of the late 1970s ended, however, the elk population began to soar at an exponential rate, from the 93 individuals recorded in the park's 1988 census to 254 individuals in 1994.[58] PRNS staff drafted an environmental assessment in August 1992 that considered three possible responses to the increasing numbers: direct control (i.e., rangers culling, as was the park's practice at that time for controlling nonnative deer), no action (with the expectation that the population would then increase to "as many as four hundred"), or creation of additional fenced elk reserves within the Seashore (an area near Limantour Estero was suggested). The possibility of removing the fence at Tomales Point and allowing the elk to disperse into the pastoral zone was explicitly rejected due to anticipated harms: "Fence damage, Johne's disease and increased competition for forage could impact the ranchers who lease land from the National Park Service. These ranchers have an expectation that the Seashore will not subject them to impacts not in existence at the time the lease and other agreements were negotiated."[59]

A year later, in October 1993, an independent scientific advisory panel concluded that earlier estimates of carrying capacity had been artificially low, due to impacts from recent cattle grazing. The panel advised that the sustainable size for the herd should be 346, in line with a range analysis conducted the same year.[60] The rapid growth rate of the herd, however, alarmed some community members, who worried about needing to "avert a crisis" if the population continued to expand and possibly overshot the 350 mark. The *Point Reyes Light* quoted Sierra Club member Anne West of Inverness at saying at one of the public meetings, "Please, above all, don't let the Park Service do nothing. We're ahead of the problem, but that does not mean pushing [the land's] carrying capacity to its limits."[61]

The scientific panel also indicated that while a passive, natural-regulation approach to management was possible, allowing the herd to reach a dynamic equilibrium with its surrounding plant community, the consequences of such a policy would be periodic swings of population size, up and down. In the downswings, reproduction would decrease and mortality would increase, which meant that the visiting public might see malnourished elk or dead and dying animals. The remaining elk could also have increased impacts on vegetation and soils in dry years.[62] On the other hand, if a series of good years pushed the population to a higher level, park management might be forced into more active interventions, including the removal of individual animals, culling by agency staff, allowing public hunting, translocating elk away from Tomales Point, or injecting females with contraceptives.

After the El Niño years of 1995–97 yielded higher than average rainfall, herd size again expanded: to 380 in 1996, 465 in 1997, and 549 in 1998.[63] Researchers from state and federal agencies as well as the University of California, Berkeley, now documented higher survival rates for both adults and calves than observed in the Owens Valley herd, and found no instances of predation on calves, despite numerous coyote sightings in the area. They concluded that the population showed little evidence of natural regulation and was likely to overshoot the area's carrying capacity during prolonged periods of drought, causing population die-backs. They also recommended a target population size for Tomales Point of 350 animals—yet the population size had already expanded beyond that limit.[64] The Seashore was in need of a new approach.

ANOTHER RANCH CLOSED DOWN

Despite official support for the remaining ranches from PRNS administration, another peninsula dairy, the Horick (D) Ranch, ended operations after permit holder Vivian Horick was killed in a car accident on April 7, 1998, having spent her entire life on the ranch. Her father, Bill Hall, a dairyman from Petaluma, had purchased D Ranch in 1934, and moved his family there in 1936 from N Ranch, where they had been renting. The NPS purchased the property from his widow, Alice Hall, in December 1971, and she retained a twenty-year RUO. Neighbors describe the old ranch in its heyday as being "the prettiest ranch on the Point," with a stately palm tree growing beside the old Victorian ranch house. Alice's daughter Vivian and her husband, Rudolph Horick, ran the dairy until Rudolph died in 1980, after which Vivian ran it herself; her sons describe her as a "matriarch—she did everything, she ran the show."[65] The RUO expired in 1991, but Sansing offered Horick a five-year special use permit, after requiring some improvement projects prior to its issuance, such as whitewashing the barn and repairing fences.[66] The permit was issued in November 1991, and renewed in 1996 for another five-year term.

After Horick's unexpected death, the NPS extended the ranch's existing permit for six months in May 1998 to provide both park staff and Horick's heirs time to consider the future use of the ranch, as Vivian had not listed any successors on her original permit.[67] Superintendent Neubacher wrote to the family of plans to have the ranch's rental value independently appraised as soon as possible, after which they could discuss options.[68] When members of the CAC, who took a field trip to the ranch that same month, expressed concern that the ranch might be taken out of operation, Neubacher said that they had "no plans to end the Horick family's special use permit."[69] But Horick's three heirs initially disagreed over who would take over the permit and the ranch operation. In July of that year, the Horick Trust named a local veterinarian, Dr. John Zimmerman, to oversee the day-to-day operations, and agreed to sell the dairy cattle to convert the ranch to a heifer operation.[70] Through the remainder of 1998, the three siblings quarreled among themselves and also with the former foreman of the ranch about ownership of specific cattle and equipment.

Vivian Horick's special use permit was extended twice more in 1999, but, according to the NPS, with no resolution to the legal problems. Finally, after eighteen months without a clear outcome, the NPS decided that enough time had passed. At a CAC meeting in October 1999, Assistant Superintendent Frank Dean announced that after the Horick family had missed multiple deadlines in the process of getting a new special use permit, the NPS was cancelling the remainder of their mother's permit, effective November 1, 1999.[71]

Two of the heirs, Todd and Roger Horick, tell a different story—as do the archives. While no one disputes that there had long been a rift among the siblings, and despite the ongoing disagreement with their sister Carol, the two brothers decided early on that Todd would take over the ranch's operation. Todd remembers that during the CAC members' May 1998 visit, "they were satisfied we would be able to continue the family operation."[72] They began having regular meetings and phone conversations with Superintendent Neubacher, and report his telling them that the NPS "wanted a Horick to take over the lease of the D Ranch."[73] The brothers assert that Neubacher told them not to worry about the permit, except for five hundred acres out on the bluffs, which the NPS wanted to take out of grazing (although no reason was given), and that if they cleaned up the ranch, "it would go in your favor." Todd had not worked on the ranch for eleven years at that time, but spent several months mowing, cutting thistle, fixing troughs, repairing plumbing and electrical problems in pumps and barns, hauling away dumpsters of material, and cleaning up the buildings. The Horicks estimate that they spent fifty to sixty thousand dollars on this effort.[74] The superintendent told their sister's lawyer that he felt Todd was doing a good job with the cleanup.[75]

The Horick brothers also recall Neubacher stating that it would "point you in the right direction" if they sold the family's dairy herd and replaced it with beef cattle, since the latter might cause fewer impacts in terms of erosion and water pollution. The dairy herd was sold in August 1998, with the assistance of Dr. Zimmerman; Todd remembers that when he met with Neubacher to report the progress, he was told "not to worry so much, everything was okay, and the ranch was on the back burner." Disputes continued among the siblings, sometimes heatedly, until December, at which point Neubacher extended the permit through March 1, 1999.[76]

Then in January 1999, the familial impasse was broken: the siblings and their lawyers agreed to conduct a bidding process among themselves, within the structure of the Horick Trust, to decide who would take on the permit. They wrote to the superintendent asking whether the independent appraisal Neubacher had described the previous May had been completed, and for details on the acreage and capital investment expectations likely to accompany the permit, to inform the bids.[77] In the spring of 1999, the various parties corresponded through their attorneys, with the Horicks pressing Neubacher for the appraisal but receiving no written response.[78] According to the archives, there were a few telephone calls between individual attorneys and the superintendent, but the Horicks remained in the dark about the appraisal or how the permit might proceed.[79]

In May, the superintendent requested a number of repairs to be completed by the Horick Trust by August 30 in response to a public health inspection and extended the existing special use permit a third time, through June 1999; his letter stated that they were still completing the appraisal, but could not determine the final rental rate until they negotiated with the future tenant. He set a new deadline of June 30 for resolving the issue.[80] The NPS archives contain no response from the attorneys, but also no appraisal. The Horicks seem to have been caught in a Catch-22; they could not settle the estate among themselves financially until the NPS gave them permit terms based on the appraisal, but the NPS insisted on knowing who the prospective permittee would be before sharing information on the lease terms. And the NPS does not appear to have ever conducted the ranch appraisal, despite the frequent references over a year's time to its being in progress.

In late July, the NPS sent a bill for overdue rent for March through June, and on August 16, 1999, Neubacher notified the Horick Trust attorney via registered mail that, because the NPS never got a response to their May 18 letter, they were ending their arrangement with the Horick family. By his recollection, Todd was told on a conference call with Neubacher and Assistant Superintendent Frank Dean that the permit would be terminated and that they would have to leave the ranch for good. Initially given only thirty days to vacate, Todd was given more time after protesting. He says that Neubacher also told him over the phone not to go to the newspapers with the news, "as it would make things worse for [the Horicks' case]."[81]

Historian Richard White has written eloquently on the relationship between history and memory, arguing that the cold facts of history are the enemy of the stories and sense of identity that are the craft of memory, which can include forgetting or suppressing certain details and rearranging sequence of events.[82] While it is impossible to confirm conversations that took place in person or over the phone, rather than in writing, and there is a good likelihood that the Horick brothers' memories of this difficult time are not completely accurate, there is no question that they feel betrayed by the NPS—that they believed at the time they would be allowed to continue operating their mother's ranch, and were only denied the permit after spending considerable time, effort, and expense cleaning it up.[83] Neubacher's letters never acknowledged the siblings' agreement regarding their process for resolving the situation, and in public his administration blamed the family's alleged inaction as the sole cause of the NPS ending the permit.[84]

After PRNS cancelled the Horick permit, the fate of the ranch was not clear. The park could lease the land to another rancher for grazing, let it go fallow, or use buildings adaptively for some other purpose. At the same October 1999 CAC meeting at which the NPS announced the permit cancelation, Neubacher assured the commissioners that that the historical elements of the ranch would not be neglected: "Over the next year we are bringing a team of experts in to look at the cultural landscape . . . . We will then present an overall plan of attack. I think we just need to be patient."[85] Two years later, when a local community member volunteered to organize some of the neighboring families to help maintain the ranch's buildings, the superintendent declined, saying, "We're working on the older historic house. We've put a new roof on it. It'll be painted this year . . . we're committed. It just takes time."[86] At the time, there was no evidence of any work being done on the historic house; Neubacher may have confused it with the hay barn, where an added-on section had collapsed, and work had been done to stabilize the original structure.

FIGURE 16. House on the Horick Ranch, with old family photograph and Kevin Lunny walking past the building, Point Reyes National Seashore, 2008. Photograph by author.

At no point did anyone at PRNS suggest that another dairy operator or beef rancher move into the buildings and actually work the ranch. Portions of the D Ranch pastures have since been leased to the neighboring ranches, Nunes and Spaletta. Vivian Horick's former residence, a modern house above the main ranch complex, is currently used for staff housing, and the Seashore's historic carpentry crew has stabilized and painted the hay barn and horse barn. The other buildings, however, including the 1870s-era Victorian house and creamery, described by former PRNS historian Dewey Livingston as "the last of their kind," were left to deteriorate, with broken windows left open to the foggy, salty wind (see figure 16).[87] At some point in 2007 or 2008, the creamery collapsed. The palm tree next to the house has now died as well, and the remains of the creamery have been removed, leaving only a blank space on the landscape between an old utility pole and a storage tank.

FREE-RANGING ELK ON THE POINT

Recently, yet another management controversy has erupted at Point Reyes, involving both the tule elk and the old Horick ranch. A conflict stemming from the presence of free-ranging tule elk moving into the pastoral zone has put increasing pressure on the working landscape. This issue has its roots in the same late-1990s period, under Superintendent Neubacher's management, when many other changes to the Seashore's landscape were occurring. In May 1997, PRNS staff made a presentation before a CAC meeting on the over-large size of the elk herd on Tomales Point. According to the local paper, when commissioners learned that about a hundred calves had been born the previous year, "that was a wake-up call" because they suddenly had "a 33 percent increase in population."[88] The presentation listed four options for addressing herd size: immuno-contraception, chemical sterilization, relocating "surplus" elk to the Limantour Spit, or the shooting of "excess elk" by rangers. The PRNS superintendent told the group "I see no easy solutions to the management of the elk . . . . But it's important to create a long-term plan."

A year later, PRNS produced its 1998 *Tule Elk Management Plan and Environmental Assessment*. By that time, the Tomales Point herd size had grown to approximately 550 individuals; the statewide population of tule elk was at 3,200. As one of its five objectives, the plan listed the establishment of a free-ranging elk herd on the Seashore's natural areas by 2005.[89] This was not a new goal; Phillips had advocated for a free-ranging herd throughout the Seashore in a 1976 socioeconomic study, and the 1993 scientific advisory panel made a similar recommendation.[90] Yet the 1998 plan took this first step toward establishing a herd that would not be restricted by a separation fence. Wildlife biologist Dale McCullough attributes the proposal to a general belief at the time that local agriculture was fading economically; the elk, which are prone to wandering long distances, would spread across the landscape as ranches were phased out.[91] Historian Peter Alagona, in his analysis of endangered species policy across California, has found that, in many cases, "the emergence of a consensus [among wildlife managers] about historical trends, *rather than the strength of the evidence supporting it,* was really what mattered in shaping subsequent management decisions."[92] That seems to have been the case at Point Reyes, since no ranches in the Seashore had shown any sign of going out of business of their own accord. And rather than prepare for the possibility of conflict between elk and ranches, at least in public, Superintendent Neubacher insisted, "our plan [for the free-ranging herd at Limantour] clearly does not promote elk in agricultural lands."[93]

Implementation did not take long: five months after the *Tule Elk Management Plan and Environmental Assessment* was signed, in

December 1998, forty-five healthy elk were relocated via helicopter from Tomales Point to a 25-acre fenced range just north of Coast Camp on the west side of Inverness Ridge, to be quarantined and monitored for six months. Of this original group, twenty-two were later euthanized and studied after extensive testing for the organism that causes Johne's disease.[94] Several of the remaining females, but not all, were given immuno-contraceptives. At the time, some residents expressed concern that the relocated elk would spread to private property on the east side of Inverness Ridge; several ranchers explicitly asked that the area be fenced in.[95] These requests were for naught: on June 1, 1999, Seashore staff released twenty-eight elk—eighteen females, three bulls, and seven calves born in the enclosure—from their quarantine holding pen into the Phillip Burton Wilderness area near Limantour Estero.[96]

Each released adult animal wore a uniquely identifiable radio transmitter collar designed to facilitate tracking and early detection of mortality.[97] Four days prior to their release, PRNS's principle wildlife scientist, Sarah Allen, had said that the radio collars would help "ensure they do not roam too far away," and confirmed that "wanderers will be retrieved or possibly killed."[98] The 1998 management plan contained similar themes, stating that "the Park Service has a responsibility to be a good neighbor to adjacent and nearby landowners."[99] The Plan's Alternative B would have allowed elk the freedom to range throughout the Seashore, but that alternative was explicitly rejected, based on anticipated negative impacts on land uses and agriculture, as well as the short-term nature of the solution.[100] The Limantour area was chosen for the elk relocation because it had "large acreage in natural zones with buffers from major highways, ranches, and lands outside the Seashore."[101] The plan specified that PRNS should be prepared to offer compensation for any damage the elk caused, and that the herd would be managed adaptively as new situations arose.[102]

At some point over the next year, several tule elk unexpectedly appeared at the former Horick (D) Ranch, located opposite their Limantour release area on the western shore of Drakes Estero. Ranchers first noticed two elk cows in the pastoral zone near Drakes Beach in the summer of 2000; by their recollection, an additional male and female elk, both fitted with radio collars on their necks, appeared that autumn.[103] PRNS staff later informed them that the elk likely swam across Drakes Estero, but it is unusual for female tule elk to wander such distances on their own.[104] Tule elk breed in harems, so that a single dominant bull controls the females' movements for much of the year; while single males or bachelor groups commonly roam far distances, females only rarely do so—although longer movements are sometimes seen in recently translocated animals that have not yet established home ranges.[105]

In contrast, PRNS records of radio collar tracking indicate that the first two female elk were in the Horseshoe Pond drainage at the Horick Ranch on June 29, 1999, just four weeks after their release. During the autumn rut, one was recorded near Limantour, then returned to the Horick Ranch by February 2000, where it calved the following summer. The two females and the calf then remained at the Horick Ranch until they were sighted together again at Limantour in August 2001, and back at the Horick Ranch a month later, in the company of an uncollared male.[106] These records show the elk first arriving at least six months before any ranchers recall seeing them, and only one day before the then-in-effect lease extension for the Horick family was due to expire. They also appear to have traveled repeatedly between Limantour and the Horick Ranch— the only parcel not then under lease around the shoreline of Drakes Estero, and after Neubacher had specifically excluded five hundred acres from the former Horick Ranch permit, out on the coastal bluffs, from cattle grazing—without anyone noticing their movements, either across land or water. But regardless of when these animals arrived near Drakes Beach, they established a new, third herd of tule elk in the Seashore.

The PRNS *Annual Report* for 2001 noted that, "Since their release, the new herd [at Limantour] has been carefully monitored to ensure animals remain within Seashore boundaries, *do not interfere with cattle ranches within the park,* and are not shedding the organism that causes Johne's disease."[107] Despite Allen's statement to the press before the elk were released, the wandering individuals were not retrieved—and not surprisingly, like the original population at Tomales Point, the free-ranging herds at both Limantour and the Horick Ranch began to increase. By early 2016, the herd sizes were 130 and 95, respectively.[108] (See map 5 for the geographic extent of the three herds.) A 2010 study of their population dynamics estimated that, without intervention, both herds would likely increase to approximately four hundred individuals by 2018. The same study predicted that, without a plan, future conflicts between ranch owners and park management would be inevitable.[109]

PARK INACTION IN RESPONSE TO CONFLICTS

Just as predicted, problems soon arose over the new free-ranging tule elk herds. In 2005 at least one rancher asked PRNS staff about improved

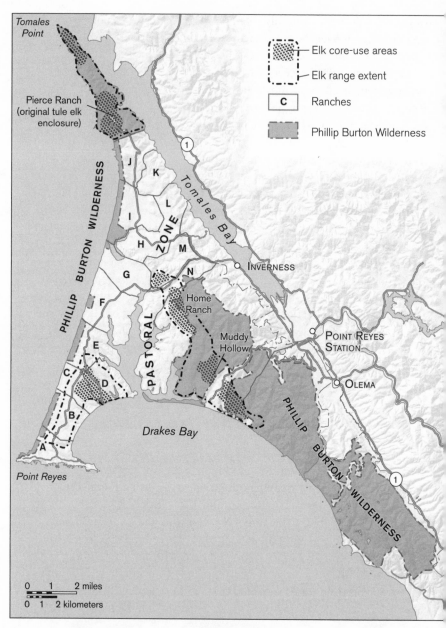

Legend:

- Elk core-use areas
- Elk range extent
- C Ranches
- Phillip Burton Wilderness

Tomales
Point

Pierce Ranch
(original tule elk
enclosure)

PHILLIP BURTON WILDERNESS

J
K
L
I
H
M
G
N
F
Home
Ranch
E
Muddy
Hollow
C
D
B
A

PASTORAL ZONE

TOMALES ZONE

Tomales Bay

1

INVERNESS

POINT REYES
STATION

OLEMA

PHILLIP BURTON WILDERNESS

1

Drakes Bay

Point Reyes

0 1 2 miles
0 1 2 kilometers

MAP 5. Map of current tule elk population ranges, Point Reyes National Seashore.
Courtesy of Ben Pease.

fencing to keep elk out of pastures. In 2008, the park held a meeting to discuss a fencing proposal, but nothing came of it. In 2010, the Spaletta family, which leases historic C Ranch plus a small portion of the former Horick or D Ranch, wrote to new PRNS Superintendent Cicely Muldoon, asking that free-ranging elk be moved off of their leased pastures: the elk were not only eating their cattle's forage, which had to be replaced with expensive hay, but were also damaging fences and irrigation systems.[110] Holes in the fencing left by the elk allowed the dairy cows to stray from their proper pastures, potentially to breed at the wrong time or with the wrong bull. At least three heifers were gored by bull elk during the breeding season, two of which died. The letter documented over thirty thousand dollars spent by the Spalettas in response to elk damage.[111]

Elk have caused problems on seven of the eleven remaining Seashore ranches. Most of the ranches are formally registered as organic operations, a status which comes with limitations on how much supplemental feed the cattle can receive in a given year. California agricultural regulations require that organic operations pasture their cattle for a certain percentage of the year, and must moreover leave a specific amount of residual dry matter (i.e., uneaten pasture) on the landscape.[112] Herd sizes are finely tuned to these requirements, so that the operation remains sustainable and does not overgraze the pastures. The wild card of elk on their lands puts dairies at risk of losing their organic certifications, which would cause them to go out of business completely.

PRNS staff began recording observations of the Drakes Beach elk herd in September 2010, noting a herd of roughly forty adult animals moving back and forth across C, D, and E Ranches on a daily basis.[113] The following January, the Spalettas' special use permit renewal contained a reference to a draft ranch unit plan, a document they had received a copy of six months earlier with no opportunity for input or suggestions.[114] This plan presumed the presence of free-ranging elk on the leased pastures. It also contained new language regarding "wildlife friendly fencing," and newly stipulated that livestock were no longer protected from wildlife. While the draft ranch plan was never finalized, in June, the Point Reyes Seashore Ranchers Association (PRSRA) sent Muldoon a letter, asking that the superintendent attend their next meeting to address the issue of elk migration into the pastoral zone, and specifically asking that PRNS address this issue for the ranchers as a group. Twenty-three PRSRA members, each a leasee at PRNS, signed the letter.[115]

Muldoon demurred, responding that the park would continue to discuss the issue "one on one with each ranching family as part of ongoing ranch plans and permit negotiations." She justified this on the basis on the Federal Advisory Committee Act (FACA), which requires that meetings with advocacy groups be advertised in the *Federal Register*. A representative from PRNS—in fact, Muldoon—would attend the PRSRA meeting, but only to provide information and listen to ranchers' views.[116] At the meeting, which took place on July 11, Muldoon talked about experimental fencing (lowering fences so that elk wouldn't damage them), but insisted that she could not discuss overall policy with the group. Moreover, any new plan or policy to remove elk from the pastoral zone would require environmental assessment under NEPA, above and beyond that undertaken for the 1998 *Tule Elk Management Plan*.[117]

The assertion that FACA prohibits NPS staff from meeting or communicating with the Ranchers' Association as a group is incorrect, because the statute does not apply to "meetings initiated with or by nongovernmental organizations." And contrary to the claims about NEPA review, the environmental impacts of moving the elk to the designated wilderness area near Limantour, regardless of what part of the Seashore they came *from*, had already been studied, in the 1998 *Tule Elk Management Plan*, the result being a formal "finding of no significant impact." The plan had analyzed possible impacts of relocating elk that wandered onto nearby private lands; the question for NEPA is whether environmental impacts have been analyzed, not whether the elk would be moved from public or private property.

The only concrete outcome of the meeting between PRSRA and NPS staff was some elk "hazing," chasing animals away from ranches. This is predictably ineffective, because the startled elk simply return after a day or two. In the press, NPS staff insisted that they couldn't have known the elk would move into the pastoral zone when they established the free-range herd at Limantour, and that, "by working on solutions with individual ranchers rather than the ranchers as a group the park is trying to avoid having to elicit a longer and more costly process, such as an environmental impact statement."[118] An anonymous rancher cited the precedent set several years earlier when NPS staff removed a "rogue" elk that repeatedly returned to the L and M Ranches, stating that Superintendent Neubacher had in effect promised to "manage these infringements if and when they begin negatively impacting ranching operations."[119] "What in the world," the rancher asked, "has changed?"[120]

In September 2011, the PRSRA wrote to Muldoon again, this time arguing that unless the NPS implemented its own elk management plan to retrieve elk that did not stay within their designated range, multigenerational ranching at Point Reyes would end. "We cannot believe that this is your intent," the letter said, "but your failure to enforce the PRNS elk management policy virtually ensures this outcome."[121] The Ranchers Association also wrote to Senator Dianne Feinstein, asking for help in removing elk from the pastoral zone and getting the NPS to enforce existing elk management guidelines.[122] Marin County Supervisor Steve Kinsey also wrote to Feinstein, asking that she alert Secretary of the Interior Ken Salazar to the serious nature of this ongoing problem and to the need for more effective measures to manage the elk.[123] In response, Feinstein requested a review of NPS actions, "to ensure that they are both compliant with the Elk Management Plan and protects [sic] the rights and property of ranching lessees."[124]

In response, the Secretary of the Interior reaffirmed that the NPS actively supports dairy and beef operations at Point Reyes, but repeated the agency's previous assertion that the 1998 management plan did not address the issue of elk in the pastoral zone. Nor did Salazar address the issue of the leasees' property rights, promising only that the NPS would "work with" ranchers to "address their concerns, preserve the unique ecological and cultural landscape of the Point Reyes peninsula, and continue to demonstrate that working ranches can be successful within the context of a national park."[125] NPS staff cite this letter from Salazar as evidence that they cannot legally relocate elk from the pastoral zone.[126]

In recent interviews, PRNS staff have stressed that the 1998 management plan did not specifically anticipate elk wandering into the pastoral zone. David Press, a NPS wildlife specialist, said in 2013, for example, that the plan offers no guidance "if [elk] end up in areas of the park where they were not expected to roam."[127] Yet tule elk's tendency to be drawn toward pastoral lands has been documented across California for more than a hundred years, and was unquestionably well understood by the scientists and NPS staff working on the 1998 plan. Both Judd Howell, formerly with the Biological Resources Division of the USGS but now retired and working as a private consultant, and Dale McCullough of the University of California, Berkeley, also now retired, confirm both that it was generally understood at that time that the elk were likely to migrate from Limantour into the pastoral zone, and that NPS staff were specifically aware of this.[128] It remains unclear why a more straightforward discussion of the species' attraction to agricultural lands and the possibility that they could invade the pastoral zone was not included in the 1998 plan, although Howell speculates that the omission was intentional, an attempt to avoid any political uproar by "kicking the can down the road."[129] Yet the NPS has interpreted the absence of specific language about managing elk on leased ranch lands to mean that they can do nothing about the problem, despite the plan's inclusion of authority to manage adaptively "as new situations arose."

After a further year of inaction, the Ranchers Association sent another letter to Superintendent Muldoon, demanding in September 2013 that the Seashore stop neglecting the problem. "It is time," the ranchers wrote, "for the Seashore to comply with its own Elk Management Plan and permanently relocate this herd back to the Limantour wilderness area where it belongs."[130] This letter triggered a series of meetings with local elected officials, including Marin County Supervisor Steve Kinsey, California Assemblyman Marc Levine, and Congressman Jared Huffman, all of whom urged the NPS to act. In some cases, the officials asked for short-term relief for the ranchers negatively affected by the elk, in addition to long-term solutions.

PRNS subsequently announced that a new ranch comprehensive management plan process would begin in Spring 2014, ostensibly in response to Secretary Salazar's November 2012 memo; Salazar's directive forced the Drakes Bay Oyster Company to cease operations (more on this in the next chapter), but also directed NPS to "pursue extending permits for the ranchers within those pastoral lands to twenty-year terms."[131] Among other issues, the new planning process aimed to examine different options for managing the tule elk. The NPS conducted scoping meetings in June 2014 and two additional public workshops that November; a draft of the new plan is not expected until late 2016, after several delays in schedule. In the meantime, park management has not changed its approach to the elk. Most ranchers are now operating on only one-year permit extensions, as the NPS has insisted that it cannot renew any special use permits until the new planning process is complete.

WHAT IS WILD?

In his classic book *Wilderness and the American Mind,* Roderick Nash explored the etymology of the word "wild," which originally meant self-willed or uncontrollable. Wild, he wrote, conveys the "idea of being lost, unruly, disordered, or confused . . . ungoverned or out of control."[132] Things, creatures, or places that are truly wild can be chaotic and unpredictable. As a society, however, we are often deeply uncomfortable with actual wildness—despite the many threads of environmentalism that sing its praises. Peter Alagona points out that, even though the grizzly bear became extinct in California by 1930, *images* of grizzly bears are nearly ubiquitous across the state. The species is idealized as a magnificent representation of wildness and made into an allegory of ecological decline—yet most agree daily life is much easier without having to look over your shoulder for an actual chaparral bear.[133] The same discomfort surfaces when calls to reintroduce predator species are dismissed as unrealistic, even when they are badly needed to stabilize overpopulations of other species, such as deer. And when existing predators turn up unexpectedly in back yards they are often relocated or shot, to limit any possibility of harm to us or our property (including livestock and household pets). We love the *idea* of the wild, but prefer to avoid the unpredictability of wildness. Similarly, we struggle to get our heads around a rare or endangered species that adapts itself to modern life, or that is being helped along by humans. Perhaps these situations fall too far outside our idealization of these creatures; often, such circumstances are interpreted as making them "less wild."

At stakeholder meetings for the new ranch management planning process in November 2014, several environmental advocates called for an "unmanaged tule elk herd" at PRNS. Yet an unmanaged population of tule elk has not existed anywhere in California since the 1870s. When the elk were first reintroduced to Tomales Point, it was understood that they *needed* to be kept separate from the ranches, hence the ten-foot-high elk fence. Since then, however, the Seashore has taken to extremes the notion that "the pastoral and the wild can coexist": it has concluded that they must coexist on the same acres. This management approach ignores the reality of ranching, particularly of dairying, which relies on organized fields and fences, keeping animals of different genders and ages separate, and carefully monitoring and controlling their impacts on the landscape.

In April 2015, the *San Francisco Chronicle* reported a drop in the Tomales Point tule elk herd size from 540 in 2012 to only 286 in 2014, a drop the paper attributed primarily to the state's ongoing drought. A representative of the Center for Biological Diversity was quoted as saying, "The loss of nearly half the Pierce Point elk herd highlights how important it is that the Park Service not cave to commercial ranchers who want free-roaming Point Reyes elk fenced in."[134] Yet the likelihood of population decline in the face of California's current intense drought was clearly anticipated decades ago by scientists.[135] According to McCullough, recently reported "die-offs" of elk do not involve huge numbers of adult animals suddenly dropping dead from starvation or thirst. He estimates that at least half the losses, maybe more, are due to lack of replacement, where calves are either not being born or are not surviving their first month.[136] The absence of replacement calves is harder to "see" than corpses dotting the landscape, which is perhaps why news of the die-off was treated so sensationally.

But the article is also a sign of the tension between natural resources management and the working landscapes at Point Reyes. The Tomales Point herds, because their movement onto ranches is limited by the elk fence, are actually fairly "natural" ecologically, in that their population size is controlled in part by the seasonal availability of vegetation (i.e., abundant grass in the winter, during the rainy season, but almost nothing near the end of the dry season), and will fluctuate up and down in response to annual changes in vegetation availability—as long as the public is willing to accept the down-cycles in numbers. Conversely, the herds at Limantour and Drakes Beach are much less natural, inasmuch as they are being artificially supported by the ranchers' fertilized fields and managed water supplies.[137] Their numbers are not constrained in any meaningful way by "natural" resources and will, therefore, be governed one way or another by what mangers do. As McCullough puts it, "There is nothing to stop the expansion of these southern herds except human interference."[138]

The herd living on the former D Ranch, spilling over on a regular basis onto working leased ranches, are nevertheless romantically referred to by the NPS and environmentalists as "free ranging," while the animals in the Tomales Point Wilderness Reserve are described as "fenced in" or "enclosed"—this despite the fact that the latter herd has access to roughly two-and-a-half times as much acreage as the former. As a practical matter, then, how freely elk at Point Reyes can move within a more or less limited space seems to be the key factor in determining whether they are considered wild or not, even though they are *all* living in a landscape that has been substantially tamed and controlled since the middle of the nineteenth century. Yet the ideal of a national park as a wild, relatively untouched place full of wild animals is putting increasing pressure on the long-term viability of the ranches.

Historically, state officials have accepted the need for human management of the tule elk. There is nothing particularly exceptional about this. Lots of other large mammals across the West are similarly managed through hunting or culling, including the iconic "wild" bison in

Yellowstone National Park. Demand for tule elk hunting tags across California is enormous. Of the twenty-two locations around the state that support tule elk populations, hunting is allowed at eighteen of them (eight in the Owens Valley alone); in 2014, hunters submitted more than 33,000 applications for only 316 tags.[139] Over the past fifteen years, however, the NPS has ignored the necessity for active elk management at Point Reyes, instead operating under an intentionally "hands off" policy, portraying the elk as wild animals, and therefore proper residents of a wilderness area.

All of these management dilemmas raise the question: What is the "natural" landscape of Point Reyes, after such a long history of human occupation and use? The native grasses disappeared centuries ago. Before the cattle were brought in, it is believed that large ungulates like the tule elk—along with active management by the Miwok, who routinely burned the landscape to retain open pastures—kept the peninsula grasslands fairly open and free of brush, yet current natural resource managers tend to see ranches as "incompatible use," and apparently find the process of working around the ranches frustrating. In contrast, the result for the landscape of the *lack* of ranch management in the southern wilderness area has been a decrease of open grasslands, which are gradually being replaced by coastal scrub and coniferous forest—neither of which represents good habitat for tule elk.[140] The PRNS staff are not aiming their management toward some idealized pre-Columbian ecosystem; one staff member described their approach as, "We just look at what is there today, what's happened historically, and ask, what do we want tomorrow and how can we achieve that?"[141] Yet the heavy emphasis on managing for natural resources often overshadows the working landscape in the Seashore, reshaping it into an increasingly nature-dominated place.

By focusing visitors' attention on certain aspects of the area's history and resources, interpretation plays a key role in defining the Seashore to the public. Current interpretative materials at PRNS call attention primarily to natural resources. Exhibits at the main Visitor Center at Bear Valley are overwhelmingly predominated by presentations on Point Reyes geology, ecology, and wildlife; panels on all human history are relegated to a single corner, with information on the Coast Miwok, early European exploration of the coast, and the ranching history clumped together. There is no mention of other nonranch agricultural uses of the land, or the old hunting lodges that once were scattered across the peninsula; these cultural uses of the land have been entirely omitted from the official PRNS history. The Visitor Center at Drakes Beach has exhibits only on marine ecology and Sir Francis Drake; the lighthouse interpretive center focuses strictly on the lighthouse's own history.

A different path to land management can be found right outside the park's boundaries. The Marin Carbon Project, which aims to increase carbon sequestration and soil health on working ranches by treating the pastures with a one-time application of compost, is a prime example of how natural resource management and cultural protection can be balanced in preserving the working landscape. In partnership with researchers from the University of California, Berkeley, and the Natural Resources Conservation Service, the organization began testing its approach to "carbon farming" in 2013 on three West Marin ranches, and plans on expanding to twenty more.[142] Their efforts could be used as part of local and regional climate mitigation strategies, as well as a way to enhance ecosystem functioning and improve the ranches' productivity and economic viability. Natural resource improvements do not need to come at the expense of local land uses.

FIGURE 17. Boat at Drakes Bay Oyster Company dock, Point Reyes National Seashore, 2013. Photograph by author.

CHAPTER 7

The Politics of Preservation

This book opened with the question of why was there an oyster farm in a national park, and what its closure might mean for understanding these kinds of protected places. The answer to the first part of the question is easy; the second is more complicated. The oyster farm pre-dated the Seashore, and fit well with its original purposes as a recreational facility, a place for visitors to learn about shellfish and enjoy a picnic by the shore. Commercial oyster beds were first established in nearby Tomales Bay as far back as 1911; in the San Francisco Bay, they appeared even earlier, in the 1800s.[1] Drakes Estero was planted with oyster seed imported from Japan in the early 1930s, as were several other tidelands areas up and down the California coast, including Morro Bay and Elkhorn Slough.[2] In 1935, David C. Dreiser obtained shellfish leases from the State of California for both Drakes and Limantour Esteros—the first incarnation of the Drakes Bay Oyster Company (DBOC). After the allotments changed hands several times, Charles Johnson purchased the rights in 1960, two years before the Seashore was established, and called his business Johnson's Oyster Company. In 1965, the State of California transferred ownership of the esteros to the federal government, but retained mineral and fishing rights; the same year, Johnson negotiated a "trade" with the NPS, giving up access to 344 acres in Limantour Estero in exchange for 70 acres adjoining his existing oyster beds in Schooner Bay.[3] He additionally owned five acres on shore, carved out of the old N Ranch, that housed the oyster processing facilities. The NPS acquired this parcel in 1972, but Johnson retained a forty-year RUO, one of the longest in the Seashore, set to run out in 2012.

Early NPS planners saw the oyster farm as an asset to the proposed Seashore. At the first public hearings in 1960, George Collins, NPS regional chief of recreation and planning, specified that "existing commercial oyster beds—which we saw yesterday as we flew around there, a very important activity—and the cannery at Drakes Estero, plus three existing commercial fisheries, would continue under national seashore status because of their public values."[4] The NPS's 1961 *Land Use Study and Economic Feasibility Report* similarly stated that "the culture of oysters is an interesting industry which presents exceptional educational opportunities for introducing students to the field of marine biology," and suggested that adding a restaurant specializing in fresh oysters would be "another recreation attraction [in] the proposed seashore."[5] When the California State Assembly passed a joint resolution that same year supporting the park proposal, they specified that the bill should contain "provisions safeguarding the legitimate interests of residents, ranchers, and fishermen in the proposed park area."[6] Commercial fishing and oystering seemed fundamentally compatible with the national seashore concept.[7]

Throughout the wilderness hearings and discussions of the mid-1970s, most participants in the debate considered the oyster farm a land use that could coexist within wilderness. As chapter 4 has documented, the tidelands of Drakes Estero were categorized as "potential wilderness," instead of being included within the full wilderness designation, because of the State's retained rights. As recently as 1993, the PRNS *Statement for Management* defined the tidelands as areas where the federal government did not own full fee title: "Since Service policy requires ownership of all rights before designation of wilderness, none of these lands were designated as wilderness. Most were classified as 'potential' wilderness additions."[8] The same document noted that the oyster farm's operation was "interesting to visitors and seems to be well-accepted by the public," and included among its management objectives "to monitor and improve mariculture operations, particularly the Oyster Farm operation in Drakes Estero, in cooperation with the Department of Fish and Game."[9]

Given this fairly innocuous history, why did the oyster farm become such a flashpoint of controversy? This question cannot be understood as an isolated incident; the battle over DBOC is ultimately a microcosm of the paradoxes of preservation that this book has explored. The power of the ideal of the national park as uninhabited natural scenery, made even stronger by an increasingly purist conception of wilderness as a place where humans can only be visitors who do not remain, makes a working landscape like Point Reyes an uneasy fit inside a national park unit. Since the late 1990s, environmental groups have been increasingly successful at pushing the NPS to manage Point Reyes according to a vision of nature that excludes residents and their traditional land uses, which are cast as inherently incompatible with national park management. The mechanism for this pressure has primarily been the park's planning processes, accompanied by assertions that, because the Seashore is publicly owned, private users *ought* to be removed—that the public interest is only served by natural resources management and recreation access, overlooking the intentions expressed when the park was first established to include the working landscape, and downplaying any public interest in the area's history or community continuity.

ACCUSATIONS OF ENVIRONMENTAL HARMS

On May 8, 2007, Superintendent Don Neubacher along with his top scientist, Dr. Sarah Allen, made public claims at the Marin County Board of Supervisors meeting that Drakes Bay Oyster Company caused environmental harm. At the heart of their presentation was the assertion that the presence of the oystering operation adversely affected local populations of harbor seals, but that allegation was only one event in what would become an extended series of environmental criticisms directed at DBOC. Starting with the NPS report *Drakes Estero: A Sheltered Wilderness Estuary*, first released in September 2006, Seashore staff faulted oyster cultivation for harming fish diversity, eelgrass coverage in the estero, sediments, and even water quality—accusations that environmental advocates amplified in the local and national press.[10] None of these claims have stood up to scientific scrutiny; instead, the NPS has been found by independent review panels to have misrepresented the data. When the National Research Council (the public policy arm of the National Academy of Sciences) was called upon to review existing science on the estuary in 2009, their panel of experts concluded that "there is a lack of strong scientific evidence that shellfish farming has major adverse ecological effects on Drakes Estero at the current levels of production and under current operational practices," and that NPS had "in some instances selectively presented, overinterpreted, or misrepresented the available scientific information on potential impacts of the oyster mariculture operation."[11] The panel felt that the Drakes Estero report "did not present a rigorous and balanced synthesis of the mariculture impacts" and "gave an interpretation of science that exaggerated the negative and overlooked potentially beneficial effects of the oyster culture operation."[12] Before addressing the question of why the NPS would do this, it is useful to convey the extent of the NPS's misconduct.

The claims that caused the most uproar involved alleged harm to harbor seals. During the 2007 pupping season, NPS staff and wilderness advocates described seals fleeing into the water from their sandy haulouts when disturbed by the oyster boat or workers. Yet observation records for the two years prior did not contain any substantiated instances of DBOC-caused disturbances. On April 26, PRNS lead scientist Allen published an article in the *Point Reyes Light* describing a huge drop in the seals' presence at pupping sites—and later that same afternoon, filed a trip report that detailed the first documented observation of the oyster boat disturbing seals.[13] DBOC's own records, however, contradicted these claims; at the time of Allen's observed disturbance, its workers had already clocked out. Further, Allen described the disturbance as being caused by a "white boat," but the oyster farm's only white boat was being repaired that day for engine trouble.[14] Two NPS volunteers documented a second disturbance the following Sunday afternoon, April 29, but DBOC employees only rarely worked on Sundays, and there was no record of them having clocked in that day.[15]

A week later, on May 5, PRNS staff installed the first of two automated cameras that took photographs of the estero's channels every sixty seconds from dawn to dusk during the three months of seal pupping season, catching in their field of view both the main seal haul-out site and DBOC boat traffic and oyster beds. Between 2007 and 2010, the cameras took over three hundred thousand digital photographs. Frame-by-frame analysis of these photographs by NPS staff, in the form of hand-written datasheets, revealed that none of the images conclusively showed disturbances caused by oyster workers.[16] PRNS withheld the existence of the camera program from both the public and the National Research Council panel that reviewed the science surrounding mariculture in 2009, and only acknowledged the photographs in 2010, when Corey Goodman, asked by Marin County Supervisor Steve Kinsey to review the scientific claims about the oyster farm, discovered reference to them in a report footnote.[17] PRNS quickly dismantled the cameras, and the same staff who had set them up year after year (and frequently visited the cameras throughout the season, to change batteries, memory cards, etc.) disavowed the program's findings as not meeting Interior's standards for a scientific product, because "the collection of these photos was not based on documented protocols and procedures."[18]

In 2011, the Interior Department's Office of the Solicitor responded to complaints about PRNS's behavior, concluding that NPS employees had committed scientific errors and appeared to have acted improperly. Their malfeasance included "incomplete and biased evaluation" of evidence, "blurring the line between exploration and advocacy through research," and withholding relevant, material, and necessary research and data from DBOC and the National Academy of Sciences.[19] The Solicitor's report found five NPS officials and scientists guilty of violating the NPS Code of Scientific and Scholarly Conduct, and concluded, "NPS, as an organization and through its employees, made mistakes which may have contributed to an erosion of public confidence."[20] Instead of being punished, however, three of the five soon were promoted to more prestigious positions elsewhere within the NPS, including Superintendent Neubacher, who became superintendent of Yosemite National Park.

From the moment of Neubacher's initial claims about the seals, the Lunny family, owners of DBOC, fought back. At DBOC's request, Senator Dianne Feinstein called a meeting between PRNS staff, the Lunnys, and local elected officials in July 2007. She urged the superintendent to work with the oyster company to come to an agreement on a special use permit, to be issued when the company's RUO expired in November 2012.[21] Over the next four years, numerous outside entities reviewed the NPS's reports and behavior, including the Interior Department's Inspector General in 2008, the National Research Council in 2009, and both the Marine Mammal Commission and Interior's Office of the Solicitor in 2011. After each review, Feinstein urged the Secretary of Interior to find a solution to the problem, and to acknowledge that "agency scientists acted as if they were advocates with no responsibility to fairly evaluate the scientific data."[22] Yet Neubacher claimed to lack jurisdiction to issue a new special use permit, despite a clear renewal clause in the RUO, citing a 2004 Interior Office of the Solicitor's opinion that wilderness law required the oyster operation to end.[23] In an attempt to work around this claim, Feinstein passed a rider through Congress in 2009 (titled Section 124) that placed authority to make the decision directly in the hands of the Secretary of the Interior, rather than the superintendent, and specified that the decision be made "not withstanding any other provision of law."[24]

In response to Feinstein's legislation, the NPS initiated a two-year environmental review process, producing a full EIS to consider four permit-issuance scenarios and the possible environmental impacts of each. This lengthy process contrasts starkly with the approach taken for converting all of the ranches' ROUs to special use permits in the early 1990s, which involved neither environmental impact statements nor less extensive environmental assessments.[25] This made sense intuitively, because by issuing the special use permits, the NPS simply allowed existing uses on the landscape, many of which have been occurring for generations, to continue—extending ongoing practices rather than causing any new environmental impacts. Similarly, when ranches have been taken *out* of operation in recent years, the NPS conducted no environmental impact analysis of any kind, despite these transformations involving sometimes-extensive changes in land use and management.

The NPS had conducted environmental review of the oyster operation once before, in an environmental assessment (EA) completed in 1998, when previous owner Charley Johnson proposed upgrading the operation's facilities. That EA considered four options: no action; construction of a new processing facility (identified as the NPS's preferred alternative); rehabilitating the existing structures with no new construction; and a fourth alternative, removing the entire operation, that was explicitly rejected.[26] Its analysis found that not only would continuing the operation have no negative environmental impacts, even *expanding* the onshore facilities would have no negative environmental impacts.[27] When Neubacher was asked by the local newspaper whether Johnson would be allowed to continue past 2012, when his RUO would expire, the superintendent replied, "As long as there's public support for oyster farming and there's no negative impacts to the environment, it may be permitted to continue. Really, it's up to the public to decide [during] future masterplan development."[28] In June 1999, Neubacher sent the Johnsons a letter stating that they were clear to proceed with their final project design.[29] The project only fell through because Johnson could not secure the necessary financing.

Why, then, did issuing a permit for DBOC to continue operating ten years later require formal environmental review? The new permit would not have involved any changes to DBOC's operations, nor any changes to the ecosystem of Drakes Estero, which had supported oyster cultivation for over eighty years. The ROU document itself contained reference to possible renewal via special use permit once the forty-year term passed.[30] Nothing had changed substantially about the oyster farm or its surrounding environment since the prior environmental assessment; if anything, DBOC had improved its environmental compliance record since the Lunnys took over from Johnson on January 1, 2005. The primary concerns about the oyster farm voiced in public comments on the 1998 EA centered on Johnson's spotty compliance with health and safety codes, as well as a desire to improve the visual quality of the operation. In 2003, the California Coastal Commission issued a cease and desist order to the Johnsons, requiring removal of some unpermitted developments, such as a mobile home and a second story on the cannery, and improvement of the wastewater treatment system. When the Lunnys decided to purchase the oyster business and Johnson's remaining ROU, they believed that they could improve the operation's environmental compliance, and overcome "the old embarrassment" that Johnson had come to represent.[31] They recounted being initially warned by Superintendent Neubacher against making the purchase, because of the farm's onshore environmental problems—but that he called them back the next day to support the idea: "He understood that our family had a great relationship with the park, that we were good stewards of our ranch and that we would take care of Drakes Estero. The park service had long supported the continuation of agriculture in the seashore, and had routinely renewed ranchers' leases. We thought that if we fixed up the farm, as Don [Neubacher] wanted, that the park would renew that lease, too—and that an important part of the agricultural fabric of our community would be saved."[32]

In their first two years of operation of the renamed Drakes Bay Oyster Company, the Lunny family invested nearly a million dollars of mostly borrowed money for cleanup and upgrades, as well as compliance with the Coastal Commission's cease and desist order; they recalled, "We truly believed the park would be relieved."[33] Yet soon after the Lunnys' purchase of the farm, Tomales Bay Association President Ken Fox warned Kevin Lunny of a meeting he had attended in January 2005 at the Seashore's Red Barn: a group of environmental leaders had met with Superintendent Neubacher, who opened the meeting by saying that, while Lunny was a good operator, the NPS felt its "hands were tied" by legal requirements and so the permit for onshore oyster operations would terminate in 2012 and not be renewed. Fox urged Lunny to "watch your back," because one of the activists had urged the group to make things difficult for the oyster farm—"we should make it so that he [Lunny] doesn't even *want* to continue past 2012"—and that the superintendent remained silent, but did not disagree.[34]

Two months later, on March 28, Neubacher sent DBOC a letter informing the Lunnys that "no new permits will be issued" after 2012.[35] In the years since, many people have asked why the Lunnys did not simply give up at that point. By the time they received the letter, they had already invested nearly three hundred thousand dollars in their cleanup efforts. Plus, the superintendent had reversed himself in his conversations with them once already, and they firmly believed, perhaps naively, that once they could show that the operation could be run cleanly and sustainably, he could be convinced of its merit.

ENVIRONMENTAL REVIEW OF THE OYSTER FARM

The draft EIS regarding DBOC, released for public comment in September 2011, was unusual in more ways than one. It had an unconventional structure; the "major federal action" under review was the issuance of the special use permit, rather than the continuation or closure of the farm. Alternative A—the "no action" alternative, which ordinarily would provide a baseline of existing impacts against which the action alternatives could be compared—would be to issue no new permit and thereby close DBOC down. This meant that the baseline did not reflect existing conditions, but a hypothetical circumstance in which the oyster farm had somehow already been removed. The differences between Alternatives B, C, and D, which all kept the oyster farm open, were minimal to the point of being almost meaningless—they varied only by harvest intensity, a variable beyond DBOC's control and instead dependent on environmental and climate conditions that vary from year to year.[36] Alternatives B, C, and D additionally all added new requirements or restrictions on DBOC's management; no proposed alternatives left the existing operations unchanged. That is the explicit purpose of requiring a "no-action" alternative, according to the Council on Environmental Quality's guidelines for the National Environmental Policy Act (NEPA) review process that governs federal environmental assessments.[37] By not providing any baseline reflecting the existing environmental impact of DBOC's current operation, and by identifying the issuance of a permit as the action under evaluation, rather than closure of the farm and its associated changes to the environment, the draft EIS distorted the primary purpose of NEPA analysis.

Environmental impact statements typically indicate a government agency's "preferred" alternative, to give the public a sense of what direction a given agency is likely to take. The NPS's draft EIS on the DBOC permit did not do this. Furthermore, in describing the environmental review process to a local reporter, PRNS representatives said "that they will rely largely on *public opinion* to determine whether oyster cultivation should continue, as it has for over nine decades, in the estero."[38] This again runs contrary to the intent of the environmental review process, which requires a careful consideration of the science of the alternatives and their impacts, rather than a popularity contest or "vote."

As a scientific document, the draft EIS was overly selective in its use of available information. It dismissed the findings from the National Research Council's 2009 report as irrelevant to the environmental review process. It also did not acknowledge the agency's seal disturbance records nor the huge collection of photos from PRNS's "secret camera" program, both of which showed that the vast majority of seal disturbances were caused by other types of wildlife, such as birds or coyotes, and by recreationists—hikers passing on shore, or kayakers paddling in too close. Instead, the draft EIS concluded that DBOC's continuing presence would have long-term moderate negative impacts on seal populations, while disturbances from recreationists were only described as minor.[39] The document also warned of possible negative impacts to birds and eelgrass from the continued presence of the oyster farm, in contrast to the Academy's conclusions of a lack of harm, and did not analyze possible benefits that the filter-feeding oysters might be providing to the ecosystem.

Some of the analysis lacked causal data to support its estimated impacts. For instance, the draft EIS claimed that DBOC's operations might adversely impact California red-legged frogs, listed as threatened under the Endangered Species Act—but the estero is brackish, not the freshwater habitat this species requires.[40] The document cited possible impacts on the frogs as a result of "degradation of a relatively small proportion of critical habitat" stemming from presence of the oyster farm structures taking up potential nonbreeding habitat, and the potential for increased mortality from vehicle strikes along the farm's half-mile long access road, but provided no data to support these conclusions. The executive summary of the draft EIS stated, "Agencies are not required to engage in speculation or analyze indirect effects that are highly uncertain (CEQ 1981 Q18 [48 Fed. Reg. 18027])"—yet the impacts analysis did exactly that, with heavy emphasis on potential, unsubstantiated impacts.[41]

In certain instances, the data itself turned out to be misleading. For example, analysis of impacts to the Seashore's soundscapes utilized "representative sound levels" caused by the oyster boat's outboard engine and onshore equipment, rather than direct measurements. Yet estimates for the boat's noise were based on a measurement of jet skis taken in New Jersey in 1995, and those for the oyster tumbler, which was powered by a twelve-volt battery, were based on a 400 HP cement mixer.[42] Senator Feinstein wrote about this to Interior Secretary Salazar in 2012, stating that she was "frankly stunned that after all the controversy over past abuse of science on this issue, Park Service employees would feel emboldened to once again fabricate the science in building a case against the oyster farm."[43] When an outside consulting firm made its own direct measurements, they found noise levels hundreds of times lower than those listed in the draft EIS.[44]

The document did not address issues related to maintaining the Seashore's working landscape. The report dismissed the question of impact to cultural resources on the basis on finding no historic integrity in the built structures at DBOC, ignoring the continuation of land use as a historically significant part of the cultural landscape.[45] Yet the oyster farm was the last vestige of what historically had been a thriving maritime economy, after the NPS closed the fish dock near Chimney Rock, previously the only working fish dock located directly on the Pacific Ocean between Half Moon Bay and (probably) Eureka, after George Nunes, the fisherman who ran it, died in 2005. Given its long history and its continuing function in controlling water nutrient levels, the oyster operation played a key role in the ranching cultural landscape. Removal of DBOC and its maricultural production could have major negative repercussions for the surrounding ranches, which are part of a nominated historic district, as well as for the wider agricultural community and economy of West Marin. As suggested in a comment letter from the University of California Agricultural Extension, current county, regional, and state policies, particularly those contained in the 2007 *Marin Countywide Plan*, support the continued existence and viability of both mariculture and agriculture at PRNS. Extension representatives wrote, "The future of agriculture on PRNS lands is an issue that strongly affects the very nature of Marin County. The Marin Countywide Plan's goals, policies and programs support the tenet that continuing a strong agricultural presence on PRNS lands is critical to Marin County agriculture as a whole. In fact, *enhancing* local agriculture, rather than simply protecting or maintaining it, is the basis for many of the Countywide Plan policies."[46]

The report's claims regarding the DBOC's impacts on visitors were similarly belied by the public comments. A large number of comments detailed visitor enjoyment of the oyster farm, yet the loss of that public benefit was not considered an adverse impact under Alternative A. Conversely, public enjoyment of the estero was portrayed as threatened by alleged noise generated by DBOC.[47] In response, the owners of the three largest kayaking tour companies to serve PRNS wrote a comment letter stating that they had never received any complaints from their clients or employees about the estero's "soundscape," and that oyster boats were rarely seen in action; when they did encounter boats, DBOC employees had always been very respectful of their presence, making sure not to disturb the kayakers or wildlife in any way. They added, "We feel that the above section of the DEIS does not accurately represent our experience of Drakes Estero or Drakes Bay Oyster Company and infers that we have stated these complaints to the park or others when we have not. Nor have we been contacted directly by the park for feedback on our experiences concerning either Drakes Estero or Drakes Bay Oyster Company."[48]

Finally, the draft EIS did not analyze the disproportionate economic impacts of closing down DBOC on people from a very underrepresented minority community—nearly all of DBOC's employees were Hispanic, and many had been employed there for decades, with specialized expertise not easily transferred to other forms of local employment. The draft EIS dismissed the environmental justice issue on the basis that the affected employees represented only 0.01 percent of Marin County's population.[49] This approach overlooked this disproportionately negative impact by calling it "regionally minimal"—but it represented a major adverse impact both on the lives of individuals as well as on the local Hispanic community of Inverness.

Of the twelve possible "impact" categories analyzed, the draft EIS concluded that the oyster farm caused major negative impacts in two: sound and wilderness. In response to questions about the scientific quality of the document, the NPS asked the National Research Council to review the draft EIS. The Council's panel concluded that most of the impact categories had moderate to high levels of uncertainty, and that for many of them, equally reasonable alternate conclusions could be reached from looking at the same data. It further criticized the draft EIS for lacking a true baseline against which all of the alternatives could be measured, and recommended defining impact intensities in ways that could be more clearly related to the magnitude of both adverse and beneficial effects.[50]

One of the most troubling things, in retrospect, about the draft EIS, is how one could only see its flaws if one *already knew* the science, or the situation. To a general member of the public, it undoubtedly seemed alarming, full of warnings of environmental repercussions if the oyster farm were allowed to continue, and apparently based in science and history. The message ought to have been counterintuitive, since the estero had so often been described as the most pristine estuary on the West Coast, despite the eighty-year presence of oyster cultivation—yet it also seemed to fit a well-understood narrative, that commercial operations are incompatible with environmental quality, and that private land uses inherently do not "belong" in a public park. The selective use of some information but not all—both the ecological data and the Seashore's history—created an overall impression of thoroughness and thoughtfulness, and fit with common assumptions about what a national park is "supposed" to be, while masking the document's shortcomings.

Under pressure from Senator Feinstein and in response to public comment on the draft EIS, the NPS agreed to have an independent expert analyze their photographs for the final EIS. They asked the U.S. Geological Survey (USGS, another Interior agency) to take on the task; it in turn contracted with Dr. Brent Stewart from Hubbs SeaWorld Research Institute to analyze the photos. In May 2012, Stewart filed his report concluding that he found "no evidence of disturbance" of the seals by the oyster boats in the NPS photos.[51] Yet the USGS in its report described two instances of disturbances in "association" with the presence of oyster boats, and in the final EIS, the NPS took this misquote a step further to imply causation, something Stewart had explicitly ruled out, and to project moderate adverse impacts to seals if the oyster farm were to continue operating.[52] Stewart later told a reporter that he was "not interested in being a whistle-blower," but felt that his findings had been altered and used to reach conclusions that his analysis directly contradicted.[53]

And after this lengthy, expensive, and contentious planning process, the revised final EIS was released for public review only late in the day on November 20, 2012, about eighteen hours before Interior Secretary Salazar was scheduled to visit DBOC in person. Coming only two days before the Thanksgiving holiday weekend, and only ten days before the Secretary's decision on DBOC's permit had to be completed, this timing minimized any possibility of public review and comment on the document's revisions before the decision deadline of November 30.

Ultimately, Salazar's decision, issued on November 29, 2012, used nebulous language. The decision relied on assertions of harm from the EIS document and appealed to wilderness law and policy as compelling his decision to deny the permit, while also claiming that Feinstein's Section 124 exempted Interior from completing the same EIS or complying with NEPA.[54] Specifically, Salazar wrote:

> Although there is scientific uncertainty and a lack of consensus in the record regarding the precise nature and scope of the impacts that DBOC's operations have on wilderness resources, visitor experience and recreation, socioeconomic resources and NPS operations, the D[raft] EIS and F[inal] EIS support the proposition that the removal of DBOC's commercial operations in the estero would result in long-term beneficial impacts to the estero's natural environment. Thus while the DEIS and FEIS do not resolve all the uncertainty surrounding the impacts of the mariculture operations at Drakes Estero, and while they are not material to the legal and policy factors that provide the central basis for my decision, they have informed me with respect to the complexities, subtleties, and uncertainties of this matter and have been helpful to me in making my decision.

In a footnote, the Secretary clarified, "My decision today is based on the incompatibility of commercial activities in wilderness and not on the data that was asserted to be flawed."[55] The final EIS document was not made publicly available for the legally required thirty-day protest period, nor was an associated record of decision ever filed with the EPA, so the environmental review process remains incomplete. Given Interior's argument that Section 124 exempted the agency from any legal requirement to comply with environmental regulations, and the lack of time for public review, it is not clear why the EIS process was undertaken at all.

OYSTERS, RANCHING, AND THE PLANNING PROCESS

The claims of environmental harm to Drakes Estero have proved to be unfounded, and this book has documented why the argument that the oyster farm's operations were inconsistent with wilderness is flawed. But the larger mystery remains: Why was the NPS so intent on closing down the oyster farm, to the point that it falsified scientific data and spent millions of dollars completing a misleading EIS that it was not legally required to perform? The answer may reflect back on the long trend, also documented in this book, of deemphasizing and eroding the working landscape at Point Reyes, steadily recreating the Seashore to better match the national park ideal.

As one point of evidence, consider a statement allegedly made by Superintendent Neubacher on the ranches' fate, reported by Marin

Agricultural Land Trust cofounder, former California Coastal Commissioner, and noted botanist Phyllis Faber, who served on the Board of Trustees for the Point Reyes National Seashore Association (PRNSA). In 2006, she attended a holiday event for PRNSA and had a conversation with Neubacher. According to Faber, Neubacher told her that the Park Service had a "plan" to get rid of the Seashore's working ranches, starting with the closure of the oyster farm in Drakes Estero. Once it was gone, the Park Service would stand by as environmental groups brought lawsuits against the surrounding ranches, claiming their operations were degrading water quality in designated wilderness. The ranchers, whose means are modest, would have no choice but to shut down— thus bringing the 150-year ranching tradition at Point Reyes to an end.[56]

Although Faber's account must be considered hearsay, Neubacher's "plan" is eerily reminiscent of the one written by the National Parks Conservation Association (NPCA) in response to the first PRNS wilderness proposal in 1971. That plan called for designating nearly the entire peninsula as wilderness, requiring that the oyster farm *and* the working ranches shut down. None of the other conservation groups advocating for wilderness at the time supported the NPCA's plan, but a similar scenario played out two decades later at Channel Islands National Park, where NPS staff worked in concert with other agencies plus the NPCA and Sierra Club to force a historic ranch on Santa Rosa Island to close in 1998, for alleged violations of the Clean Water Act and Endangered Species Act.[57]

Under Superintendent Neubacher's leadership, the Point Reyes National Seashore initiated a formal process to update its 1980 general management plan (GMP), starting with a "notice of intent" published in the Federal Register in October 1997.[58] The notice directed that "comments on the scoping of the proposed GMP/EIS should be received no later than January 31, 1998," and that public scoping sessions would be announced. It went on to anticipate a draft GMP/EIS in Spring 1999, with the final document to be completed early in 2000. Before becoming superintendent at PRNS, Neubacher had spent four years working on a GMP for the Presidio in San Francisco, so he had extensive experience with this sort of large planning project.

The first comment letters actually arrived before the planning process formally began; the earliest is stamped as received at PRNS on September 10, 1997, over a month prior to publication of the notice of intent.[59] This suggests that the NPS alerted certain environmental groups of their intentions to begin the planning process ahead of the rest of the public. Furthermore, the first comments received were all "form letters," prewritten with identical text, sometimes even down to the font type; at least ten of these letters were sent in, all received by May 1998. Many of these letters arrived from out of state, and they all contained only one request, urging the NPS not to renew any grazing leases when they came due: "The needs of the people who love Point Reyes for what it offers in the form of hiking, camping, interpretive centers, wildlife viewing, [and] research, must come ahead of commercial ranching operations which offers nothing to anyone except for those who maintain these operations."[60] Very similar language also appeared in a later comment letter from Beula Edmiston's former organization, the Committee for the Preservation of the Tule Elk, from its new representative Bruce Keegan.[61]

A second form letter of boilerplate text, longer and more subtly worded than the first, began appearing in November 1999; one paragraph asserted that, "with 13 operating ranches, there are potential conflicts between natural and cultural resource management," giving an example of "runoff from ranching harming salmon and steelhead runs and the water quality in Tomales Bay." An identical sentence appeared in the NPCA's official comment letter, dated November 30, 1999. Their organization's letter also requested that the NPS look into the impacts of fishing and oyster farming, in particular "the appropriateness of current fishing in coastal wilderness, and the impacts of Johnson's Oyster Farm on surrounding park resources."[62]

Despite the published deadline for scoping comments in January 1998, initial public scoping meetings were not held until October 1999, in Point Reyes Station, San Rafael, and Santa Rosa. In all, over the two-year period, the NPS received over a hundred comment letters, and released a summary report of the public's input in January 2000.[63] After some community members raised questions as to why the planning process did not include the northern district of the GGNRA (Olema Valley and Lagunitas Loop), also managed by PRNS, an additional scoping meeting was held on February 29, 2000, and the public comment period was reopened for forty-five days. A newsletter included a new schedule for the planning process, with a draft document to be released in Summer 2001, and the final GMP/EIS slated for publication in Spring 2002; after the additional scoping meeting, this schedule was again revised to produce a final plan in Fall 2002.[64]

Nothing then occurred until PRNS sent out a second newsletter late in 2003, identifying five management "concepts" as "preliminary ideas for the General Management Plan."[65] These seemed to parallel the usual planning alternatives required by standard environmental review processes—a no-action "concept" was offered, continuing current management unchanged, plus four action "concepts"—yet the newsletter described these concepts, each accompanied by a map of possible management zones and actions, as "preliminary, and we need your input to fully refine future alternatives and management zones." The concepts themselves represented a range of vague approaches, each promising increased emphasis on a different area of management, from natural resources to visitor experience to sustainable agriculture—yet the language in several concepts implied that continuing agricultural uses at current levels, as a form of protecting the cultural landscape, was incompatible with resource preservation and restoration. Three of the four action "concepts" proposed expansion of wilderness and natural areas, along with a reduction of working agriculture.

The NPS invited additional public comment, either in writing or via a single scoping meeting held at the Red Barn at PRNS headquarters on January 14, 2004. Over 120 people crammed into the small space, with additional people spilling out the doorway. The *Point Reyes Light* published a proposed "Concept 6," written by a local group called Marin-Watch, suggesting the enhancement of cultural *and* natural resource restoration and preservation through sustainable agriculture, modeled on the Cuyahoga Valley National Park's "Countryside Initiative" plan.[66] The newsletter and public meeting triggered twenty comment letters from organizations, and nearly a thousand letters, faxes, and emails from individuals.[67] The NPS projected that a draft GMP/EIS would be available for public review in late 2005 or early 2006.

But then nothing was released. The twice-revised estimates for publication of a draft document came and went. In the summer of 2008 PRNS produced another newsletter, promising that the draft plan would be released late that year and foreseeing a final GMP and record of decision in 2009.[68] But no draft plan has ever been published. As of February 2016, the Seashore's website still describes the GMP as "under development," and states that "Once the draft document is released, and after public review and comment, a final revised GMP/EIS will be available. Public workshops regarding the draft document will be announced."[69] Nineteen years have now elapsed since the initial Notice of Intent was published, and thirty-six years since the original GMP of 1980.

According to Gordon White, chief of integrated resource management at PRNS, efforts to update GMPs have slowed in the NPS systemwide; funding has apparently dried up, and some in the agency question whether they are really needed. Unlike the U.S. Forest Service and the Bureau of Land Management, both of which are required by federal law to produce management plans for their lands, the NPS is not legally obligated to update its plans. The agency's national policy states that GMPs should be updated every ten to fifteen years, but adds, "If conditions remain substantially unchanged, a longer period between reviews would be acceptable."[70] White states that Point Reyes staff struggled to craft plan alternatives that were not simply a continuation of the status quo, and were uncertain why they were writing a huge plan if it was just to keep the same management course.[71]

Yet despite this lack of official articulation of the vision and long-term direction for the Seashore, since the late 1990s PRNS has undertaken a number of major planning projects and changes to the landscape. In addition to the campaign to oust the oyster farm, the long-standing permit at the Horick Ranch was cancelled in 1999 and the family evicted after the unexpected death of Vivian Horick. At Rancho Baulines, Mary Tiscornia was evicted in 2000; although the NPS claimed to need the space for an environmental education center, that plan has not come to pass. The main ranch house has been used primarily to house NPS staff, and the surrounding landscape has been largely taken over by fire-prone weeds. The nearby Hagmeier Ranch was converted to adaptive NPS use as the new Pacific Coast Science and Learning Center in 2000. The Giacomini Ranch wetlands restoration project removed another historic dairy operation and converted its pastures to tidal wetlands, and dredge material from the project was used to fill and close all the ranchers' rock quarries on the Point—projects all conducted, except for the wetland restoration, without environmental review. Nonnative axis and fallow deer were eliminated, based on claims that they were outcompeting native deer, amidst enormous local uproar and protest, and tule elk were relocated from Tomales Point into the wilderness area around Limantour, from which they have been spreading onto the pastoral zone. All of these changes have been undertaken piecemeal, instead of as part of a clearly stated plan for the Seashore's management, and all have nibbled away at the working landscape.

PRESSURE FROM ENVIRONMENTAL GROUPS VIA PUBLIC COMMENT

During the GMP scoping processes in the late 1990s and early 2000s, the *only* mentions of the oyster farm came in the NPCA's comment, cited above, plus one additional letter from the Sierra Club Marin Group. In both of these letters, questions about the oyster farm's authorization or continuance were tied explicitly to similar questions about ranching.

In the latter, authored by then-chapter president Gordon Bennett and dated February 20, 2004, the organization raised a series of points about grazing. Bennett urged a study of "congressional intent" in allowing the presence of beef and dairy ranching and oystering in the PRNS. He suggested that PRNS should "commission an exhaustive legal analysis (including actual legislation, testimony before committees, floor statements, and committee reports) to see whether Congress intended existing beef, dairy, and oyster operations to be permanent or temporary within the PRNS management areas (PRNS and the north District of GGNRA). We urge that this legal analysis determine with a reasonable degree of certainty the extent to which there may exist any legal obligation on the Park Service to renew or extend leases for these existing agricultural or maricultural operations."[72] The letter went on to question whether Congress intended to allow "expansion" of agriculture into diversified areas such as row-cropping, orchards, chickens et cetera, asked for numerous studies on habitat impacts of grazing, and questioned the importance of the Seashore ranches to the larger West Marin agricultural economy. It lastly asked for a "more definite timetable" should scientific studies determine that ranching activities ought to be decreased or relocated within the park.

The timing of the Sierra Club's letter requesting clarification of legal obligations associated with agriculture and mariculture is particularly notable, as the Department of the Interior field solicitor's legal opinion— which first articulated the argument that the oyster farm must close when its RUO expired in 2012 due to wilderness law, and which became the primary basis for the NPS's claim to not have the legal jurisdiction to issue a permit—was written six days later, dated February 26, 2004.[73] The solicitor did not conduct the exhaustive analysis that the Sierra Club suggested; his opinion only reviewed the language in the 1976 wilderness legislation itself, Interior's argument at the time that the tidal areas were inconsistent with wilderness, and one House report.[74] The timing, however, suggests that PRNS's switch to insisting that the oyster farm must close was prompted at least in part by pressure from representatives of environmental groups, and that it was indeed related to those groups' long-standing ambitions to remove active ranching from the Seashore.

After the draft EIS for the oyster farm was released for public comment in 2011, several environmental groups went into action again to generate targeted responses. The document triggered a staggering 52,473 public comments, roughly 90 percent of which were identical form letters solicited by emails sent to members of four national environmental groups: Sierra Club, NPCA, the Natural Resources Defense Council (NRDC), and the National Wildlife Federation (NWF). The emails asked the recipients to help protect "the only marine wilderness on the West Coast," and provided a prewritten form letter that was sent electronically when the recipient pressed an online button.[75] None of the emails mentioned the formal environmental review process, nor did they provide links to the draft EIS or suggest that the recipients should read the document itself. Only twenty minutes after the NPS made the raw comment data available on their website, the local Environmental Action Committee of West Marin (EAC) issued a press release declaring that 92 percent of the public input supported shutting the oyster farm down.[76] A year later, the final EIS declared the form letters nonsubstantive and set them aside from the analysis, but the statistics dominated much of the press's coverage of the process, creating a strong impression of huge public outcry—yet most of the people who commented had likely never seen the EIS document.[77]

POLITICIZATION OF THE OYSTER FARM'S LEGAL BATTLE

The numerous assertions of environmental harm caused by the oyster farm were never substantiated. In March 2012, Senator Feinstein wrote to Interior Secretary Salazar, "The Park Service has falsified and misrepresented data, hidden science and even promoted the employees who knew about the falsehoods, all in an effort to advance a predetermined outcome against the oyster farm. It is my belief that the case against the Drakes Bay Oyster Company is deceptive and potentially fraudulent."[78] Similarly, three former legislators directly involved with transferring the Point Reyes tidelands from the State in 1965, passing the additional legislation with funding for land acquisition in 1970, and establishing the wilderness area in 1976, wrote the Secretary a letter in August 2011 urging continuance of DBOC via issuing a new permit, arguing that Congress had never intended to end the oyster operation. They concluded their letter by noting the deliberate misrepresentations of science by the NPS, and suggested that some elements of the NPS seemed to have "a secret agenda for some years to drive out not only the oyster farm, but the privately-leased ranches as well."[79] One of them, former California

FIGURE 18. U.S. Congressman Pete McCloskey and California Assemblyman Bill Bagley in Tomales, CA, 2011. Photograph by author.

Assemblyman William T. Bagley, quipped to a local paper, "Legislative intent is lasting. It is very difficult to change history if the authors of that history are still alive."[80] Former Congressman John Burton, author of the 1976 wilderness bill, stated that "the issue of what to do with the oyster farm wasn't even under contention. Several things were grandfathered in, and aquaculture—oyster culture—was one of them. If this had been a big issue, then trust me, I would have remembered it."[81]

Yet the environmental groups' insistence that the mariculture operation conflicted with wilderness law apparently held greater sway. Secretary Salazar cited wilderness policy as his primary reason for deciding in late November 2012 to deny issuing a special use permit thereby closing the oyster farm down. His letter specified, "Although Sect. 124 grants me the authority to issue a new SUP [special use permit] and provides that such a decision would not be considered to establish any national precedent with respect to wilderness, it in no way overrides the intent of Congress as expressed in the 1976 act to establish wilderness at the estero. With that in my mind, my decision effectuates that Congressional intent."[82]

Opponents of DBOC often quoted from *House Report 94–1680*, referring to the phrase "efforts to steadily continue to remove all obstacles" to eventual wilderness conversion. Interior Field Solicitor Mihan cited the same sentence in his 2004 letter to Superintendent Neubacher as indicating a congressional mandate to "convert potential wilderness, i.e. the Johnson Oyster Company tract and the adjoining Estero, to wilderness status as soon as the non conforming use can be eliminated."[83] Interior Secretary Salazar similarly cited it in his 2012 letter denying DBOC a special use permit, as the primary indication of congressional intent. Neither Mihan nor Salazar referenced any of the copious legislative history that identified the State's reserved rights as the obstacle, rather than the oyster farm's operations. The original legislators' more recent assertions as to their intent were similarly ignored.

Only a few days after the Secretary's decision, lawyers working pro bono on DBOC's behalf filed a complaint for declaratory and injunctive relief in federal district court.[84] The Lunny family could not afford a legal battle with the NPS on their own, and so relied on donated legal assistance. The Lunnys' legal team filed on December 21 for a preliminary injunction that would allow the oyster farm to remain open while the court considered the merits of their case.[85] The legal hurdle for obtaining such an injunction is to show likelihood of being able to prevail in court.

Their main argument stated that the Interior Secretary had acted in an arbitrary and capricious manner, and thus violated the Administrative Procedures Act as an abuse of discretion, when he ordered DBOC to close. His November 29 letter insisted that he could not issue a permit because to do so would run counter to wilderness policy, yet Feinstein's Section 124 law gave the Secretary authority to issue a permit "notwithstanding any other law"—a phrase which he had invoked to excuse the NPS from completing its environmental impact review process—hence he should not have felt constrained by wilderness law.[86] The legal team also asserted that the final EIS had misrepresented the NPS's own findings about harbor seals and used false data for noise measurements, which violated NEPA. Their third argument stated that the NPS had violated both the Administrative Procedures Act and NEPA by not engaging in a notice-and-comment rulemaking before declaring Drakes Estero "full" wilderness, which it did by publishing a notice in the Federal Register on December 4, 2012—despite the State's remaining reserved rights.[87]

The subsequent legal case was complicated by the involvement of Cause of Action, a Washington, D.C.-based nonprofit organization, which donated pro bono legal support to the Lunnys' case. Cause of Action, which describes itself as "advocates for government accountability," had received funding from a number of conservative backers, allegedly including the notorious Koch brothers, and was headed by a former staffer from one of the Kochs' charitable foundations.[88] Environmental advocates fanned the flames of right-wing conspiracies, suggesting that DBOC's lawsuit represented a plot to weaken wilderness protections or privatize public lands nationally—describing it as a "precedent-setting land-grab effort."[89] The headlines began to overshadow the legal case, as the oyster farm was accused of "working lockstep with pro-corporate lawyers and legislators on an agenda to open up the voters' beloved national parks and wilderness areas to industry and commercialization."[90] Then in May 2013, Cause of Action sent a letter and Freedom of Information Act (FOIA) request to the Public Broadcasting Service, following a report aired on *NewsHour* about DBOC's legal case, the language of which implied that the Lunnys knew about and supported the FOIA request.[91] The Lunnys did not, and soon after, DBOC severed its connection with Cause of Action.[92] But the damage had been done; the oyster farm's nonpartisan dispute with the NPS had effectively been transformed in the public's eye into a greatly simplified left-wing cause against conservative forces.

On February 4, 2013, the federal district court denied DBOC's motion for an injunction, but did not address the merits of the legal team's arguments; rather, the court stated that it did not have jurisdiction over the Interior Secretary's discretionary decision. DBOC's lawyers immediately appealed to the Ninth Circuit, and were joined by a joint amicus brief from a number of agricultural and sustainable food organizations and individuals, including Alice Waters from the storied Chez Panisse Restaurant, arguing that the public interest would be best served by keeping DBOC open and operating.[93] Two of the three judges on the Ninth Circuit panel agreed with the district court, stating that the court "lacked jurisdiction to review the Secretary's ultimate discretionary decision whether to issue a new permit," and that by enacting Section 124, "Congress did nothing more than let the Secretary know his hands were not tied."[94]

A dissent written by the third judge, Paul Watford, forcefully disagreed, arguing that, by passing Section 124, Congress "sought to override the Department of the Interior's misinterpretation of the Point Reyes Wilderness."[95] In his dissent, Watford cited much of the same legislative history documented in this book. He stated that both the

Field Solicitor's opinion in 2004 and the Interior Secretary's decision in 2012 erroneously concluded that in 1976 Congress considered the oyster farm's operation to be an obstacle to wilderness protection, and criticized the court's majority opinion:

> What does the majority offer in response to this analysis? Some hand waving, to be sure, but nothing of any substance. Most tellingly, the majority never attempts to argue that the Interior Department's interpretation of the Point Reyes Wilderness Act was correct. Nor could it make that argument with a straight face given the Act's clear legislative history, which the majority never attempts to address, much less refute. The majority thus has no explanation for Congress's inclusion of the notwithstanding clause in §124 other than the one I have offered: that it was included to override the Department's misinterpretation of the Point Reyes Wilderness Act.[96]

The Lunnys filed one last appeal to the Supreme Court, arguing that their case involved issues of administrative law with broad national implications, but the Supreme Court declined to hear their case in July 2014. The environmental advocates who had worked for years opposing the oyster farm hailed the decision; Amy Trainer, head of the Environmental Action Committee of West Marin (EAC), having earlier characterized the legal case as a "desperate move full of desperate arguments," announced that "good government prevailed and iconic Point Reyes remains protected as long-intended by Congress."[97] Jerry Meral, former deputy secretary of California's Natural Resources Agency, told a reporter, "I think [Lunny] was treated fairly . . . [The Park Service] intended to close him down, they did everything they could to close him down, and eventually they did close him down."[98] Despite the years of asserting environmental harms in the estero, the NPS now characterized the main issue as being wilderness policy; PRNS Outreach Coordinator Melanie Gunn stated that, "Science will always be debated, like climate change . . . . But the law and policy of the Wilderness Act is very clear."[99]

REMOVAL OF THE DRAKES BAY OYSTER COMPANY

At Point Reyes, representatives of the NPCA and Sierra Club, as well as the locally based EAC, employed the national park ideal, with its emphasis on pristine wilderness, as a crowbar with which to pry the oyster farm out of the park landscape. Collectively, they warned that allowing DBOC to continue past the 2012 expiration of its reservation of use would "overturn" the estero's wilderness designation.[100] Yet the wilderness status of Point Reyes was never actually in danger: Drakes

Estero's designation as potential wilderness meant that the estuary had been *managed as wilderness* ever since, with the sole nonconforming elements of maintaining the oyster rack structures, which long pre-dated the designation (and the park itself), and allowing the use of an outboard motor to propel the harvest boat. While the Wilderness Act prohibits commercial operations in wilderness areas, DBOC's "com-mercial operation" itself was on shore, on land that is historically part of the Seashore's pastoral zone, and which was not included in the wil-derness designation.

And while both esteros (Drakes and Limantour) plus Abbotts Lagoon have now all been formally converted from potential wilderness to "full wilderness," nearly five thousand acres of potential wilderness remain at Point Reyes, in the form of the quarter-mile tidal strip along much of the Point's coastline. There are no uses of this tidal zone that violate the 1964 Wilderness Act; only the State's reserved rights remain, as they do in the other wetland areas. It is not clear why these submerged acres remain as potential wilderness, given the argument that DBOC's leased tidelands were required by law to be converted to full wilderness as soon as the ROU expired.

DBOC was part of a long history of fishing and mariculture in West Marin, and many visitors maintained traditions of hiking the estero or kayaking its waters and then gathering around a picnic table to cele-brate with a plateful of oysters. For them, there was no either/or between sustainable agriculture and the wild. As a point of comparison, there was little difference between DBOC's use of the potential wilderness of the estuary and that of commercial kayak tours. Both provided a service to paying customers that relied on the natural ecosystem, without inputs or manufacturing. Oysters thrive in the same wild conditions that other organisms do, and DBOC workers, in guiding people's use of the estu-ary's shellfish resource, were providing a service similar to that of a kayak guide who shows visitors to the most spectacular spots. DBOC's onshore shop hosted as many as fifty thousand visitors a year, making it a major recreation destination, particularly as local food increasingly became a tourist draw for the West Marin area. An oyster even tastes wild, bringing the sharp brininess of the sea to our mouths along with a deep appreciation of place, like the idea of terroir in winemaking.

A more direct parallel exists between the oyster operation and live-stock grazing, a practice specifically grandfathered in, with appropriate restrictions and oversight, by the 1964 Wilderness Act. In ranching, the cattle or sheep move across the landscape, finding their own forage in the meadows of a wide variety of public lands, and convert something inedible to humans, grass, into useful meat, milk, and fiber. Human guardians lightly guide their movements, but the animals do the work themselves. Similarly, oysters hang around in the water column and filter out the nutrients they need, turning something inaccessible to us into (delicious) edible protein—the oyster racks and bags are akin to fences, and the oyster workers like cowboys who check on the bivalves' progress and safety, and occasionally round them up, without diminishing the wild ecosystem upon which they depend.

A strong sense of reciprocity existed between the oyster operation and its surroundings, in a relationship that transcended one-way extraction. While few reliable studies have been done on oysters' role in the Drakes Estero ecosystem, oysters are generally recognized in the scientific literature as being *beneficial* to their ecosystems, where they maintain or restore water quality and provide habitat to other species.[101] Up and down the Atlantic coast of the United States, oyster cultivation is currently being employed to help restore damaged estuaries and local maritime economies. From Drakes Estero, DBOC sustainably produced roughly 40 percent of California's oysters, all sold exclusively within the San Francisco Bay Area; importing an equivalent quantity of this high-demand product from other locations, such as Seattle, Chesapeake Bay, or Japan, comes with a greatly increased carbon footprint. In addition, oyster workers were "eyes on the water" for recreationists on the estuary, and frequently helped or even rescued kayakers stuck in the mud or having other trouble. In recent years DBOC donated many tons of oyster shell to other Bay Area restoration efforts—including projects to restore native oysters to the San Francisco Bay and to improve nesting habitat and success rates for the threatened western snowy plover. Since only oysters shucked onsite contributed to this stockpile of oyster shell, and DBOC was California's last operating oyster cannery (all other oyster farms sell oysters only in their shells, rather than shucked and packed in jars), this resource has disappeared along with the company.

The wildness of the estero and the passive production of oysters were not as mutually exclusive as the rhetoric sometimes suggested. Point Reyes represents the future challenges for American parks management, as we will increasingly need to reconcile the presence of humans, both in the past and looking forward, in the natural world, finding new ways for people and parks to coexist and complement one another. As environmental historian William Cronon has stated, "If we wish to preserve wild nature, then we must permit ourselves to imagine a way of living

FIGURE 19. National Park Service demolition of Drakes Bay Oyster Company shack, Point Reyes National Seashore, January 2015. Courtesy of David Briggs.

in nature that can use and protect it at the same time."[102] PRNS could have embraced DBOC as an integral part of the Point Reyes experience, as an example of relying on nature for something in addition to recreation. The presence of an oyster operation could have underlined for visitors the interdependence that has always existed on the peninsula between nature and humans. In a letter to Senator Dianne Feinstein in October 2012, author Michael Pollan wrote, "There are deep roots to the hostility of environmentalism toward agriculture . . . . An 'all or nothing' ethic that pits man against nature, wilderness against agriculture, may be useful in some places, under some circumstances, but surely not in this place at this time." The Lunnys' oyster farm, he wrote, "stands as a model for how we might heal these divisions."[103]

But instead, DBOC has closed, ending its last harvests in December 2014. Less than three months later, a local newspaper reported, "On the windswept shore of Drakes Estero, it's as if eighty years of oyster farming have been erased overnight."[104] The former location of the oyster sales shack, cannery, and wooden docks have been "scraped clean" by a government contractor. Standing on the shore now, instead of seeing pristine wilderness, one sees heavy machinery preparing to remove the decades-old oyster racks. The Lunny family has been driven nearly to bankruptcy, and roughly thirty full-time employees, men and women who had worked on the estuary for fifteen, twenty, or even more than twenty-five years, have been put out of work, and for those who also lived onsite, put out of their homes as well. A long tradition of cultivation has vanished—in exchange, more or less, for a label, since the estero was *already* managed as wilderness. By insisting on a more "pure" vision of wilderness at Point Reyes than was ever imagined by the legislators who made the designation, or by the authors of the original Wilderness Act, environmental advocates have sacrificed the relative wild for an idealized one.

FIGURE 20. Schooner Creek, at the top end of Drakes Estero, Point Reyes National Seashore, 2013. Photograph by author.

Point Reyes as a Leopoldian Park

The closure of Drakes Bay Oyster Company may have captured most of the headlines in recent years, but the oyster farm's struggle is only one part of a continuing story of landscape change at Point Reyes National Seashore, where preservation management has devalued and eroded the working landscape in favor of a more wilderness-like vision of the future. This runs counter to societal trends that increasingly value sustainable agriculture and a more intimate connection with the natural world through cultural use and engagement. It is also part of an agency-wide, century-old trend within the NPS of downplaying residents and working land uses in parks where they still exist, and generally working to remove those elements in favor of the more stereotypical natural scenery. Not only is NPS ownership and management of these places an assertion of power and control, but it reshapes these lived-in landscapes according to its own policies and practices, "building in" expectations that residents and working farms somehow do not belong.

At the same time, management of PRNS is increasingly out of step with changes to the agricultural economy at the local scale, changes to public lands management at the national scale, and changes to heritage preservation at the global scale. While the Seashore's leadership has taken ranches out of operation and shuttered the oyster farm, the rest of West Marin has turned into an amazing hub of food production and consumption, truly a mecca of modern sustainable agriculture. Most of the area's dairies and beef ranches have gone organic and/or primarily grass-fed. There has been a huge increase in artisanal cheese-making, for instance, as part of an overall diversification of small-scale local agriculture, including production of more vegetables, chickens, honey, et cetera. Organic production of all kinds of agricultural goods has moved from being a fringe activity to being the economic mainstream of West Marin, as part of a broader nationwide movement. In particular, partly in reaction to the 2002 federal definition of "organic" and the increasing presence of industrial-scale production of organic products, many West Marin producers have focused more on emphasizing local and seasonal production. The area's proximity to the high-end metropolitan market of the larger Bay Area, which seems increasingly willing to pay the higher prices of premium agricultural products, has given West Marin a new boost of confidence that food production is not doomed to disappear, as has been presumed by some to be almost inevitable since the 1950s.[1]

The agricultural operations on PRNS and GGNRA lands represent a substantial portion of this economy, contributing 17 percent of Marin's overall agricultural production and 17 percent of its agricultural land base.[2] The 2007 Marin Countywide Plan lists among its core goals the protection of the area's working agricultural landscapes, and development of greater community food security by increasing the availability and diversity of locally produced foods.[3] Grazing on these lands also provides important ecosystem services, such as managing nonnative weedy species and reducing fire danger. And most of the ranching families have historic connections to the land that go back through generations, helping to anchor the overall community's sense of identity and place.

At the other end of the scale, at the international level, world heritage management policy is increasingly articulated as the protection of both cultural and natural resources and values, emphasizing local uniqueness and community input into management. For example, UNESCO's 2009 *World Heritage Cultural Landscapes: A Handbook for Conservation and Management* identifies six guiding principles in its management framework, including the core recognition that people associated with cultural landscapes are primary stakeholders for stewardship, and that successful management must be inclusive, transparent, and shaped through dialogue. Living cultural landscapes are shaped by the interaction between people and their environment, the report asserts, thus management needs to focus on guiding change to retain and respect this relationship.[4]

In contrast, the NPS's website describing cultural landscapes focuses mostly on the process of documenting such places, but does not mention inclusion of the landscape users or other local community members in decisions about management or direction.[5] Elsewhere across the United States, partnerships between public agencies and private land users are becoming more commonplace, and these kinds of community-based collaborations are increasingly cited as offering new possibilities for land and species conservation.[6] For instance, in September 2015, Secretary of Interior Sally Jewell stated that collaborative efforts of both public and private stakeholders had so significantly reduced threats to the rare greater sage-grouse, that listing it under the Endangered Species Act was not necessary. A Colorado newspaper hailed the decision as "an example of the federal government acknowledging that different stakeholders, not just government agencies, could find their own solution to protect the bird and its habitat."[7] Yet in a recent edited collection of essays and case studies regarding collaborative conservation of working landscapes, across both public and private holdings and involving a myriad of government agencies and nonprofit organizations, the NPS barely shows up in the index.[8]

As this book has shown, the American conception of parks and wilderness has historically tended to dichotomize landscapes into natural and wild versus inhabited and somehow degraded; back in 1964, geographer David Lowenthal wryly observed, "It is no accident that God's own wilderness and His junkyard are in the same country."[9] Historian Christopher Conte argues that this park ideal has too often been "exported" to other parts of the globe, often merging with the "authoritarian legacy of colonialism, resulting in places controlled by bureaucracies rather than local people."[10] Mark Dowie's powerful book *Conservation Refuges* documents numerous clashes in protected areas around the globe between efforts to protect biological diversity on the one hand, and indigenous peoples' residence and use rights on the other—starting with the example of the Miwok people in Yosemite National Park.[11] Gary Nabham calls this a "cultural parallax," as romanticized conceptions of pristine wilderness make indigenous influences on the landscape invisible to scientists and bureaucratic managers.[12] In a review of the popularity of the term "rewilding" over the past two decades, Dolly Jørgensen finds that the concept is vague and fuzzy at best, and frequently "disavows human history and finds value only in historical ecologies prior to human habitation."[13]

Here in the United States, wilderness advocates need to recognize that federal wilderness designation is not *necessary* for a place or an ecosystem to be wild and ecologically healthy, and in fact greater acceptance of the ways in which humans and wild nature can coexist is needed as we face the challenges of climate change. The ultimate irony of efforts to increase habitat protection for the adorable San Joaquin Valley kit fox, for example, is that a stable and growing population is living in urban Bakersfield, even while the species is struggling in the other, more "wild" parts of its range. Yet these city dwellers are more or less invisible to most conservation efforts, except as a source of additional genetic diversity for their cousins living in nature reserves.[14] Similarly, work documenting the habits of urban wildlife—a colleague recently told me about how automatic cameras at intersections in Chicago, intended to catch drivers running red lights, have revealed urban coyotes waiting for the traffic lights to change, as they have learned that it is easier to cross on the green—is forcing us to rethink our categories as some animals increasingly adapt themselves to changing conditions.[15] The importance of broadening our conceptions of wildness reaches back to the work of Aldo Leopold, who observed in a 1925 essay that "wilderness exists in all degrees, from the little accidental spot at the head of a ravine in the Corn Belt woodlot to vast expanses of virgin country . . . . Wilderness is a relative condition. As a form of land use it cannot be a rigid entity of unchanging content, exclusive of all other forms."[16]

CUYAHOGA VALLEY COUNTRYSIDE INITIATIVE AS A MODEL

Pockets of support for working landscape protection do exist within the NPS, particularly the work of the Conservation Study Institute, which in 2007 put out a booklet titled *Stewardship Begins with People* that highlighted both David Evans's ranch and Kevin Lunny's oyster operation as exemplars of sustainable agriculture at Point Reyes.[17] Certain park units within the system, like Ebey's Landing in Washington and Boxley Valley in Arkansas, are making great strides in working with local communities to maintain and encourage the working landscape. However, this approach has been slow to catch on with Point Reyes managers. While a good deal of cultural landscape inventory work has been completed at PRNS, nominations to the National Register of Historic Places remain stalled, so it is unclear whether this inventory work will translate into any change in management.

There is still time to reverse this trend. Reprivatizing the ranches on the Point Reyes peninsula is probably not feasible at this point, even with conservation easements to helping to bring prices down—as was done at Boxley Valley at the Buffalo National River to restore its agricultural community—as land values in Marin County are among the highest in the nation. However, PRNS could easily follow the model of Cuyahoga Valley National Park (CVNP) in Ohio, which went through a controversial acquisition process by the NPS in the 1970s (detailed in chapter 3), but now is implementing a Countryside Initiative to bring agricultural use back to former rural communities in that park after decades of absence. Starting in 1999, CVNP has worked with a non-profit cooperating partner, the Countryside Conservancy, to reestablish working agriculture by rehabilitating historic farms and farmland and offering long-term leases, up to sixty years in length, via a competitive proposal process, with the goal of resurrecting and then maintaining the former rural character of the valley.[18] In its 2015 call for new proposals from prospective leasees, CVNP even referenced the importance of parks as lived-in places:

> Farming in a national park (or any other park) is a most unconventional idea in America. Americans tend to perceive parks as places to visit, not live in—regardless of whether it is a Yellowstone-like wilderness, or a manicured metropark. That is not the case in many other parts of the world. In Great Britain, for example, over 10 percent of the English landscape is located within the boundaries of a national park—over 90 percent of that is privately owned, and most of it is in farms. In Great Britain, farming in the boundaries of national parks is considered the only practical way to maintain the openness, beauty, and diversity of the countryside.[19]

One of the key elements of the Countryside Initiative program is the use of long-term leases; as stated in its 2015 request for proposals:

> Prior to these current authorizations, use of NPS lands for agricultural purposes has been limited to Special Use Permits (SUPs) covering periods of one to five years. Although short-term SUPs are intended to prevent or limit serious damage to park lands, ironically, *they act as a negative incentive to basic land stewardship*. It is economically infeasible for farmers to undertake costly long-term land care programs, which can take years or decades to implement, since they have little assurance of a reasonable return on their investment. The leasing authority now available for the Countryside Initiative resolves this inherent dilemma.[20]

As of this writing, there are now nine farms operating within the CVNP, with two more new leases anticipated in early 2016, most of which are organic and using modern sustainable practices to produce goats, chickens, turkeys, lamb, wine, vegetables, berries, and other products.

Point Reyes could establish a similar relationship with Marin Agricultural Land Trust, which has a long-established track record of working well with local ranchers to conserve their lands, or some other community-based nonprofit focused on cultural landscape protection and management. The nonprofit could serve as an intermediary partner, negotiating lease terms, working with ranchers to maintain land management practices, and generally getting the ranchers out of the direct tenant/landlord role with the NPS. It would also be essential to provide a clearer, more permanent avenue for community collaboration, viewing the ranchers and other locals as stakeholders in Seashore management and planning, distinct and separate from the general visiting public. The NPS needs to recognize that residents have a different relationship to place than do visitors, and particularly that working the land, especially over generations, creates a unique connection that should be respected and incorporated into management practices.

One perennial problem for long-term residents of parks is the changeover in management staff; as soon as ranchers establish a working relationship with one, another turns up, often with a different approach or management style, and often with very little knowledge of the landscape history. For instance, recent range management specialists at PRNS have rarely lasted in their positions longer than a year or two, making for a constant shifting of terms and approaches for the longtime resident ranchers. While turnover in a government agency is unavoidable, the NPS should have long-term measures in place—such as better training for their staff on the historical development of the ranchers' relationship to this landscape, and clear guidelines to help shape staff interactions with park residents—that could help compensate for the inevitable change in staffing that occurs in national agencies.

In addition to forming this kind of management partnership, PRNS should revise a number of its policies to encourage and strengthen long-term agricultural viability. These recommendations are not new; they were clearly articulated in 2009 in letter from Senator Feinstein to PRNS and the Seashore Ranchers Association, and then expanded in a report written by the University of California Cooperative Extension office that same year. Feinstein wrote: "What came through loud and clear at these meetings were three things: first, that Special Use Permits which allow you to operate at Point Reyes need to be issued for longer periods of time than five years. Second, that many of you would like the opportunity to diversify your operations in an effort to stabilize your income. And third, it was very apparent that the National Park Service

FIGURE 21. Cattle on the Lunny (G) Ranch, Point Reyes National Seashore, 2008. Photograph by author.

needed to do a better job of communicating with ranchers and facilitating communications among interested groups in the West Marin area."[21]

The UC Cooperative Extension report added detail to these recommendations, suggesting not only granting longer permit terms, but also formalizing agricultural diversification through the permitting process, thereby giving ranching families more flexibility to raise a variety of products and respond to changing market demands. It also recommends restoring agricultural uses to some lands that have been taken out of production, as part of a wider embrace of the working landscape component of the Seashore. There is absolutely a need for a clearer process for dealing with ranch succession to family members or other community members, so that continuation of ranching will not come into question in cases when permittees retire or pass away. Improved communication and commitment to utilizing ranchers in resource management roles are also needed, to genuinely recognize the value and dedication of these working families to the Seashore as their home and livelihood.[22]

THE IMPORTANCE OF SECURE LAND TENURE

Ultimately, the case of Point Reyes raises the question, why does ownership matter? Private land with public encumbrances, such as conservation easements, grants users most of the same rights and restrictions as public lands with private leases. Why does it matter which side of this divide the ownership remains on? While this book does not advocate for reprivatizing lands already in public ownership, history has shown that ownership makes a huge difference, since an easement is part of the deed, and any changes must be agreed to by both parties—whereas NPS leases or special use permits are constantly in danger of revision, or even cancellation if the NPS so chooses, unilaterally. Full-fee ownership gives the agency too much latitude to determine land uses or abruptly change management direction, and this has not been good for the working landscape at Point Reyes.

Giving the agency such one-sided control over the landscape is particularly problematic when the land use in question is one that does not seem to "fit" into our society's current idealization of parks and wild spaces. Despite the origins of Point Reyes National Seashore in aiming to provide recreation access in a pastoral setting, and the importance put on supporting local agriculture through the design of the Seashore, ranching has long been only tolerated by NPS staff, not embraced. The agency's long history of positing private ownership and use as antithetical to park management suggests that this may be slow to change. Explaining to a local reporter the letter he and two other former legislators sent to the Interior Secretary to plead the oyster farm's case, Congressman Pete McCloskey observed that the NPS "secretly feels very uncomfortable about administering privately leased lands." But, the reporter went on, "he believes that that discomfort—or the beliefs of those who would like to see the entire seashore become 'pure' wilderness—do not justify a misreading of the laws he authored."[23]

This discomfort may link back to the general sense, discussed in chapter 2, that all public access to public lands be *equal* somehow—there's almost an element of jealousy in the fervor with which some advocates seem to be trying to sever the relationships that the ranching families have with the Point Reyes landscape. This bias against private users has deep historical roots. For example, when Congress first set Yosemite Valley aside in 1864 as a public park, it was quite isolated from most of the country's population, and did not become a major tourist draw until some years later. Yet park managers quickly took the approach that small operators within the parks were not acceptable, as they posed too much of a threat to the protection of such symbolic grandeur in perpetuity—there was too much danger of a repeat of Niagara. Landscape designer Frederick Law Olmsted, writing to the California State Legislature just after the park was established, hailed the Yosemite grant as precluding the possibility "that such scenes might become private property and . . . their value to posterity be injured."[24] His report emphasized that private landowners should not be allowed to monopolize the scenery, making it less accessible to the masses, or degrading its beauty for future users. Two private land claims in Yosemite Valley actually went before the Supreme Court in 1872; the claimants lost, with the Court asserting that the claimants *should have known* that the "remarkable features of the place" would cause the government to treat these lands differently than ordinary agricultural lands.[25] Public interest in protecting an unchanging symbolic landscape took precedence over private claims.[26]

Conte's analysis of national parks in developing countries reveals a similar dynamic, in which "an interest group claims a moral imperative to define wilderness, pushes for its adjudication, and then seeks to determine how local groups fit into the new landscape. The impositions come from outside the local area, and those who know the land through labor and experience lose control of landscapes they had a hand in creating. Wilderness advocates then celebrate park designations uncritically, equating their own aesthetic choices with the appreciation of nature: they champion naturalness in the form of remote areas free from signs of human habitation."[27]

It is striking how similar this language is to some of the arguments made against the PRNS residents today. During the oyster controversy, a frequent refrain from wilderness activists was "A deal's a deal," implying that because the public owned the land, the oyster farm was obligated to leave.[28] Similarly, Neal Desai, associate director of the NPCA's Pacific regional office, asserted that "For more than thirty years, the public has waited for nature—not business plans and motorized uses—to be the estero's heartbeat."[29] In the context of the more recent tule elk discussions, environmental advocates have characterized the Point Reyes ranches as "subsidized," accused them of having "sweetheart deals," and implied that they are attempting to eliminate wildlife from the Seashore.[30]

Yet the presence of the still-operating ranches does not inhibit any one else's access to the coastal grandeur of Point Reyes. And in the case of the ranchers within the Seashore and GGNRA lands, it is their work and their ancestors' work that have *made* the place in so many ways. Walking across a pasture with one or another of these people, the kinds of stories they tell, the sense of responsibility and linkage that they feel to the land are so different from those of a recreational visitor, even someone who returns to hike the same trail again and again, year after year. One type of relationship is not necessarily *better* than the other, but the NPS must find a way in its management policies to acknowledge and value that difference in a substantive manner.

Because of this 150-year-old belief that private interests are fundamentally incompatible with public parks, it is essential, if the working landscape at Point Reyes is to be maintained over time, that the ranchers' tenure be strengthened legally. In his November 2012 statement denying DBOC its permit, Secretary Salazar added an offering to the remaining ranchers on the Seashore: "Because of the importance of sustainable agriculture on the pastoral lands within Point Reyes, I direct that [NPS] pursue extending permits for the ranchers within those pastoral lands to twenty-year terms."[31] This was followed by a letter to Interior's western regional director in late January 2013, authorizing the issuance of these new longer-termed permits.[32] Yet no permits have yet been issued; instead, two months after the Ranchers Association's September 2013 letter to the Seashore complaining about elk in the pastoral zone, the NPS announced that it was embarking on yet another new planning effort, a Comprehensive Ranch Management Plan, which is estimated to take at least another two years to complete.[33] Questions about what triggered this new planning effort at this particular time, rather than when the Secretary directed the permits to be issued, have gone unanswered, but it came suspiciously soon after the ranchers' complaints, and after local elected officials began pressing the NPS on the ranchers' behalf. The NPS conducted scoping meetings in June 2014 and two additional public workshops that November; a draft of the new plan is not expected until late 2016, and in the meantime there is no change to the park's management of the tule elk. Most ranchers are now operating on only one-year lease extensions, as the NPS has insisted that it cannot renew any special use permits until the new planning process is complete.

While longer-term leases would be welcomed by the ranchers, the specific language would also need to be more concrete than in the past. After all, quite a few ranches have had agricultural leases with five-year terms and three options to renew (totaling twenty years)—yet the language in these agreements is almost verbatim to that in Mary Tiscornia's lease at Rancho Baulines, where the NPS asserted they had no obligation to renew. The language in her lease read: "Six months prior to the expiration of this lease, this lease or a similar lease including changes in conditions negotiated with the leasee *will be issued* for an additional five-year period. If an additional five-year lease is not entered into prior to the expiration of the existing lease, the leasehold interest will terminate and removal conditions of this lease will apply."[34] A group of leases renewed in 2003 have nearly identical language, followed by an additional sentence, "Notwithstanding the foregoing, Lessor shall have no obligation to offer a subsequent lease to Leasee if Leasee breaches any of the provisions of this Lease or if Leasee terminates this Lease prior to its Expiration Date."[35] Special use permits have similar language, but instead of stating that a similar lease "will be issued," they simply say that "Prior to the Expiration Date of this Permit, this Permit, or a similar Permit, *may be offered* to Permittee for an additional period."[36]

The argument made by the NPS in Tiscornia's case, that "if there is to be a renewal of the lease, it must be 'entered into' by mutual agreement," but that if the agency decided not to renew, it was under no legal obligation to do so, makes all of the renewal clauses meaningless— despite a common understanding in commercial leasing that renewal clauses are a benefit to the tenant that allow the *tenant* to decide whether to renew or not.[37] For now, all permit renewals or extensions are on hold while the Seashore undertakes its new Comprehensive Ranch Management Plan effort. The ranchers are stuck in limbo, uncertain of their long-term prospects. The NPS needs new policy to give them clearer certainty and greater security.

Another suggestion for Point Reyes management would be to reinstitute the Citizens Advisory Commission, which served as a sounding board and public forum for both North Bay park units until 2002, when it "sunseted" out of the original GGNRA founding legislation's requirements. Geographer Megan Foster has found that, despite its functioning as something of a "rubber stamp" for NPS decisions, the CAC played a vital role as an intermediary, providing an open forum for discussion of park-related issues and interaction between local residents and NPS managers.[38] In a similar vein, political scientist Leigh Raymond has concluded that Grazing Advisory Boards set up in the 1930s on lands managed by the Grazing Service (later the Bureau of Land Management), rather than representing a "captured agency," actually served to balance local and federal influence over range policy, an earlier precursor to today's efforts at community-based conservation.[39] There needs to be clearer recognition by park managers that while local land users are no longer owners, they retain a different kind of knowledge of, and relationship with, the landscape that both parties aim to protect and sustain. They should truly be brought into management conversations as partners.

As part of this policy shift, Interior and the NPS should also reverse their decision to close down the oyster farm. It represents an absolutely pointless loss, valuing semantics ("the first marine wilderness on the West Coast") over on-the-ground ecological and cultural reality—and many are concerned that it may lead to further closures of ranches in the future. Any departures from Point Reyes' working landscape are a tragic loss to this vibrant area's sustainable agriculture and distinctive character. Drakes Estero with the oyster farm in place, as a designated potential wilderness supporting production of a sustainable, local food source, represented a rare opportunity for the public to experience wilderness and working landscape together, *seeing* how attached and interwoven they are, how each depends on the other. It truly was the core of Point Reyes, both geographically and in its role as a Leopoldian park, showcasing agriculture and wilderness side by side, celebrating the ways in which humans *can* coexist with the natural world.

A NEW OUTLOOK ON PARKS MANAGEMENT

In his 2010 book *Uncertain Path: A Search for the Future of National Parks,* long-time NPS naturalist and historian William Tweed called for a fundamental redefinition of the agency's core mission and management goals. With climate change already bringing ecological shifts to the High Sierra he so loves, Tweed argued that park management should respond to evolving ecosystems and shifting public expectations for the parks. The Park Service must learn to accommodate change, he forcefully declared. It's no longer enough to just preserve parks as islands of stability: "Today, the idea that parks can protect forever everything significant within their boundaries reflects a mind-set somewhere between denial and fantasy."[40]

What he failed to acknowledge is that this idea has *always* been fantasy—that preservation itself causes changes (ecological change, visual change, changes in use, etc.) that the NPS has never managed to acknowledge or see its own hand in. The removal of Native peoples and their active landscape manipulations from parks was just the start of the many transformations wrought by park management. Photographs of Yosemite Valley from 1864 to the 1930s to 2016 reveal substantial ecological shifts, from open meadows and oak woodlands to closed coniferous forest. Anthropogenic change in the Sierra is not something new brought on by climate change or air pollution drifting in from the Central Valley's exploding development; anthropogenic change, first through

Native peoples' active engagement with and manipulation of the landscape, later through park management practices, has constantly been reshaping these places.

Tweed also stressed the importance of public education, particularly the need to persuade visitors that the Park Service simply cannot fulfill its promise of an unchanging park system: "If 'unimpaired for future generations' must be abandoned, a new vision and a new set of values must be offered."[41] His book underlines the need for the Park Service not only to stop making that promise, but also to stop expecting it from itself. Until the Park Service acknowledges that preservation is not neutral, that *nothing* stays the same forever, this call for a new mission will fall flat. The NPS has simply been in denial of its own impacts on the landscape, and while Tweed's call for a change of mindset is a wonderful and welcome step, it does not go far enough to reconcile the agency with its long history of being an agent of change.

Furthermore, the agency needs to further broaden this message to recognize that ideas about what parks "should" be are changing; in other words, that the controversies and conflicts at Point Reyes are a bellwether of sorts, that they indicate changing ideas about the environment, parks, and public recreation, as well as our relationship with nature and wilderness. What has been called the "Yellowstone model" of natural preservation is limited in terms of what it can offer us, particularly as we increasingly recognize that humans have influence on every aspect of the natural world, that we are not distinct and separate from it.[42] In the conservation literature, authors from a variety of backgrounds call for "a middle ground, to reconcile the twin impulses of pragmatism and purity in nature preservation."[43] Point Reyes could be just such a middle ground. There is no need for a zero-sum game of agriculture versus wilderness; there is room, and an essential role, for both. The NPS managers at Point Reyes need to get beyond that opposition, and actually go back in time to the original vision for the Seashore, as a Leopold-inspired place that offers protection of natural resources and public recreation while embracing and collaboratively supporting its working agricultural landscapes.

FIGURE 22. Working landscape, on the Barinaga Ranch, Marshall, CA, with Tomales Bay and Tomales Point in the distance, 2011. Photograph by author.

Epilogue

Just as I was working on the last revisions to this book, in early 2016, a new lawsuit was filed against the National Park Service at Point Reyes, this time by a trio of environmental groups: the Resource Renewal Institute (headed by Huey Johnson, former director of the Trust for Public Land, a national-scale land trust organization, and resident of Marin County), the Center for Biological Diversity (CBD), and a group called the Western Watersheds Project, based in Idaho. Their suit demands that the ongoing ranch management planning process be suspended until the thirty-six-year-old PRNS General Management Plan can finally be updated with studies of the environmental impacts of grazing.[1] This would halt the renewal process for permits, including the one-year extensions all the ranches are currently operating under—which could functionally end all ranching in the Seashore. Local and national newspapers are now full of the groups' accusations about "'welfare ranchers' who take advantage of subsidized grazing rates they pay on park grounds," and accusations that the ranchers generated the dispute about free-ranging elk to "derail" the species' restoration.[2] Johnson told the San Jose *Mercury News* that if their suit is successful, the coalition plans to press for removal of cattle from all national park units; in the same article, Ken Brower, affiliated with Earth Island Institute and son of former Sierra Club leader David Brower, weighed in as saying that while many people admire ranchers as icons of the West, "they really shouldn't be in a national park."[3]

Readers of this book will not be surprised: indeed, Phyllis Faber, cofounder of the Marin Agricultural Land Trust, Peter Prows, the lead lawyer for Drakes Bay Oyster Company, and I predicted this kind of outcome in a *Point Reyes Light* article two years ago.[4] The CBD has been saber-rattling over the tule elk issue since late 2014, but as this book has shown, there has been a steady drumbeat of pressure from various environmental organizations to eliminate the ranches for decades—not only since the late 1990s in comment letters on planning processes, but in some ways all the way back to the early wilderness hearings of the 1970s. This history of Point Reyes management reveals a long pattern of downplaying, ignoring, and sometimes actively pushing the working landscape to the margins in this park where founding legislation intentionally accommodated agriculture.

The CBD is fairly well known across the West for its use of litigation to force federal managers to prioritize species protections; they have likened their own legal tactics to psychological warfare.[5] In contrast, several key local environmentalists who had long been campaigners against agricultural land uses at the Seashore have apparently had a change of heart lately: for instance, Gordon Bennett—the former Sierra Club representative whose 2004 comment letter questioned congressional intent for continuing agricultural uses at Point Reyes and appeared to trigger the Interior Solicitor's legal opinion that wilderness law required DBOC to close in 2012—is now speaking out against the new lawsuit and publicly advocating for the ranch management planning process to proceed.[6] The Environmental Action Committee of West Marin (EAC), under new leadership, has joined Bennett in his opposition to the lawsuit, as have representatives of the NPCA—two of the most vociferous opponents of the oyster farm only a few years ago. Whether this is a genuine change of perspective or a calculated political move aimed at producing a compromise (such as, fewer ranches instead of none) is hard to gauge at this point, but I am hopeful that they are genuine in their call for "a renaissance of ranching at Point Reyes, and a model for sustainable ranching nationwide."[7]

Elsewhere in the news, an armed standoff at the Malheur National Wildlife Refuge has just simmered down, with the arrest of Ammon Bundy and most of his instigators, but soberingly with the loss of one person's life. Geographer Peter Walker from the University of Oregon (a graduate school classmate of mine!) bravely attended many of the town hall meetings in Harney County during the standoff and reported to the world via Facebook. Bristling with firearms, the occupiers, none of whom were local, asserted that federal ownership of the Refuge was unconstitutional, called for the land to be "returned" to the county level of government, and urged neighboring ranchers to tear up their grazing permits as a form of resistance. The militants failed to inspire the people of Harney County to rise up and join their insurgency; not a single local grazing permit was destroyed, and there were numerous counterprotests by county residents and the local Paiute tribe, making it clear that the militants did not speak for them, and demanding that they leave.

Yet across the West there have been murmurings of understanding, not of the Bundy group's tactics, but of the pinch that many western public lands users find themselves in, subject to ever-shifting federal management practices, and feeling hounded by environmental absolutists who push for reduction or removal of their livelihoods from the landscape. I have heard mutterings around West Marin, concerns that the new lawsuit against the NPS at Point Reyes could inspire Bundy-type militants to show up here next, and some of the advocates pushing the new lawsuit have explicitly attempted to link Point Reyes ranchers to the Oregon standoff in the press.[8]

At the same time, commentators like Nancy Langston, whose environmental history of the Malheur provides essential background to understanding that region of Oregon, have praised recent collaborative planning processes in that area, in which local ranchers, Paiute tribal members, and conservation groups, working together through an organization called the High Desert Partnership, have assisted the U.S. Fish and Wildlife Service in crafting a new fifteen-year management plan for the Refuge.[9] Langston writes, "This requires a great deal of flexibility, which addresses a core frustration ranchers have had with federal lands. Decisions had become bogged down in rules, environmental regulations, and litigation, causing even the smallest of changes to take years to implement. Collaborative plans devolve much of the day-to-day decision-making to local participants, allowing for responsiveness to changing ecological conditions."[10] A former Forest Service supervisor writing in *Time* magazine struck a similar note: "The heart of the matter is that the human dimensions of managing public lands are just as important as the physical and biological dimensions."[11]

Given this recent context, I feel the need to state clearly that this book does not advocate for privatization of Point Reyes, nor of any other public lands—yet neither would I encourage absolute federal control of working landscapes, as it is simply not a good fit at this time. As Brower's comment about the Point Reyes lawsuit suggests, the old ideal of what a national park "ought" to be—pristine natural scenery empty of human habitation but open for recreational uses—remains a powerful force. It presumes that there is only one "green" relationship possible between humans and the natural world: one of standing back in awe and wonder. It also implicitly argues that private use rights on public lands *necessarily* erode the public's interest, just by being there—regardless of management outcomes. And it overlooks not only the value of the ranches at Point Reyes as historic and cultural resources, which the NPS is obliged to protect, but also the deep and lasting relationship the ranching families have with the landscape itself, their work over generations inscribed into the current appearance and health of the land.

The idealization of nature as pristine and empty of human habitation is not unique to the national parks and wilderness, but it is often expressed in its most absolute and intolerant forms in debates over the management of these kinds of public spaces. Partly this is a form of what Franz Vera calls "shifting baselines," the ways in which the meaning of specific words—like "park," like "wilderness"—shift over time to obscure past relationships (both ecological and cultural) and impose new meaning on the landscape.[12] Certainly this has been the case at Point Reyes, where the pragmatic vision of wilderness that held sway in the 1970s, which considered the existing oyster farm to be compatible with wilderness protection, has been replaced by an assertion that wilderness required its demise. Yet the landscape itself contains traces of those older meanings, if you dig back far enough.

The NPS itself is beginning to change, with some of its leaders calling for a greater focus on integrated stewardship, as well as on "deepening public engagement and establishing ever-more-meaningful connections" between parks and the communities they serve.[13] Geographer David Lowenthal has advised the agency that parks and wilderness areas "must begin to exemplify, rather than be set apart from, the everyday terrain of our ordinary places of work and play, travel and repose."[14] By building on the insight of Aldo Leopold, recognizing that the wild and the pastoral can not only coexist but also strengthen each other, Point Reyes could be a powerful model of this evolving stewardship approach.

Collaborative organizations like the Southwest's Malpai Borderlands Group, as well as many others, have shown again and again in the last two decades that the strongest way forward on public lands is to accept resident land users as allies and partners, rather than opponents, to work together toward increased ecological *and* economic stability—and to take these peoples' knowledge of and connection to the land seriously.[15] This approach has worked in Harney County, Oregon. Conversely, when absolutist environmental organizations sling lawsuits at the NPS that explicitly aim to end ranching at Point Reyes, they are bringing the legal equivalent of the rifles and threats of the Bundy militants to the local community. My hope is that the history in this book can help the Point Reyes ranchers, the NPS, and environmental interests to leave behind the threats and start to see a way forward to a more collaborative partnership in public land management—to a more Leopoldian vision for Point Reyes National Seashore.

Notes

INTRODUCTION

1. Memo from Secretary of Interior Kenneth Salazar to Director, National Park Service, regarding Point Reyes National Seashore / Drakes Bay Oyster Company, dated November 29, 2012.

2. NPS, *Drakes Estero: A Sheltered Wilderness Estuary.* This report was revised and reissued a number of times, dated September 2006, April 1, 2007, May 8, 2007, and May 11, 2007; a version dated July 27, 2007 was not released publicly.

3. Transcript of Marin County Board of Supervisors meeting, May 8, 2007; video available www.marincounty.org/depts/bs/meeting-archive. The item of consideration at the meeting was a proposed letter to be sent by the board to Senator Dianne Feinstein, asking for her assistance resolving DBOC's current lease, finalizing cleanup and other requirements after acquisition of the operation from Johnsons Oyster Company in 2005; the board voted to send an amended version of the letter.

4. John Muir, 1912, "Hetch Hetchy Valley," in *The Yosemite* (Madison: University of Wisconsin Press, repr. 1986), 262. Rolf Diamant, "From Management to Stewardship: The Making and Remaking of the U.S. National Park System," *George Wright Forum* 17, no 2 (2000): 31–45.

5. Peter F. Cannavò, *The Working Landscape: Founding, Preserving, and the Politics of Place* (Cambridge, MA: MIT Press, 2007), 220.

6. National Park Service Act of 1916, 39 Stat. 535.

7. Steven H. Corey and Lisa Krissoff Bohem, eds., *The American Urban Reader: History and Theory* (New York: Routledge, 2011), 207–8.

8. For an overview of the residents and their use of the land that became Central Park see Roy Rosenzweig and Elizabeth Blackmar, *The Park and the People: A History of Central Park* (Ithaca, NY: Cornell University Press, 1992), 59–91.

9. Aldo Leopold, "The Farmer as a Conservationist," first published in *American Forests* (1939), reprinted in *For the Health of the Land: Previously Unpublished Essays and Other Writings*, eds. J. Baird Callicott and Eric T. Frey-fogle (Washington, DC: Island Press, 1999), 161.

CHAPTER 1. LANDSCAPES, PRESERVATION, AND THE
NATIONAL PARK IDEAL

1. Peter F. Cannavò, *The Working Landscape: Founding, Preserving, and the Politics of Place* (Cambridge, MA: MIT Press, 2007).

2. Peirce Lewis, "Axioms for Reading the Landscape: Some Guides to the American Scene," in *The Interpretation of Ordinary Landscapes: Geographical Essays*, ed. Donald W. Meinig (New York: Oxford University Press, 1973), 12.

3. Paul Groth, "Frameworks for Landscape Study," in *Understanding Ordinary Landscapes*, eds. Paul Groth and Todd Bressi (New Haven, CT: Yale University Press, 1997), 1.

4. Barbara Bender, "Introduction: Landscape—Meaning and Action," in *Landscape: Politics and Perspective*, ed. Barbara Bender (Oxford: Berg, 1993), 3; emphasis is mine.

5. Kenneth Olwig, "Landscape: The Lowenthal Legacy," *Annals of the Association of American Geographers* 93, no. 4 (2003): 871–77, 873.

6. Henri Lefebvre, *The Production of Space* (Oxford: Basil Blackwell, 1991), 44.

7. Henri Lefebvre, *The Critique of Everyday Life* (London: Verso, 1991).

8. Timothy Mitchell, *Colonizing Egypt* (Cambridge: Cambridge University Press, 1988), 14.

9. Kenneth R. Olwig, "Sexual Cosmology: Nation and Landscape at the Conceptual Interstices of Nature and Culture; or What Does Landscape Really Mean?" in *Landscape: Politics and Perspective*, ed. Barbara Bender (Oxford: Berg Press, 1993), 307.

10. Don Mitchell, *The Lie of the Land: Migrant Workers and the California Landscape* (Minneapolis: University of Minnesota Press, 1996), 28.

11. William Cronon, "Rails and Water," in *Nature's Metropolis: Chicago and the Great West* (New York: W.W. Norton, 1991), 72.

12. Noel Castree, "The Nature of Produced Nature: Materiality and Knowledge Construction in Marxism," *Antipode* 27, no. 1 (1995): 12–48, 15.

13. Kenneth R. Olwig, "Recovering the Substantive Nature of Landscape," *Annals of the Association of American Geographers* 86, no. 4 (1996): 630–53, 640.

14. Robert Pahre, "Introduction: Patterns of National Park Interpretation in the West," *Journal of the West* 50, no. 3 (2011): 7–14.

15. Raymond Williams, *The Country and the City* (New York: Oxford University Press, 1973), 124–25.

16. David Lowenthal, *The Past Is a Foreign Country* (Cambridge: Cambridge University Press, 1985), 14.

17. Laura A. Watt, Leigh Raymond, and Meryl L. Eschen, "On Preserving Ecological and Cultural Landscapes," *Environmental History* 9, no. 4 (2004): 620–47.

18. Note that some things were deliberately preserved as part of burial rituals or tombs, going back thousands of years, and many specific objects with religious symbolism or meaning have been protected, but large-scale "preservation for preservation's sake—because an object is old, regardless of its religious or aesthetic content" is not seen in any ancient societies. E. R. Chamberlin, *Preserving the Past* (London: J. M. Dent and Sons, 1979), ix.

19. David Lowenthal, "Introduction," in *Our Past before Us: Why Do We Save It?*, eds. David Lowenthal and Marcus Binney (London: Temple Smith, 1981), 10.

20. Joseph Sax traces the beginnings of preservation as a state act to the French Revolution, as a way of stabilizing the sense of French culture during turbulent political times, in his 1990 article "Heritage Preservation as a Public Duty: The Abbé Grégoire and the Origins of an Idea," *Michigan Law Review* 88, no. 5 (1990): 1142–69.

21. Peirce Lewis calls this "successful proxemics"—we become accustomed to certain dimensions of our living spaces, and so we don't like them to change, as the new spaces or arrangements make us feel uncertain and uncomfortable. Lewis, "The Future of the Past: Our Clouded Vision of Historic Preservation," *Pioneer America* 7, no. 2 (1975): 1–20, 13.

22. Kevin Lynch, *What Time Is This Place?* (Cambridge, MA: MIT Press, 1972), 105.

23. Lowenthal, *The Past*, 384—also described in Lynch, *What Time*, chapter 2.

24. Lowenthal, *The Past*, 243.

25. Gillian Tindall, *The Fields Beneath: The History of One London Village* (London: Granada/Paladin, 1980), 116.

26. Lowenthal, *The Past*, xvii.

27. David Lowenthal, "The American Way of History," *Columbia University Forum* 9, no. 3 (1966): 27–32.

28. Yi-Fu Tuan, "Perceptual and Cultural Geography: A Commentary," *Annals of the Association of American Geographers* 93, no. 4 (2003): 879–80. Cannavò, *Working Landscape*, 6.

29. For example, an organization called the Society for Creative Anachronism, which stages reenactments of medieval life, describes its goal as "to recreate the Middle Ages not as it was but as it should have been, doing away with all the strife and pestilence." Lowenthal, *The Past*, 363.

30. Edward M. Bruner, "Epilogue: Creative Persona and the Problem of Authenticity," in *Creativity/Anthropology*, eds. Smadar Lavie, Kirin Narayan, and Renato Rosaldo (Ithaca, NY: Cornell University Press, 1993), 324; emphasis in original. Note that this is the opposite of progress, an Edenic narrative where the original innocence is lost, and the rest of history is a constant struggle to reattain that original state of perfection.

31. Even today as ethnographers are trying to move away from this idea of a single "authentic" past or original state, tourism increasingly demands it, seeking it out worldwide. Bruner, "Epilogue," at 324.

32. Dydia DeLyser, "Good, by God, We're Going to Bodie! Landscape and Social Memory in a California Ghost Town," PhD diss., Syracuse University, 1998.

33. Wallace Stegner, "The Best Idea We Ever Had," *Wilderness* 46, no. 160 (1983): 4–13. The full quote is, "National parks are the best idea we ever had. Absolutely American, absolutely democratic, they reflect us at our best rather than our worst."

34. Alfred Runte, *National Parks: The American Experience,* 2nd ed. (Lincoln: University of Nebraska Press, 1987).

35. This and additional presidential quotes, from Eisenhower to George W. Bush, can be found at www.nps.gov/nama/historyculture/presidential-quotations-about-national-parks.htm, accessed March 10, 2015.

36. Tourism is rarely if ever considered a "use" of these lands, even though it clearly is. See Hal Rothman, *Devil's Bargains: Tourism in the Twentieth-Century American West* (Lawrence: University of Kansas Press, 1998).

37. Roderick Nash, *Wilderness and the American Mind,* 3rd ed. (New Haven, CT: Yale University Press 1982), 23–24.

38. For more extensive discussion of the conception of wilderness, see generally Nash, *Wilderness,* also Max Oelschlaeger, *The Idea of Wilderness: From Prehistory to the Age of Ecology* (New Haven, CT: Yale University Press, 1991), and J. Baird Callicott and Michael P. Nelson, eds., *The Great New Wilderness Debate: An Expansive Collection of Writings Defining Wilderness from John Muir to Gary Snyder* (Athens: University of Georgia Press, 1988).

39. J. Baird Callicott and Michael P. Nelson, "Introduction," in Callicott and Nelson, eds., *The Great New Wilderness Debate,* 4.

40. Nash, *Wilderness,* 51, notes that Romantic appreciations "are found among writers, artists, scientists, vacationers, gentlemen—people, in short, who did not face wilderness from the pioneer's perspective."

41. William Cronon, "The Trouble with Wilderness; or, Getting Back to the Wrong Nature," in *Uncommon Ground: Toward Reinventing Nature,* ed. William Cronon (New York: W. W. Norton, 1995), 76.

42. Mark David Spence, *Dispossessing the Wilderness: Indian Removal and the Making of the National Parks* (New York: Oxford University Press, 1999), 12.

43. Runte, *National Parks,* 11.

44. Nash, *Wilderness,* 67.

45. John F. Sears, *Sacred Places: American Tourist Attractions in the Nineteenth Century* (New York: Oxford University Press, 1989). This creation of a national identity as represented by the natural landscape "demanded a body of images and descriptions of those places—*a mythology of unusual things to see*—to excite people's imaginations and induce them to travel" (3).

46. For extended discussions on Niagara and its cultural meanings see Patrick McGreevy, *Imagining Niagara: The Meaning and Making of Niagara Falls* (Amherst: University of Massachusetts Press, 1994), and Elizabeth McKinsey, *Niagara Falls: Icon of the American Sublime* (Cambridge: Cambridge University Press, 1985).

47. Elizabeth McKinsey, "An American Icon," in *Niagara: Two Centuries of Changing Attitudes, 1697–1901,* ed. Jeremy Elwell Adamson (Washington, DC: The Corcoran Gallery of Art, 1985), 89; William Irwin, *The New Niagara: Tourism, Technology and the Landscape of Niagara Falls, 1776–1917* (University Park: Pennsylvania State University Press, 1996), xix.

48. Irwin, *New Niagara*, 18. Sears finds that many tourist accounts from Niagara express disappointment, boredom, or irreverence, the viewers having been distracted from the view by the variety of business operations surrounding it. The commercial nature of the experience reduced the grand natural scenery to something more like a carnival or freak show; Sears, *Sacred Places*, 18.

49. As part of his 1831 trip to America, de Tocqueville visited the falls, and warned that it would soon be spoiled by development. Runte, *National Parks*, 6.

50. Richard White, "The Conquest of the West, in *"It's Your Misfortune and None of My Own": A New History of the American West* (Norman: University of Oklahoma Press, 1991).

51. William H. Truettner, "Ideology and Image: Justifying Western Expansion," in *The West as America: Reinterpreting Images of the Frontier 1820–1920,* ed. William H. Truettner (Washington, DC: Smithsonian Institute, 1991).

52. Patricia Hills, "Picturing Progress in the Era of Westward Expansion," in Truettner, ed., *The West as America*, 101, quoting from Gilpin, 1860, *Mission of the North American People: Geographical, Social and Political;* all emphases in original.

53. Hills, "Picturing Progress," 102.

54. Spence, *Dispossessing the Wilderness*, 28–29, emphasis in original. Also Cronon, "The Trouble with Wilderness," 79: he describes wilderness as "flight from history."

55. Samuel Bowles, *Across the Continent* (Springfield, MA: Samuel Bowles and Company, 1865), 226–27.

56. Clarence King, *Mountaineering in the Sierra Nevada* (1872), as quoted in Runte, *National Parks*, 21.

57. Spence, *Dispossessing the Wilderness*, 102.

58. Jen A. Huntley, *The Making of Yosemite: James Mason Hutchings and the Origin of America's Most Popular National Park* (Lawrence: University Press of Kansas, 2011).

59. Many Europeans thought the tree slices were a hoax, as they simply could not imagine that any tree could actually reach such an enormous size. See Jared Farmer, *Trees in Paradise: A California History* (New York: W. W. Norton and Co., 2013).

60. As quoted in Joseph H. Engbeck, Jr., *The Enduring Giants* (Berkeley: University Extension, University of California, 1973), 77.

61. Alfred Runte, *Yosemite: The Embattled Wilderness* (Lincoln: University of Nebraska Press, 1990), 18.

62. According to Runte, the motives pushing the legislation through Congress were primarily business-related, yet these remained mostly hidden in the public debate.

63. Yosemite Land Grant, 13 Stat. 325 (1864).

64. It was in fact fully two years before California agreed to take over management of the park. The park was effectively expanded in 1890, to accommodate increasing numbers of tourists, by creating a ring of "reserved forest lands" surrounding the original grant; this expansion was encouraged by the Southern Pacific Railroad, as well as agricultural interests hoping for greater watershed protection. California eventually receded Yosemite Valley back to federal

ownership, and combined it with the surrounding protected forestlands, to form the current park boundary in 1906. See Runte, *Yosemite*, 45–56, 83.

65. Runte, *National Parks*, 29.

66. Charles W. Cook, "The Valley of the Upper Yellowstone," *Western Monthly* 4 (1870): 61, as quoted in Runte, *National Parks*, 35.

67. The tribes in the area actively managed the landscape, and in particular used fire to manage the forests, opening up "broad savannas" to attract game, ease travel, and encourage growth of certain kinds of plants, as well as to keep favorite sites clear of brush and insects. Spence, *Dispossessing the Wilderness*, 43–44.

68. See Mark Daniel Barringer, *Selling Yellowstone: Capitalism and the Construction of Nature* (Lawrence: University of Kansas Press, 2002) for more detail on the role of concessionaires, beginning with the railroad company, in creating and shaping Yellowstone National Park.

69. *The Yellowstone Park: House Report 26 to accompany H.R. 764*, 42nd Congress, 2nd Session, House Committee on the Public Lands (February 27, 1872), at 1–2; emphasis is mine.

70. An Act to Set Apart a Certain Tract of Land Lying Near the Headwaters of the Yellowstone River as a Public Park, 17 Stat. 32 (March 1, 1872).

71. Richard West Sellars, *Preserving Nature in the National Parks: A History* (New Haven, CT: Yale University Press, 1997), 19.

72. The national parks served by railroads were: Mount Rainer by the Northern Pacific; Glacier by the Great Northern; Yellowstone by the Northern Pacific, Union Pacific, and C.B.&Q.; Bryce-Zion by the Union Pacific; Grand Canyon by the Atchison, Topeka & Santa Fe; Crater Lake and Yosemite by the Southern Pacific.

73. Mitchell, *The Lie of the Land*, 18.

74. Spence, *Dispossessing the Wilderness*, chapter 8.

75. John Brinckerhoff Jackson, *Discovering the Vernacular Landscape* (New Haven, CT: Yale University Press, 1984), x.

76. www.nationaltrust.org.uk/what-we-do, accessed Dec. 27, 2014.

77. Barbara Bender, "Stonehenge: Contested Landscapes," in *Landscape: Politics and Perspective*, ed. Barbara Bender (Oxford: Berg, 1993), 269.

78. Bender, "Stonehenge," 271.

79. Williams, *The Country and the City*, 120.

80. See Rolf Diamant et al., *Stewardship Begins with People: An Atlas of Places, People, and Handmade Products* (Woodstock, VT: Conservation Study Institute, National Park Service, 2007), for a number of examples of parks that aim to protect working landscapes.

81. Peter Jackson, *Maps of Meaning: An Introduction to Cultural Geography* (London: Unwin Hyman, 1989), 59.

82. Sellars, *Preserving Nature*, 290.

83. See Denis Cosgrove, *Social Formation and Symbolic Landscape* (London: Croon Helm, 1984), Barbara Bender, ed., *Landscape: Politics and Perspective* (Oxford: Berg, 1993), and Mitchell, *Lie of the Land*, for starters.

84. See Williams, *The Country and the City*, and Richard White, "'Are You an Environmentalist or Do You Work for a Living?' Work and Nature," in

Uncommon Ground: Toward Reinventing Nature, ed. William Cronon (New York: W.W. Norton, 1995).

85. James C. Scott, *Seeing Like a State: How Certain Schemes to Improve the Human Condition Have Failed* (New Haven, CT: Yale University Press, 1998); this work has been particularly applied to the NPS by James W. Feldman, *A Storied Wilderness: Rewilding the Apostle Islands* (Seattle: University of Washington Press, 2011).

86. Joseph L. Sax, "Do Communities Have Rights? The National Parks as a Laboratory of New Ideas," *University of Pittsburgh Law Review* 45, no. 3 (1984): 505.

87. See National Park Service, *Cultural Resources Management Guideline,* NPS-28 (Washington, DC: National Park Service, 1997); Robert R. Page, Cathy A. Gilbert, and Susan A. Dolan, *A Guide to Cultural Landscape Reports: Contents, Process and Techniques* (Washington, DC: National Park Service, 1998); and Richard Westmacott, *Managing Culturally Significant Agricultural Landscapes in the National Park System* (Washington, DC: National Park Service, 1998).

88. Arnold Alanen and Robert K. Melnick, "Introduction: Why Preserve Cultural Landscapes?" in *Preserving Cultural Landscapes in America,* eds. Arnold Alanen and Robert K. Melnick (Baltimore: Johns Hopkins University Press, 2000), 7.

89. For example, Alston Chase, *Playing God in Yellowstone: The Destruction of America's First National Park* (San Diego: Harcourt Brace Jovanovich, 1984) and Sellars, *Preserving Nature.*

90. See Christopher Conte, "Creating Wild Places from Domesticated Landscapes: The Internationalization of the American Wilderness Concept," in *American Wilderness: A New History,* ed. Michael Lewis (New York: Oxford University Press, 2007), 223–41.

91. Emma Marris, *Rambunctious Garden: Saving Nature in a Post-Wild World* (New York: Bloomsbury, 2011), particularly chapter 2, "The Yellowstone Model."

CHAPTER 2. PUBLIC PARKS FROM PRIVATE LANDS

1. For more detail on local Miwok history, see Kent Lightfoot and Otis Parrish, *California Indians and Their Environments: An Introduction* (Berkeley: University of California Press, 2009); also the website for the Federated Indians of Graton Rancheria, www.gratonrancheria.com/ourpeople.htm.

2. Email from Kent Lightfoot to Michael Newland, in author's collection, dated Jan. 21, 2015: "The environment we are seeing at Point Reyes today is probably characterized, in large part, by a highly transformed vegetation that is the product of both the termination of Native burning and more recent fire suppression policies." For more detail, see R. Scott Anderson, Ana Ejarque, Peter M. Brown, and Douglas J. Hallett, "Holocene and Historical Vegetation Change and Fire History on the North-Central Coast of California, USA," *The Holocene* 23, no. 12 (2013): 1797–1810; also see report by R. Scott Anderson, *Contrasting Vegetation and Fire Histories on the Point Reyes Peninsula During the*

Pre-Settlement and Settlement Periods: 15,000 Years of Change (Flagstaff, AZ: Center for Environmental Sciences and Education, Quaternary Sciences Program, 2005).

3. Dewey Livingston, *Ranching on the Point Reyes Peninsula: A History of the Dairy and Beef Ranches within Point Reyes National Seashore, 1834–1992,* Historic Resource Study, National Park Service, 1993. See also the excellent collection of historic photographs in Carola DeRooy and Dewey Livingston, *Point Reyes Peninsula: Olema, Point Reyes Station, and Inverness,* Images of America (Charleston, SC: Arcadia, 2008).

4. Livingston, *Ranching on the Point Reyes Peninsula,* 37.

5. Hal K. Rothman, *Preserving Different Pasts: The American National Monuments* (Urbana: University of Illinois Press, 1989).

6. Richard White, "'Are You an Environmentalist or Do You Work for a Living?' Work and Nature," in *Uncommon Ground: Toward Reinventing Nature,* ed. William Cronon (New York: W. W. Norton, 1995), 171–85, 173.

7. C. B. Macpherson, *Property: Mainstream and Critical Positions* (Toronto: University of Toronto Press, 1978), 5.

8. See John Locke, "Of Property" (1689), reprinted in Macpherson, *Property,* 17–30. Note that Locke's conception of self-ownership only extended to white men; women and people of other ethnicities, such as Native Americans or black slaves, were not considered to hold the same rights.

9. Eric T. Freyfogle, *The Land We Share: Private Property and the Common Good* (Washington, DC: Island Press, 2003), 5.

10. A similar principle regarding use was present in the 1841 Preemption Act, the Timer Culture Act, the Timber and Stone Act, and in numerous grants of federal land to states or railroad companies, to name a few. See Samuel Trask Dana and Sally K. Fairfax, "Conservation from the Broad Arrow Policy through the Disposition of the Public Domain," in *Forest and Range Policy: Its Development in the United States,* 2nd ed. (New York: McGraw Hill, 1980).

11. Sally K. Fairfax, Lauren Gwin, Mary Ann King, Leigh Raymond, and Laura A. Watt, *Buying Nature: The Limits of Land Acquisition as a Conservation Strategy, 1780–2003* (Cambridge, MA: MIT Press, 2005), 16.

12. Leigh Raymond and Sally Fairfax, "Fragmentation of Public Domain Law and Policy: An Alternative to the 'Shift to Retention' Thesis," *Natural Resources Journal* 39, no. 4 (1999): 649–753.

13. Frederick Law Olmsted, "The Yosemite Valley and the Mariposa Big Trees: A Preliminary Report" (1865), reprinted in *Landscape Architecture* 43, no. 1 (1952): 12–25, 21.

14. See Paul Herman Buck, *The Evolution of the National Park System of the United States* (Washington, DC: U.S. Government Printing Office, for the National Park Service, U.S. Department of the Interior, 1946), 9, and Joseph Sax, *Mountains without Handrails: Reflections on the National Parks* (Ann Arbor: University of Michigan Press, 1980), 10.

15. See Harvey Mayerson, *Nature's Army: When Soldiers Fought for Yosemite* (Lawrence: University of Kansas Press, 2001).

16. See generally Samuel P. Hays, *Conservation and the Gospel of Efficiency: The Progressive Conservation Movement, 1890–1920* (Cambridge, MA: Har-

vard University Press, 1959); also Samuel Trask Dana and Sally K. Fairfax, "Conservation in Practice and Politics: The Golden Era of Roosevelt and Pinchot, 1898–1910," in *Forest and Range Policy: Its Development in the United States*, second edition (New York: McGraw Hill, 1980), 69–97.

17. Ethan Carr, *Wilderness by Design: Landscape Architecture and the National Park Service* (Lincoln: University of Nebraska Press, 1998), 55. While scenic quality had originally been one of the considerations in selecting forest reserves in the 1890s, by the arrival of Gifford Pinchot in the 1900s, scenic preservation was only one of an array of multiple uses being considered in resource management.

18. Of particular concern were Gifford Pinchot's efforts to get the Interior parks transferred into his new, more utilitarian Forest Service in the Agriculture Department. See Hays, *Conservation and the Gospel of Efficiency*, at 196–98.

19. Horace M. Albright and Robert Cahn, *The Birth of the National Park Service: The Founding Years, 1913–33* (Salt Lake City: Howe Brothers, 1985), chapters 1–2. Note that Mather had made his fortune in borax, but was also an avid mountaineer and longtime member of the Sierra Club.

20. Donald C. Swain, "The Passage of the National Park Service Act of 1916," *Wisconsin Magazine of History* 50, no. 1 (1966): 4–17, 7. For more on the role of the automobile in parks tourism, see Hal K. Rothman, "Intraregional Tourism: Automobiles, Roads, and the National Parks," in *Devil's Bargains: Tourism in the Twentieth-Century American West* (Lawrence: University of Kansas Press, 1998).

21. An Act to Establish a National Park Service, and Other Purposes, 39 Stat. 535 (August 25, 1916).

22. Frederick Law Olmsted Jr., 1916, "The Distinction between National Parks and National Forests," *Landscape Architecture* 6, no. 3 (1916): 114–115, 114; emphasis is mine. Also see E. A. Sherman, "The Forest Service and the Preservation of Natural Beauty," *Landscape Architecture* 6, no. 3 (1916), 115–19. Sherman, the acting chief forester of the Forest Service at the time, wrote specifically that "National Parks are areas where economic considerations can be practically excluded" (117).

23. Swain, "Passage of the National Park Service Act," 9.

24. Jerry Frank calls this "the opportunity to build a broader constituency, which would bolster support for a national parks agency to better protect and manage America's growing constellation of parks." Jerry Frank, *Making Rocky Mountain National Park: The Environmental History of an American Treasure* (Lawrence: University of Kansas Press, 2013), 8.

25. Historian Robin Wink asserts that preservation is somehow privileged by the phrasing of the Act (see his article, "Dispelling the Myth," *National Parks* 70, nos. 7–8 [1996]: 52–53), but most other researchers disagree. Sellars identifies the primary concerns for NPS founders as "preservation of scenery, the economic benefits of tourism, and efficient management of the parks," rather than a mandate for "exacting preservation of natural conditions"; Richard West Sellars, *Preserving Nature in the National Parks: A History* (New Haven, CT: Yale University Press, 1997), 29.

26. Carr, *Wilderness by Design*, 5.

27. Standard discussions of these trends include James Q. Wilson, "The Rise of the Bureaucratic State," *The Public Interest* 41 (1975): 77–104; and Herbert Kaufman, "Emerging Conflicts in the Doctrine of Public Administration," *American Political Science Review* 50, no. 4 (1956): 1057–73.

28. Linda Flint McClelland, *Building the National Parks: Historic Landscape Design and Construction* (Baltimore: Johns Hopkins University Press, 1998), 8.

29. McClelland, *Building the National Parks*; emphasis is mine. In fact, the American Society of Landscape Architects recommended, among other things, that the 1916 bill include "an advisory board composed of landscape architects and an engineer, whose services would be called upon whenever landscape questions in existing parks or proposals for new parks were considered." This suggestion was dropped from the eventual bill.

30. Carr, *Wilderness by Design*, 6.

31. Sellars, *Preserving Nature*, 49.

32. Charles B. Hosmer, Jr., *Preservation Comes of Age: From Williamsburg to the National Trust, 1926–1947*, vol. 1 (Charlottesville: University Press of Virginia, 1981), 1.

33. Carr, *Wilderness by Design*, at 85. At the time there was general concern in the popular culture about the recent influx of Southern and Eastern European immigrants, including fears that their presence was eroding national identity and patriotism.

34. Carr, *Wilderness by Design*, 72, quoting McFarland.

35. Park visitors expected facilities comparable to those found while traveling through Europe, depending "not only upon efficient highways, but also upon a comfortable inn, a clean bed, and a palatable cuisine," and the NPS was determined to provide them. Buck, *Evolution of the National Park System*, at 53.

36. In a 1922 conference, the park superintendents issued a statement that a park without proper facilities would be "merely a wilderness, not serving the purpose for which it was set aside, not benefiting the general public." Sellars, *Preserving Nature*, 63.

37. Carr, *Wilderness by Design*, 1.

38. For example, as recently as 2008, video playing in Joshua Tree National Park's visitor center described the local tribes as having "vanished into the mists of time," despite the fact that the park has three organized tribes as neighbors. Laura A. Watt, "Reimagining Joshua Tree: Applying Environmental History to National Park Interpretation," *Journal of the West* 50, no. 3 (2011): 15–20.

39. Philip Burnham, *Indian Country, God's Country: Native Americans and the National Parks* (Washington, DC: Island Press, 2000), 15. For more on Caitlin, see John Hausdoerffer, *Caitlin's Lament: Indians, Manifest Destiny, and the Ethics of Nature* (Lawrence: University of Kansas Press, 2009).

40. Mark David Spence, 1999, *Dispossessing the Wilderness: Indian Removal and the Making of the National Parks* (New York: Oxford University Press), 3. He also notes (32) that many tribes understood the early reservation treaties as designating areas where the United States guaranteed to not let its citizens invade, rather than as limiting where Indians could live, and it was the U.S. government's attempt to restrict off-reservation use that triggered the Indian Wars of the 1870s.

41. Spence, *Dispossessing the Wilderness*, 4.

42. John Muir, *Our National Parks* (1901; San Francisco: Sierra Club Books, 1991), 21.

43. Theodore Catton, *Inhabited Wilderness: Indians, Eskimos, and National Parks in Alaska* (Albuquerque: University of New Mexico Press, 1997), 9.

44. Carolyn Merchant, "Shades of Darkness: Race and Environmental History," *Environmental History* 8, no. 3 (2003): 380–94, 381–83.

45. Lafayette Houghton Bunnell, *Discovery of the Yosemite, and the Indian War of 1851, Which Led to that Event* (1880; n.p.: General Books, 2009), 132.

46. Note, too, the irony of whites disliking the idea of communally held Indian land, yet then making the areas into parks owned jointly "by the people"; see Burnham, *Indian Country*, chapter 2, for a discussion of this idea.

47. Sellars, *Preserving Nature*, 56. Also see Albright and Cahn, *Birth of the National Park Service*, chapter 6. The Lane letter was actually penned by Albright, and approved by Mather.

48. Albright and Cahn, *Birth of the National Park Service*, 69.

49. Sellars, *Preserving Nature*, 65.

50. Buck, *Evolution of the National Park System*, 45–46. The fact that many parks were created from national forest lands caused continued animosity between the NPS and the U.S. Forest Service.

51. Sellars, *Preserving Nature*, 58–60; also see Albright and Cahn, *Birth of the National Park Service*, chapters 2–5.

52. See Jenks Cameron, *The National Park Service: Its History, Activities and Organization* (Baltimore: Institute for Government Research, Johns Hopkins Press, 1922), for early detail.

53. The project was first financed by a group that included George B. Dorr, a wealthy Bostonian, and Charles W. Eliot, then president of Harvard University; see Alfred Runte, *National Parks: The American Experience*, 2nd, revised ed. (Lincoln: University of Nebraska Press, 1987), 114–15. Eliot then went on to create the first private land trust in the country, The Trustees for Reservation, in Boston.

54. Nancy Wayne Newhall, *A Contribution to the Heritage of Every American: The Conservation Activities of John D. Rockefeller, Jr.* (New York: Knopf, 1957), 49. From Newhall: "A park perfected, stretching from Frenchman Bay to the ocean itself, with roadways opening up its vistas and making its heights accessible to visitors, was the ambition of John D. Rockefeller, Jr. Quietly buying up land to supplement the areas already given to the Park, he was also thinking of how the roads could be threaded through the hills to blend with the natural features of the landscape and leave no scars. First working out on paper the curves, bridges and overpasses, and then vigilantly following construction on the spot, Mr. Rockefeller worked closely with the engineers, hiring in seasons when work was scarce on the Island and relaxing the pace when other jobs were plentiful. As each unit was completed, it was deeded to the Government as a gift from Mr. Rockefeller" (56).

55. John Ise, *Our National Parks: A Critical History* (Washington, DC: Resources for the Future, 1961) describes this original area as "a stingy, skimpy, niggardly little park of only about 150 square miles, from three to nine miles wide and twenty-seven miles long" (329).

56. Robert W. Righter, *Crucible for Conservation: The Creation of Grand Teton National Park* (Jackson Hole, WY: Grand Teton Natural History Association, 1982), 31.

57. Alice Wondrak, "Teton Dreams: Horace M. Albright, John D. Rockefeller, Jr., and the Making of the Landscape in Grand Teton National Park," Boulder: University of Colorado, 1997, unpublished manuscript on file with author.

58. Righter, *Crucible for Conservation*, 49–50.

59. Rockefeller also donated six million dollars in 1952 to the Jackson Hole Preserve for the construction of a hotel and cottages near Jackson Lake, adding to the development of the area specifically for tourists. Newhall, *Contribution to the Heritage*, 110.

60. Wondrak points out that according to Albright and Rockefeller's visions of the Tetons, "In order to allow these places to fulfill their potential as glorious natural landscapes, messy traces of historical human intervention needed to be removed; even if it had been that human intervention which had made the landscape accessible to viewers and shaped the view in the first place."

61. When the Shenandoah proposal was still competing with that for Great Smoky Mountains, North Carolina newspapers refuted this claim, and noted that fifteen thousand people would be displaced by the creation of Shenandoah with its originally planned boundary—yet this fact didn't come up at the congressional hearings. The project idea also got considerable support early on from Stephen Mather himself, who particularly wanted a park close to the nation's capital; Darwin Lambert, *The Undying Past of Shenandoah National Park* (Boulder, CO: Roberts Rinehart, 1989), 194, 202.

62. Lambert, *Undying Past*, 216. The legislation, passed in 1928, was called the Public Park Condemnation Act.

63. Durwood Dunn identifies the primary motive of promoters to be profit and national prestige; one prominent spokesman, Colonel Chapman, even went so far as to specify that greater "advertising and prestige" would result from the area becoming a national park rather than a national forest. Durwood Dunn, *Cades Cove: The Life and Death of a Southern Appalachian Community, 1818–1937* (Knoxville: University of Tennessee Press, 1988), 242–43. Also see Carlos C. Campbell, *Birth of A National Park in the Great Smoky Mountains* (Knoxville: University of Tennessee Press, 1960), written by one of the principle players in that park movement, from the promoter's perspective.

64. In the Great Smokies case, the most critical legal question was "whether one sovereignty, the state of Tennessee, could exercise the power of eminent domain to secure land for the public use of another sovereignty, the federal government." The original judge found that it could not, but the Tennessee State Supreme Court rejected that decision. Dunn, *Cades Cove*, at 249.

65. Lambert, *Undying Past*, 223. Note that Lambert was one of the first employees at Shenandoah National Park. Also see Justin Reich, "Re-Creating the Wilderness: Shaping Narratives and Landscapes in Shenandoah National Park," *Environmental History* 6, no. 1 (2001): 95–117.

66. Dunn, *Cades Cove*, xiii. Note a similar observation in Lambert, *Undying Past*, 171: "Romanticists considered them noble primitives or dangerous bar-

barians. Missionary-educators thought they were 'lost sheep.' Sociologists and social workers were sure they needed 'help.'"

67. Lambert notes the existence of a "secret list" of "aged and *especially meritous*" residents who were allowed to stay in the park until the end of their lives. The list had forty-three people on it in 1934, chosen by "personal situation and 'merit,' recommended by Will Carson and approved by Ickes." The last of these people died in 1979. Lambert, *Undying Past*, 254.

68. Dunn writes: "Although it is evident that park promoters intended to include Cades Cove within park boundaries almost from the beginning of the movement, they launched an elaborate campaign to assure cove citizens that their homes and farms would never be molested." Every major group brought to see the proposed park site was taken to Cades Cove, yet "Campbell and others reiterated over and over their promise that private homes would not be taken over by the proposed park." Dunn, *Cades Cove*, 243.

69. Dunn, *Cades Cove*, 246.

70. It is also worth noting that, unlike the other projects mentioned here, Blue Ridge Parkway was *not* a national park, but rather an area intended primarily for motoring recreation, which had lower expectations of significance and management; this may explain the divergence from the general pattern. In contrast, both Big Bend National Park (1935) and later, Virgin Islands National Park (1956) stuck to the usual model of removing residents to create an uninhabited natural landscape.

71. Harley E. Jolley, *The Blue Ridge Parkway* (Knoxville: University of Tennessee Press, 1969), 50–53. The mountain people were also considered "fertile and prolific" (51), and Jolley suggested that "such increases were proportionally much greater than the ability of the soil properly to sustain them; thus, squalor and misery continued to multiply" (52).

72. Jolley, *Blue Ridge Parkway*, 130. This perhaps reflects the NPS's greater involvement with historic resources at this time.

73. North Carolina passed a law in 1935 stating that just by mapping and posting in counties which rights-of-way lands it wanted, the state automatically became possessor of that land. Private owners could then accept an offer of compensation or file suit for damages. In the meantime, the state could proceed with construction. Similarly, in Virginia, the State Assembly gave the State Highway Commission power "to procure the land as it would for any state road," which would then be transferred to the federal government. Jolley, *Blue Ridge Parkway*, 104–5.

74. Jolley, *Blue Ridge Parkway*, 103.

75. Jolley, *Blue Ridge Parkway*, 132.

76. William Cronon, "The Trouble with Wilderness; or, Getting Back to the Wrong Nature," in *Uncommon Ground: Toward Reinventing Nature*, ed. William Cronon (New York: W. W. Norton, 1995), 80.

77. Catton, *Inhabited Wilderness*, 2.

78. William C. Everhart, *The National Park Service*, 2nd ed. (Boulder, CO: Westview Press, 1983), 22.

79. Hosmer, *Preservation Comes of Age*, 3.

80. Charles B. Hosmer Jr., *Presence of the Past: A History of the Preservation Movement in the United States Before Williamsburg* (New York: G.P. Putnam's Sons, 1965), 299.

81. Executive Order 6166, Section 2 (1933). In addition, the NPS received jurisdiction for some things it did *not* want, such as the Fine Arts Commission, the National Capital Park and Planning Commission, Arlington National Cemetery, and all District of Columbia parks and buildings; the executive order also changed the name of the agency to the Office of National Parks, Buildings, and Reservations. The NPS eventually managed to drop some of the extra responsibilities it did not want, as well as to get its original name back.

82. Hosmer, *Preservation Comes of Age*, 549–56. HABS was made into a permanent program in July 1934, with headquarters and organizing staff housed within the NPS, and the American Institute of Architects (AIA) providing field organization.

83. S. 2073, 74 Cong., 1 Sess., Stat. 49 (1935), at 666. Hosmer, *Preservation Comes of Age* notes that much debate over the bill turned on the question of whether the government would be able to confiscate property so as to create National Historic Sites (572).

84. James A. Glass, *The Beginnings of a New National Historic Preservation Program, 1957 to 1969* (Nashville: American Association for State and Local History, 1990), xiii.

85. Hosmer, *Preservation Comes of Age*, 585. Hosmer also records the breakdown of the requests: 57 percent came from Congress, 38 percent from NPS offices, and the remainder from state or private historical agencies; see 1936 letter from Cammerer to Ickes, Hosmer, *Preservation Comes of Age*, 592.

86. Barry Mackintosh, *The Historic Sites Survey and National Historic Landmarks Program: A History* (Washington, DC: History Division, National Park Service, 1984), 81.

87. Such a stance was hardly convincing, though: "When explaining the national historic landmarks program in connection with controversial sites, the Service regularly contended that landmark designation constituted a *neutral* recognition of historical importance rather than an 'honoring' of the subject involved. In reality, the idea that designation entailed a degree of honoring could not be so easily dismissed. The Service's leaflet describing the program spoke of landmarks as 'among the most treasured' tangible reminders of the nation's history. The homes of unmitigated scoundrels, however great their influence, were not made landmarks (unless justified on architectural grounds) . . . . And while the moral neutrality of a mere listing of sites might have been credible, it was difficult for the general public not to view the bronze plaque as a sign of official sanction or approval." Mackintosh, *Historic Sites Survey*, 84; emphasis added.

88. Other factors contributed to this trend as well. Hosmer notes that the "emphasis on military parks and sites that illustrated various themes in American history meant the Park Service would perpetuate the museum aspects of preservation." Hosmer, *Preservation Comes of Age*, 580.

89. See Ronald A. Foresta, *America's National Parks and Their Keepers* (Washington, DC: Resources for the Future, 1984), 131.

90. Hosmer, *Preservation Comes of Age*, 929.

CHAPTER 3. ACQUISITION AND ITS ALTERNATIVES

1. Other national parks established during the 1930s include Isle Royale (1931) and Everglades (1934), which historian Paul Sadin describe as having dominantly horizontal views, rather than the traditional, vertical mountainous landscapes of earlier parks; he suggests these set the stage for interest in seashores and lakeshores. Paul Sadin, *Managing a Land in Motion: An Administrative History of Point Reyes National Seashore* (Washington, DC: National Park Service, U.S. Department of Interior, 2007), 41.

2. For example, at Smoky Mountains, a Roosevelt executive order made an allocation in 1933, which was followed by a congressional appropriation in 1938. Also at Isle Royale, land was purchased in 1936, and then needed additional legislation to allow the land to be added to the park in 1940. Barry Mackintosh, *The National Parks: Shaping the System* (Washington, DC: National Park Service, U.S. Department of Interior, 1985), 48.

3. Sharon A. Brown, *Administrative History: Jefferson National Expansion Memorial, 1935–1980* (Washington, DC: National Park Service, U.S. Department of Interior, 1984), 27.

4. The NPS acquired its first Philadelphia property, the Second Bank of the United States, by transfer from the Department of the Treasury in 1939. Constance M. Greiff, *Independence: The Creation of a National Park* (Philadelphia: University of Pennsylvania Press, 1987), 40–41.

5. 57 Stat. 563 (July 14, 1943).

6. Anna Coxe Toogood, *George Washington Carver National Monument, Diamond, Missouri: Historic Resource Study and Administrative History* (Denver, CO: Denver Service Center, Historic Preservation Team, National Park Service, 1973).

7. Note that July 1959 also briefly became a formal cutoff date between the NPS's official definition of "old" parks, for which the policy was eventual acquisition of inholders, and "new" parks, for which the policy was prompt acquisition. See National Park Service, *Revised Land Acquisition Policy*, 44 Fed. Reg. 24790 (1979). This policy was then revised again in 1983, in 48 Fed. Reg. 21121. For a discussion of the new/old parks policy, see Joseph L. Sax, "Buying Scenery: Land Acquisitions for the National Park Service," *Duke Law Journal*, no. 4 (1980): 709–40, 714–15.

8. *Report to Accompany H.R. 5892*, Report No. 600, 86th Congress, 1st Session, Committee of Interior and Insular Affairs (1959), at 6. Note that the committee specifically cited the authority to condemn "under the provisions of the act of August 1, 1888." In his 1986 article, Steven Hemmat argued that the 1888 Act empowered the NPS "to exercise condemnation power at *all* national park units unless specifically prohibited." Steven A. Hemmat, "Parks, People, and Private Property: The National Park Service and Eminent Domain," *Environmental Law* 16, no. 4 (1986): 935–61, 938, emphasis in original. However, in 1938 the NPS specifically recommended that "all privately owned lands within existing national parks and monuments be acquired by the Federal Government, and the power of condemnation, to be used only when necessary, be extended to apply to all areas"; *Recreational Use of Land in the United States*, report prepared by the NPS for the Land Planning Committee of the National

Resources Board (Washington DC: U.S. Government Printing Office, 1938). This, along with the language in the Minuteman legislative report, suggests that the NPS believed that the 1888 Condemnation Act did *not* grant the agency broad authority to condemn.

9. At community meetings in Lincoln, MA, in 1959, NPS Director Conrad Wirth promised there would be no evictions; this led to controversy later when the agency did indeed threaten condemnation proceedings against owners uninterested in selling. See Charlene K. Roise, Edward W. Gordon and Bruce C. Fernald, *Minute Man National Historic Park: An Administrative History,* unpublished manuscript, not approved by the NPS North Atlantic Region, 1989, pp. 16.

10. Cape Hattaras was authorized by Congress in 1937, but not enough land was acquired to allow full establishment until 1953, when substantial grants from the Mellon family and the State of North Carolina provided funding for purchase. Mackintosh, *National Parks,* 56.

11. Douglas W. Doe, "The New Deal Origins of the Cape Cod National Seashore," *Historical Journal of Massachusetts* 26, no. 2 (1997): 136–56, 148.

12. *Our Vanishing Shoreline* (1955); *Seashore Recreation Area Survey of the Atlantic and Gulf Coasts* (1955); *Our Fourth Shore: Great Lakes Shoreline, Recreation Area Survey* (1959); and *Pacific Coast Recreation Area Survey* (1959), all Washington, DC: U.S. Dept. of Interior, National Park Service. Mission 66 was a ten-year campaign to dramatically expand and improve visitor services by the agency's fiftieth anniversary in 1966, and was particularly focused on roads and infrastructure as well as facilities like developed campgrounds and visitor centers; as described by former director Wirth, the program was "designed to overcome the inroads of neglect [dating from World War II] and to restore to the American people a national park system adequate for their needs." Conrad L. Wirth, *Parks, Politics, and the People* (Norman: University of Oklahoma Press, 1980), 237.

13. S. 2010, introduced by Senator Richard Neuberger (OR) would have established Cape Cod, Oregon Dunes, and Padre Island; S. 2460, introduced by Sen. James Murray (MT), chair of the Interior and Insular Affairs Committee, would have established the same three plus Indiana Dunes, Point Reyes, Channel Islands, Cumberland Island, Huron Mountains, Pictured Rocks, and Sleeping Bear Dunes.

14. While locals were initially uneasy about federal involvement, there was also concern that trends of uncontrolled development would both alter their communities' character and "kill the goose that was laying the golden egg," i.e., tourist revenues. Charles H. W. Foster, *The Cape Cod National Seashore: A Landmark Alliance* (Hanover, NH: University Press of New England, 1985), 69. The Cape's traditional fishing and agriculture industries had declined since the turn of the century, leaving land development and seasonal tourism as the dominant "industries."

15. Foster, *Cape Cod,* at 68.

16. The regulations prescribing these standards were approved in July 1962, with three underlying purposes: preserving the natural and scenic features of the coastline; providing recreational facilities for visitors; and protecting the inter-

ests of property owners who lived on the Cape prior to Seashore establishment. For greater detail see Charlotte E. Thomas, "The Cape Cod National Seashore: A Case Study of Federal Administrative Control over Traditionally Local Land Use Decisions," *Boston College Environmental Affairs Law Review* 12, no. 2 (1985): 225–72.

17. Joseph L. Sax, "Helpless Giants: The National Parks and the Regulation of Private Lands," *Michigan Law Review* 75, no. 2 (1976): 239–74 , 242.

18. The framework has been regarded as less successful at controlling development at Fire Island than at Cape Cod.

19. Holborn, on writing the legislation: "We did negotiate with them for a while. The Park Service and particularly the Interior Department's Solicitor's Office didn't feel they had any mandate to introduce this rather different type of legislation, but they didn't actually discourage us." Holborn also notes that he wasn't sure Wirth liked the idea, but that "he wanted the park, and he didn't like visible political fights." In Francis P. Burling, *The Birth of Cape Cod National Seashore* (Plymouth, MA: Leyden Press, 1979), 19.

20. Conrad Wirth stated in an interview in 1975 that he personally had "encouraged a servicewide policy of opposition to the establishment of park advisory bodies." Foster, *Cape Cod*, at 7.

21. Conrad Wirth, Study of a National Seashore Recreation Area, Point Reyes Peninsula, California (Washington, DC: National Park Service, U.S. Department of Interior, 1935), 14, as cited in John Hart, *An Island in Time: Fifty Years of Point Reyes National Seashore* (Mill Valley, CA: Pickleweed Press, 2012), 52.

22. Murphy's initial purchase was one of the two pieces originally owned by J. M. Shafter; his daughter Julia Shafter Hamilton borrowed $144,000, with her Point Reyes land as collateral, three weeks before the stock market crash, and so had to sell. Murphy kept the Home Ranch and sold the rest to M. P. "Doc" Ottinger (O Ranch), Sayles Turney (R Ranch), and Robert Marshall Jr. (T Ranch), and sold S Ranch to an investment group. He apparently then intended to subdivide his remaining land; he formed a land company called New Albion Properties, and later joined Turney and Frank Merz to form a development corporation called Limantour Lands, Inc. See D. S. Livingston, *Ranching on the Point Reyes Peninsula: A History of the Dairy and Beef Ranches Within Point Reyes National Seashore, 1834–1992* (Point Reyes Station, CA: Point Reyes National Seashore, National Park Service, 1993), 380–85.

23. DeRooy and Livingston list these as the Murphy Ranch and B, F, H, and N Ranches. During World War II, both of these newcomer groups were prevented from living close to the coast, and the lands they had rented reverted to pasture. Carola DeRooy and Dewey Livingston, *Point Reyes Peninsula: Olema, Point Reyes Station, and Inverness* (Charleston, SC: Arcadia Publishing, 2008), 43.

24. Membership was limited to 125, all of whom had to first belong to the elite Pacific-Union Club in San Francisco, and then be invited to join. DeRooy and Livingston, *Point Reyes Peninsula*, 75.

25. DeRooy and Livingston, *Point Reyes Peninsula*, 95.

26. DeRooy and Livingston, *Point Reyes Peninsula*, 99–111.

27. See Adam Rome, *The Bulldozer in the Countryside: Suburban Sprawl and the Rise of American Environmentalism* (Cambridge: Cambridge University Press, 2001).

28. *Calling for the Preparation of a Report on the Proposed Point Reyes National Seashore Recreational Area, Marin County, Calif.*, 85th Congress, 2nd Session, Committee on Interior and Insular Affairs, House Report 2463 (August 5, 1958), at 2.

29. Hart, *Island in Time*, 52; *Hearings on S. 476, to Establish the Point Reyes National Seashore in the State of California, and for Other Purposes*, 87th Congress, 1st Session (March 28, 30, and 31, 1961).

30. *Hearing on S. 2428, to Establish the Point Reyes National Seashore in the State of California and for Other Purposes*, 86th Congress, 2nd Session (April 16, 1960); the proceedings include a written report from the Marin County Planning Commission listing "uses indicating transition away from agriculture" (26).

31. *Hearings on S. 476* (1961) (testimony of Stewart Udall), at 8 and 32.

32. Interview with Boyd Stewart, in Ann Lage and William J. Duddleston, *Saving Point Reyes National Seashore, 1969–1970: An Oral History of Citizen Action in Conservation* (Berkeley: Regional Oral History Office, Bancroft Library, University of California at Berkeley, 1993), 234.

33. A 1964 independent appraisal of the Home Ranch, describing overall population increase in Marin County, specifically noted that "this population surge has barely affected West Marin and the Point Reyes peninsula. The population was almost static during the 1950–1960 period. Building activity has been relatively minor." Kermichel appraisal of the Home Ranch, Dec. 24, 1964; PRNS archives.

34. *Hearings on S. 476* (1961), at 103. Joseph Mendoza added later that "This subdivision thing is poppycock" (106).

35. *Hearings on S. 476* (1961) (testimony of Bryan McCarthy), at 89.

36. *Hearings on S. 476* (1961) (testimony of Bryan McCarthy), at 89.

37. Letter from Representative Clem Miller, dated July 3, 1961, to the chair of the House Subcommittee on National Parks, in *Hearings on H.R. 2775 and H.R. 3244, to Establish the Point Reyes National Seashore in the State of California, and for Other Purposes*, 87th Congress, 1st Session (March 24, July 6, and August 11, 1961), at 111.

38. "Chart of Subdivision Activity as of Aug. 6, 1962," from appraisal of Alamea, Bolema, and Point Reyes Land & Development Company parcels, compiled by Edward Morphy, San Rafael, Aug. 15, 1962, pp. 9; PRNS archives.

39. For example, Peter Behr referred to the 1960s as experiencing "a tremendous explosion" of development in the park; Lage and Duddleston, *Saving Point Reyes*, 115. In contrast, McCarthy argued that "this argument only came about to justify getting the dairymen out of there"; *Hearings on H.R. 2775 and H.R. 3244* (1961), at 22.

40. "Questions and Answers Concerning the Proposed Point Reyes National Seashore," draft dated Feb. 23, 1962, pp. 13; PRNS archives. A map made in 1960 of the proposed park area listed only sixty-two landowners; 95 percent of the land was represented by twenty-five ranches.

41. John Hart, *Farming on the Edge: Saving Family Farms in Marin County, California* (Berkeley: University of California Press, 1991), 28.

42. Sadin, *Managing a Land in Motion*, 57.

43. Sadin, *Managing a Land in Motion*, 74, quoting Harold Gilliam.

44. Sadin, *Managing a Land in Motion*, 72.

45. *Hearing on S.* 2428, 86th Congress, 2nd Session (April 16, 1960), at 1.

46. "Questions and Answers Concerning the Proposed Point Reyes National Seashore," undated draft, pp. 9, question 15; PRNS archives. A later draft of this document is marked "Sent to Rep. Miller," suggesting this compilation of information regarding the proposed seashore may have been authored by NPS regional planning staff, rather than by Miller's staff.

47. *Hearing on S.* 2428 (1960) (testimony of George Collins), at 13.

48. *Hearing on S.* 2428 (1960) (testimony of B. W. Broemmel, Marin County assessor), at 19.

49. *Hearing on S.* 2428 (1960) (testimony of Ed Rennington), at 186.

50. *Hearing on S.* 2428 (1960) (testimony of Waldo Giacomini), at 165.

51. Sadin, *Managing a Land in Motion*, 79–81.

52. Climate was also cited by locals as one reason why the peninsula would never be developed for year-round residential use—see *Hearing on S.* 2428 (1960) (testimony of Joseph Mendoza), at 183, for example.

53. Minutes from West Marin Property Owners Association meeting with William Grader, staff assistant to Congressman Miller, July 9, 1959, 38; PRNS archives.

54. Lage and Duddleston, *Saving Point Reyes*, 237.

55. *Hearings on H.R.* 2775 *and H.R.* 3244 (1961), at 37. Not only had the ranchers and their representatives received a copy of the Land Use Survey only that same day, they also had no details about what the proposed lease-back terms might be.

56. *Hearings on S.* 476 (1961), at 113.

57. County Supervisor James Marshall made this point at the 1960 hearings: "Most of the Point Reyes Peninsula was privately owned by two landowners until prior to World War II. These ranches have gotten out from under the yoke of the landlords and now are facing the same problem again." *Hearing on S.* 2428 (1960), at 17.

58. *Hearings on S.* 476 (1961) (testimony of Bryan McCarthy), at 86–87; see also *Hearings on H.R.* 2775 *and H.R.* 3244 (1961), at 17–19.

59. *Hearings on H.R.* 2775 *and H.R.* 3244 (1961), at 37.

60. In the 1961 Senate hearings, it is particularly evident that several senators were concerned about the locals being displaced, and suggested that the NPS had not taken this issue seriously enough. Senator Dworshak (ID) actually scolded NPS Director Wirth during his testimony for the NPS's lack of preparedness: "We waste literally hours because you people do not know what you are talking about"; *Hearings on S.* 476 (1961), at 220–21.

61. *Hearings on S.* 476 (1961) (Senator Dworshak speaking), at 221.

62. Founding member Margaret Azevedo described the Point Reyes National Seashore Foundation as a "paper organization," needed because there was "no evidence for support" for the seashore idea in Marin County, and aiming to

"create leadership and a political face for local public support where it had yet to coalesce"; Lage and Duddleston, *Saving Point Reyes*, 171. *Hearing on S. 2428* (1960) (testimony of Harold Gilliam), at 199.

63. See *Hearings on H.R. 2775 and H.R. 3244* (1961) (testimony of Conrad Wirth, NPS director), at 140. The "donut" lands at Everglades were eventually acquired by the NPS in 1975, as part of a larger effort at ecological restoration in the park.

64. See letters from AT&T, RCA, the Vedanta Society, California governor Edmund G. Brown and Oregon governor Mark Hatfield, as well as various officials from Interior; *Hearings on H.R. 2775 and H.R. 3244* (1961), at 130–38.

65. 107 Cong. Rec., part 14 (Sept. 6–13, 1961), at 18462–3.

66. 108 Cong. Rec., part 11 (July 20–Aug. 3, 1962), at 14412.

67. 108 Cong. Rec., part 11 (July 20—Aug. 3, 1962) (statement of Rep. John Kyl), at 14410–11.

68. This figure was based on testimony from the Marin County assessor in 1960, who based his estimates on 1958–59 assessed values, and said that there had been little change for 1959–60. *Hearing on S. 2428* (1960) (testimony of B. W. Broemmel, Marin County assessor), at 20.

69. National Park Service, *Land Use Survey and Economic Feasibility Report: Proposed Point Reyes National Seashore* (San Francisco: Region Four Office, National Park Service, 1961), 7.

70. *Hearings on H.R. 2775 and H.R. 3244* (1961), at 7. It is worth noting that developer David Adams purchased the Drakes Bay Pines property from Doc Ottinger in December 1959, intending to subdivide, for $350 per acre; Drakes Bay Land Co. v. United States, 275–66 (U.S. Court of Claims, April 17, 1970).

71. *Hearing on S. 2428* (1960), at 14.

72. Transcript of meeting with Bruce Kelham and Francis Hutchens (attorney), Doris Leonard, Dick Leonard, and George Collins at NPS office, June 8, 1960 (notes taken by Doris Leonard); PRNS archives, PORE 9809, box 2, folder 17. Kelham "asked that this discussion be kept confidential because of his neighbors. Said that we might hear from one of them that Mr. Kelham says that he doesn't know what is going on any more than they do. He said that this is what he always tells them, and shall continue to do so." He also specified that if the park proposal dragged on for five years or longer, his price might change.

73. *Hearings on S. 476* (1961), at 88.

74. PRNS has on file two appraisals of the 1,115-acre property, one from July 1962 assessing its value at $875,000, the other from July 1963 at $974,200; PRNS archives.

75. Lynn Ludlow, "How Delay Has Cost Thirty Million Dollars at Point Reyes," *San Francisco Chronicle/Examiner*, September 19, 1971. The article states that the Heims ranch was first appraised in 1962 at $255,565, or $225 per acre. In her oral history interview, Margaret Azevedo mentioned that the same assessor told her the Park Service had paid too much for the ranches; Lage and Duddleston, *Saving Point Reyes*, 172.

76. July 1962 appraisal of N Ranch by Edwin F. Jordan, pp. 10; PRNS archives. As Ludlow wrote, "[PRNS project manager James] Cole declines to

say if Heims was rewarded for blocking further subdivisions. But the effect on the inflation of land prices was significant"; Ludlow, "How Delay Has Cost Thirty Million." Note that Wirth had also felt the Heims sale was important to show a "break in [the ranchers'] apparently solid opposition" to selling. Letter from Wirth to Secretary of Interior, January 10, 1962, as quoted in Sadin, *Managing a Land in Motion*, 104. After the Seashore was established, Bonelli asked the NPS to purchase his parcel, but the government refused; the Court of Claims later determined this to be an illegal taking of his property. Drakes Bay Land Co. v. United States, 275–66 (U.S. Court of Claims, April 17, 1970).

77. Note that the Gallaghers had been leasing C Ranch to Ernie Spaletta and family since 1955, and the Spalettas stayed on as tenants after the purchase; F Ranch had been used to graze beef cattle, and the Gallaghers continued to lease it after selling; Livingston, *Ranching on the Point Reyes Peninsula*, 130 and 179. Congress increased the approved appropriation ceiling in 1966 by roughly five million dollars, in part to help fund acquisition of the Murphy Ranch. Public Law 89–666 (1966).

78. Total purchase price for Bear Valley Ranch was $5,725,000, or roughly $737 per acre; Sadin, *Managing a Land in Motion*, 105.

79. Lage and Duddleston, *Saving Point Reyes*, 16. Similarly, in the July 1962 discussions of H.R. 732, Representative Smith (CA) states, "In my opinion fourteen million dollars will not take care of the entire cost of acquiring the land. It may cost more than that. But we are not talking billions of dollars in this bill; we are talking about millions of dollars. This is a start." 108 Cong. Rec., part II (1962), at 14409.

80. Lage and Duddleston, *Saving Point Reyes*, 142. Similarly, in 1966, Senator Kuchel opened new PRNS hearings with a comment that "it is fair to say that the initial estimates of land values, perhaps, were too low." *Hearing on S. 1607, to Establish the Point Reyes National Seashore in the State of California and For Other Purposes*, 89th Congress, 2nd Session (July 27, 1966), at 4.

81. Gladwin Hill, "A Patchwork Park in Trouble," *New York Times*, August 5, 1969.

82. D. S. Livingston, *A Good Life: Dairy Farming in the Olema Valley: A History of the Dairy and Beef Ranches of the Olema Valley and Lagunitas Canyon* (San Francisco: Golden Gate National Recreation Area and Point Reyes National Seashore, National Park Service, 1995), 82.

83. Lynn Huntsinger, James W. Bartolome, and Carla M. D'Antonio, "Grazing Management on California's Mediterranean Grasslands," in *Ecology and Management of California Grasslands* eds. J. Corbin, M. Stromberg, and C. D'Antonio (Berkeley: University of California Press, 2015), 236–38.

84. Douglas Maloney, county counsel for Marin County, explained the situation at the 1969 hearings: "Just as all farmers in the United States each year get older as a group, so do these ranchers, and as members of their families die, the Government assesses this land for estate tax purposes at these very, very high values and these people have little or no liquid assets at all"; *Hearing on H.R. 3786 and Related Bills, to Authorize the Appropriation of Additional Funds Necessary for Acquisition of Land at the Point Reyes National Seashore in California*, 91st Congress, 1st Session (May 13, 1969), at 67–68.

85. Sadin, *Managing a Land in Motion*, 105–7; the proposed Oregon exchange was opposed by various officials in that state, including its governor.

86. Peter Behr, who was on the Marin Board of Supervisors at the time, suggests that this "earned [the Board] the undying enmity of George Hartzog"; Lage and Duddleston, *Saving Point Reyes*, 115.

87. See Hart, *Farming on the Edge*, 32–40. In the recent documentary *Rebels with A Cause*, former supervisor Gary Giacomini comments that he wasn't sure at the time that "A-60," the zone name for the minimum plot size, wouldn't be ruled as a taking, and remains unsure a court wouldn't rule so today.

88. *Hearing on S. 1607* (1966) (written proposal from Interior), at 3; emphasis added.

89. *Hearing on S. 1607* (1966), at 11, 14. Hartzog's assertion is followed by Senator Bible: "What you are saying is that if we do not receive some kind of remedial legislation you are apt to have a complete subdivision within the pastoral zone; this is what you are attempting to say?" Hartzog: "Yes, sir; that is correct."

90. *Hearing on S. 1607* (1966), at 16.

91. *Hearing on S. 1607* (1966), at 15.

92. *Hearing on S. 1607* (1966), at 15.

93. *Hearing on S. 1607* (1966), at 16.

94. The specific problem facing Murphy was the unexpected death of his wife in 1964, which resulted in an estate tax bill that he could not afford unless he was able sell his land to the NPS, which had already run out of appropriations.

95. *Hearing on H.R. 3786* (1969), at 11–12.

96. *Hearing on H.R. 3786* (1969), at 18.

97. *Hearing on S. 1530 and H.R. 3786, to Authorize the Appropriation of Additional Funds Necessary for the Acquisition of Lands at the Point Reyes National Seashore in California*, 91st Congress, 2nd Session (February 26, 1970), at 73: "The Sweet family it to be commended for its willingness to hold back the developers in order that this may be a National recreation facility. However, there is a limit to how long they can carry this financial burden by themselves."

98. Web Otis, interview with the author, San Francisco, June 3, 2009.

99. *Hearing on H.R. 3786* (1969), at 100.

100. By 1969, the county had published a revised West Marin General Plan projecting eventual development for 125,000 people, roughly triple the area's population density at the time.

101. William Duddleston, "Introduction," in Lage and Duddleston, *Saving Point Reyes*, viii.

102. *Hearing on H.R. 3786* (1969), at 34.

103. Later at the hearing, rancher Boyd Stewart confirmed this lack of speculation inside the park boundary, saying that he knew of only two cases where "land has moved," both caused by deaths in the family. *Hearing on H.R. 3786* (1969), at 63.

104. *Hearing on H.R. 3786* (1969), at 20–21.

105. *Hearing on H.R. 3786* (1969), at 47.

106. *Hearing on H.R. 3786* (1969), at 85.

107. See Lage and Duddleson, *Saving Point Reyes*, interview with Katy Miller Johnson.

108. Lage and Duddleson, *Saving Point Reyes*, ix.

109. See Lage and Duddleson, *Saving Point Reyes*, particularly the "Introduction" written by William Duddleson, former chief of staff for Representative Clem Miller.

110. Throughout the 1969–70 discussions, Boyd Stewart served as a representative of sorts for the ranchers. He was a member of the Marin Conservation League, and had experience working with recreation and environmental interests. It is also interesting that most of Stewart's ranch is not actually within the PRNS boundary, so he was not directly affected by the establishment of the park (although his ranch was shortly thereafter included in the GGNRA boundaries).

111. Interview with Boyd Stewart in Lage and Duddleson, *Saving Point Reyes*, 243–44.

112. *Hearing on S. 1530 and H.R. 3786* (1970), at 58.

113. *Hearing on S. 1530 and H.R. 3786* (1970), at 60.

114. An Act to Authorize the Appropriation of Additional Funds Necessary for Acquisition of Land at the Point Reyes National Seashore, California, Public Law 91–223 (1970).

115. Interview with Peter Behr in Lage and Duddleson, *Saving Point Reyes*, 129–30, 145. Behr stated that the owners of the Pierce Ranch were supportive of the SOS campaign; Web Otis confirmed that they donated to Behr's campaign. Web Otis, interview with the author, San Francisco, June 3, 2009.

116. Note that in Fall 1970, with ironic timing, the Marin Board of Supervisors once again became anti-development, and the Williamson Act was amended and strengthened. See Hart, *Farming on the Edge*, 32.

117. These national seashores and lakeshores are: in 1962, Point Reyes National Seashore (CA) and Padre Island National Seashore (TX); in 1964, Fire Island National Seashore (NY); in 1965, Assateague National Seashore (MD and VA); in 1966, Cape Lookout National Seashore (NC), Indiana Dunes National Lakeshore (IN), and Pictured Rocks National Lakeshore (MI); in 1970, Sleeping Bear Dunes National Lakeshore (MI) and Apostle Islands National Lakeshore (WI); in 1971, Gulf Islands National Seashore (FL and MS); and in 1972, Cumberland Island National Seashore (GA).

118. Theodore J. Karamanski, *A Nationalized Lakeshore: The Creation and Administration of Sleeping Bear Dunes National Lakeshore* (Washington, DC: National Park Service, Dept. of Interior, 2000), section 2d, available at https://www.nps.gov/parkhistory/online_books/slbe/. The NPS also proposed making a national park out of another area nearby, the Huron Mountains, considered to have much more spectacular scenery, but the very wealthy landowners there successfully fought off NPS advances. Karamanski quotes Wirth as agreeing that the area was in good hands, and concludes that "the fifty well-heeled owners of the Huron Mountains accomplished what the hundreds of less well-connected cottage owners in the Sleeping Bear area could not do—stop a national park" (section 1e).

119. Karamanski, *Nationalized Lakeshore*, section 2b.

120. Karamanski, *Nationalized Lakeshore*, section 3c; he describes the chief land acquisition officer in 1970 as "unsympathetic, unsmiling, and unrelenting."

121. Karamanski, *Nationalized Lakeshore*, "Introduction."

122. Karamanski, *Nationalized Lakeshore*, section 3c.

123. Dwight T. Pitcaithley, *Let the River Be: A History of the Ozark's Buffalo River* (Santa Fe: Southwest Cultural Resources Center, Southwest Regional Office, National Park Service, 1989), section 6, ". . . Let the River Be," available at www.nps.gov/parkhistory/online_books/buff/history/chap6.htm.

124. An Act to Provide for the Establishment of the Buffalo National River in the State of Arkansas, and for Other Purposes, Public Law 92–237 (March 1, 1972).

125. Kenneth L. Smith, *Buffalo River Handbook* (Little Rock: University of Arkansas Press, 2004), 113. A later superintendent changed course, and actually reprivatized a portion of the community of Boxley Valley, selling more than a dozen parcels back to the original owners. Jim Liles, "Boxley Valley: Buffalo National River: A National Park Service Historic District in Private Hands," *OzarksWatch* 4, no. 1 (1990): 14–18.

126. Interview with Jim Liles, assistant superintendent, Buffalo National River, conducted by Sally Fairfax, 1997, in the author's collection: "We did pretty much eviscerate the landscape of Boxley when we acquired it."

127. Gateway and Golden Gate were also the last parks established under Director George Hartzog, who had a strong personal interest and commitment to developing national parks in urban areas. It is worth noting that prior to becoming director, Hartzog had been superintendent at Jefferson Expansion National Memorial in St. Louis.

128. The earliest proposal for the GGNRA in 1970 only included two private acquisitions: the 12.5-acre Sutro Baths/Cliff House property in San Francisco, and sixteen hundred acres in Marin County slated for development as the Marincello project. When the park proposal was revised in 1972, however, it suddenly included an additional eight thousand acres of private land along the Marin coast, to meet up with the southern end of PRNS. The proposal suggested that only twenty-five hundred acres of this would need to be purchased in fee; the rest was expected to be donated. This shift is attributed to the vigorous efforts of People for a Golden Gate National Recreation Area (PFGGNRA). See James Smith, "The Gateways: Parks for Whom?" in *National Parks for the Future*, ed. Conservation Foundation (Washington DC: Conservation Foundation, 1972), 228–31. The PFGGNRA remained deeply involved with the administration of the park for decades, even though it only had two actual members, local activist Amy Meyer and then-president of the Sierra Club Edgar Wayburn.

129. Mackintosh, *National Parks*, 86. Reaction within the NPS to its own first report on Cuyahoga, released in 1971, "was largely 'raised eyebrows and a lot of jokes' about a national park featuring a flammable river." Officially, the NPS and Interior protested that Cuyahoga was only of regional interest, that it would cost too much to develop, and that its patchwork of land ownership would be impossible to administer. Ron Cockrell, *A Green Shrouded Miracle:*

The Administrative History of Cuyahoga Valley National Recreation Area, Ohio (Omaha: Midwest Regional Office, National Park Service, 1992), 80, 86.

130. An Act to Provide for the Establishment of the Cuyahoga Valley National Recreation Area, Public Law 93–555 (Dec. 27, 1974).

131. Cockrell, *Green Shrouded Miracle*, 96.

132. Public Law 93–555, Section 2(b) (1974).

133. Public Law 93–555, Section 2(c) (1974).

134. This approach makes sense if you consider the Corps' usual situation: buying land that would soon be under water, a status that makes retained rights or easements unnecessary, and use of eminent domain common. Cockrell, *Green Shrouded Miracle*, 141–42.

135. Cockrell, *Green Shrouded Miracle*, 145.

136. This left six thousand acres still to be negotiated. Cockrell, *Green Shrouded Miracle*, 137.

137. Cockrell, *Green Shrouded Miracle*, 144–45.

138. The total acreage of Redwood National Park at initial establishment was 58,000 acres, including Prairie Creek, Jedediah Smith, and Del Norte State Parks, which remained in state ownership. Only 10,876 acres of the park represent old-growth redwoods; the rest had been logged. Susan R. Schrepfer, *The Fight to Save the Redwoods: A History of Environmental Reform, 1917–1978* (Madison: University of Wisconsin Press, 1983), 156–58.

139. Redwood National Park ended up costing more like $210 million, making it the most expensive single park unit. Schrepfer, *Fight to Save the Redwoods*, 158.

140. See David Louter, *Contested Terrain: North Cascades National Park Service Complex, an Administrative History* (Seattle: National Park Service, 1998), 145–49.

141. This later drew criticism, as it was not clear how necessary full-fee purchase of the paper companies' lands really was. NPS assigned priority to lands to be acquired, but then did not use that priority system to direct acquisition, instead acquiring the land as appraisals determined values. General Accounting Office, *The Federal Drive to Acquire Private Lands Should Be Reassessed* (Washington, DC: U.S. General Accounting Office, 1979), 110–15.

142. General Accounting Office, *Federal Drive*, i. The report recommends that Interior and Agriculture jointly establish a policy for federal land acquisition that explores alternatives and gives agencies guidance as to when lands should be purchased and when protected by other means. Agencies should, it recommended, evaluate the need to purchase in existing projects, making a detailed review of alternatives. For new projects, land plans should be prepared *before* private lands are acquired, to determine what is needed to meet the project purposes, consider alternate strategies for protection, et cetera.

143. National Parks and Recreation Act of 1978, Public Law 95–625 (1978). For discussion in greater detail see John Jacobs, "Park Barrel," in *A Rage for Justice: The Passion and Politics of Phillip Burton* (Berkeley: University of California Press, 1995), 351–79. The chapter suggests, particularly at 364, that Burton assembled the enormous omnibus bill, which involved park projects in over two hundred congressional districts, in order to garner support for

separate legislation expanding Redwoods National Park (P.L. 95–250, 1978), which with its Title II compensation for out-of-work loggers was a controversial bill. Others suggested that the massive bill was constructed to punish political enemies in the House, whose districts did not receive any new park projects. Jacobs, *Rage for Justice*, 373.

144. Santa Monica Mountains National Recreation Area, *Final General Management Plan and Environmental Impact Statement* (Los Angeles: National Park Service, 2002).

145. Omnibus Parks and Public Lands Management Act of 1996, Public Law 104–333 (Nov. 12, 1996). The Presidio Trust reached financial self-sufficiency in 2013; Presidio Trust, *Milestones: Presidio Trust 2012 Year-End Report to Congress and the Community*.

146. NPS Land Resources Division Listing of Acreage, summary, Dec. 31, 2011, https://irma.nps.gov/Stats/reports/national. More than half of the Boston Harbor islands are owned by the Massachusetts Department of Conservation and Recreation: www.mass.gov/eea/agencies/dcr/massparks/region-south/boston-harbor-islands-generic.html, accessed June 9, 2016.

147. For more on Tallgrass Prairie, see Nature Conservancy website, www.nature.org/ourinitiatives/regions/northamerica/unitedstates/kansas/placeswe-protect/tallgrass-prairie-national-preserve.xml, accessed February 12, 2015.

148. See *Rebels with a Cause* website, http://rebelsdocumentary.org.

149. Hart, *Farming on the Edge*; he notes that "within months of the rezoning, three ranches north of Marshall on Tomales Bay—several thousand acres that had been in speculators' hands—were sold once more to ranchers, at prices that the ranchers could afford" (40).

150. Marin Agricultural Land Trust website, www.malt.org, accessed Jan. 21, 2014.

151. Shelley Brooks, "Inhabiting the Wild: Land Management and Environmental Politics in Big Sur," *Western Historical Quarterly* 44, no. 3 (2013): 295–317, 297. She quotes resident Nathaniel Owings, a well-known architect who became the driving force for the CMP, as saying, "We want to try and preserve this beautiful coastline without large-scale federal subsidization. Nobody wants a national park along the Big Sur coast" (302).

152. Brooks, "Inhabiting the Wild," 307, quoting from Robert A. Jones, "Control of Big Sur Prompting Identity Crisis," *Los Angeles Times*, March 30, 1980.

153. Brooks, "Inhabiting the Wild," 311.

CHAPTER 4. PARKS AS (POTENTIAL) WILDERNESS

1. Mark Woods, "Federal Wilderness Preservation in the United States: The Preservation of Wilderness?" in *The Great New Wilderness Debate* (Athens: University of Georgia Press, 1998), 136; emphasis in original.

2. *Hearings on S. 476, to Establish the Point Reyes National Seashore in the State of California, and for Other Purposes*, 87th Congress, 1st Session (March 28, 30, and 31, 1961) (testimony of Representative Clement Miller), at 70, 168.

3. Quoted in *Hearing on S. 2428, to Establish the Point Reyes National Seashore in the State of California and for Other Purposes*, 86th Congress, 2nd Session (April 16, 1960) (testimony of Brian McCarthy), at 79–80.

4. *Hearing on S. 2428* (1960) (testimony of NPS Director Conrad Wirth), at 6.

5. NPS, "Questions and Answers Concerning the Proposed Point Reyes National Seashore," dated March 1, 1960, PRNS Archives; Harold Gilliam, *Island in Time: The Point Reyes Peninsula* (San Francisco: Sierra Club Books, 1962).

6. Point Reyes National Seashore, *Statement of Management Objectives* (National Park Service, 1970), 10. PRNS files.

7. National Park Service, *General Management Plan, Environmental Analysis: Golden Gate National Recreation Area/Point Reyes National Seashore* (Denver, CO: Denver Service Center, National Park Service, 1980), 110–12.

8. William Cronon, "The Trouble With Wilderness; or, Getting Back to the Wrong Nature," in *Uncommon Ground: Toward Reinventing Nature,* ed. William Cronon (New York: W. W. Norton, 1995), 80.

9. Paul Sutter, *Driven Wild: How the Fight against Automobiles Launched the Modern Wilderness Movement* (Seattle: University of Washington Press, 2002), 13–16.

10. See Christopher W. Wells, *Car Country: An Environmental History* (Seattle: University of Washington Press, 2012), particularly "The Paths Out of Town," for more detailed discussion of the development of roads for outdoor recreation, connecting the American public with "car-friendly nature" in the national parks.

11. Sutter, *Driven Wild*, 31.

12. Sutter, *Driven Wild*, 55.

13. Richard White, "From Wilderness to Hybrid Landscapes: The Cultural Turn in Environmental History," in *Companion to American Environmental History,* ed. Douglas Cazaux Sackman (Hoboken, NJ: Wiley-Blackwell, 2010), 185.

14. Curt Meine, *Aldo Leopold: His Life and Work* (Madison: University of Wisconsin Press, 1988), 196.

15. Sutter, *Driven Wild*, 94.

16. Sutter, *Driven Wild*, 85–88.

17. Mark Harvey, *Wilderness Forever: Howard Zahniser and the Path to the Wilderness Act* (Seattle: University of Washington Press, 2005), 156–57. For more on Mission 66, see Ethan Carr, *Mission 66: Modernism and the National Park Dilemma* (Amherst: University of Massachusetts Press, 2007).

18. For more on the Echo Park controversy at Dinosaur, see Mark Harvey, *A Symbol of Wilderness: Echo Park and the American Conservation Movement* (Seattle: University of Washington Press, 1994).

19. James Morton Turner, *The Promise of Wilderness: American Environmental Politics Since 1964* (Seattle: University of Washington Press, 2012), 30; also see Harvey, *Wilderness Forever*, 203–5.

20. See Kevin Marsh, *Drawing Lines in the Forest: Creating Wilderness Areas in the Pacific Northwest* (Seattle: University of Washington Press, 2007).

21. The Wilderness Act of 1964, Public Law 88–577, 88th Congress, 2nd Session (Sept. 3, 1964).

22. It is worth noting that the NPS originally opposed the law, considering it "unnecessary," as did the Forest Service; only the U.S. Fish and Wildlife Service supported it. See Turner, *Promise of Wilderness*, 52–53.

23. Turner, *Promise of Wilderness*, 59.

24. Point Reyes National Seashore, *Wilderness Study* (Point Reyes Station, CA: National Park Service, April 1971), 4–5.

25. "Strict Preservation of Seashore Urged," *Marin Independent Journal,* Sept. 23, 1971, quoting Doug Nadeau, the "captain of the park service planning team"; the article also quotes him as saying, "The cuts cannot be obliterated so we might as well use it [*sic*]."

26. "Marin Urges Ban on Cars in Pt. Reyes," *San Francisco Chronicle,* Sept. 15, 1971.

27. "Strict Preservation of Seashore Urged," *Marin Independent Journal,* Sept. 23, 1971. The article also notes, "Almost all speakers pleaded for a citizens advisory committee to watch the park planning and development each step of the way."

28. Doug Dempster, "Point Reyes Plea: Too Small," *Sacramento Bee,* Sept. 24, 1971.

29. Point Reyes National Seashore, *Wilderness Recommendation* (Point Reyes Station, CA: National Park Service, August 1972), 12. The National Parks and Conservation Association started in 1919 as the National Parks Association, but added the words "and Conservation" in 1970 to reflect the environmentalism of the time. In 2000 it dropped "and," and is now named the National Parks Conservation Association.

30. Larry D. Hatfield, "Reyes Wilderness Argued," *San Francisco Examiner,* Sept. 23, 1971.

31. See, for instance, PRNS, *Wilderness Recommendation*, 18.

32. Comment letter from State of California Resources Agency, dated Oct. 22, 1971, printed in PRNS, *Wilderness Recommendation*, "Appendix: Hearing Officer's Report."

33. PRNS, *Wilderness Recommendation*, 13.

34. Point Reyes National Seashore, *Final Environmental Statement for Proposed Wilderness* (Western Region, National Park Service, 1974), 14–15. Note that in the mid-1970s, PRNS kept minutes of their weekly staff meetings, and these are full of details of removing fences, buildings, old fishing docks, reseeding former roads, even burying foundations and old cars in some instances. Squad meeting minutes, 1973–76, file folder code A4031, PRNS archives.

35. PRNS, *Final Environmental Statement*, 25.

36. Archeological sites within other proposed wilderness areas, such as at Mesa Verde National Park, posed an interesting conundrum due to this language—see discussion in *Hearings on H.R. 13562 and H.R. 13563, to Designate Certain Lands in the National Park System as Wilderness*, 93rd Congress, 2nd Session (March 22, 25, and 26, 1974), in which both NPS and Wilderness Society representatives concluded that "evidence of pre-historic works of man is not considered as cause to exclude areas from wilderness protection," adding

that "moreover, we feel that wilderness designation . . . will assure protection against unwise development that could threaten their archeological and natural features" (45). The NPCA argued those archeological areas should be designated potential wilderness (51), whereas the Sierra Club (55) thought they were not incompatible with wilderness.

37. An Act to Convey Certain Tide and Submerged Lands to the United States in Furtherance of the Point Reyes National Seashore, California A.B. 1024 (July 9, 1965).

38. The Drakes Estero Unit (number nine on their list) is a total of twenty-two hundred acres, including the whole estero, Drakes Beach and Headlands. Its description in PRNS, *Final Environmental Statement* includes the following: "This is the only oyster farm in the seashore. Control of the lease from the California Department of Fish and Game, with *presumed renewal indefinitely,* is within the rights reserved by the State on these submerged lands . . . . The existence of the oyster-farm operation renders the estero unsuitable for wilderness classification at present, and there is *no foreseeable termination* of this condition" (56). Emphasis is mine.

39. Memo from Assistant Secretary of the Interior Nathaniel Reed to the directors of the National Park Service and the Bureau of Sport Fisheries and Wildlife, titled "Guidelines for Wilderness Proposals, Reference Secretarial Order No. 2920," June 24, 1972. Copy on file with author.

40. Turner, *Promise of Wilderness*, 129, quoting memo from Reed to Brandborg, August 22, 1972.

41. PRNS, *Final Environmental Statement*, A-20; emphasis in original.

42. PRNS, *Final Environmental Statement*, A-51.

43. H.R. 8002 (June 18, 1975).

44. Doug Scott, *The Enduring Wilderness: Protecting Our Natural Heritage through the Wilderness Act* (Golden, CO: Fulcrum Publishing, 2004), 66.

45. Hal Rothman, *The New Urban Park: Golden Gate National Recreation Area and Civic Environmentalism* (Lawrence: University of Kansas Press, 2004), 59, quoting Commissioner Richard Bartke.

46. The subcommittee included John Mitchell (chair), Fred Blumberg, Daphne Greene, Joe Mendoza, Amy Meyer, and Merritt Robinson; their three meetings were held on July 10, July 22, and August 5, 1975.

47. "Summary of GGNRA CAC Wilderness Subcommittee Meetings," Sept 7, 1975, printed on NPS letterhead; PRNS archives.

48. "Summary of GGNRA CAC Wilderness Subcommittee Meetings," 4–5. The subcommittee also recommended that "the preamble of the wilderness legislation should clearly state the atypical nature of wilderness at Point Reyes."

49. *Hearings on S. 1093, to Designate Certain Lands in the Point Reyes National Seashore, California, as Wilderness, and on S. 2472, to Designate Certain Lands in the Point Reyes National Seashore, California, as Wilderness; to Designate Point Reyes National Seashore as a Natural Area of the National Park System, and for Other Purposes,* 94th Congress, 2nd Session (March 2, 1976). The hearing proceedings include a November 5, 1975, letter from Interior, recommending that S. 1093 be amended to cover 25,480 acres. The original president's recommendation, from November 28, 1973, had been for 10,600

acres, but since then the NPS had done additional study, and also acquired some additional land, and had come up with 14,880 acres more—making the total 25,480, with twenty acres of "potential wilderness addition."

50. *Hearings on S. 1093* (1976) (testimony of Senator Tunney), at 270.

51. *Hearings on S. 1093* (1976), at 308–9.

52. *Hearings on S. 1093* (1976) (oral testimony), at 265; (written testimony), at 269 .

53. *Hearings on S. 1093* (1976), at 270.

54. *Hearings on S. 1093* (1976), at 271.

55. *Hearings on S. 1093* (1976), at 272–73.

56. *Hearings on S. 1093* (1976), at 310.

57. *Hearings on S. 1093* (1976), at 311. Everhardt testified, "There are lands there that certainly should be kept in their condition. These lands primarily are all being recommended as wilderness. There are other areas that we feel should be developed for recreational use; and still there are other areas that have outstanding historical value. Furthermore, Mr. Chairman, we feel that the designation of the seashore as a natural area would be generally inconsistent with grazing and commercial oyster farming activities that are presently found at Point Reyes, and also an authorization for hunting at Point Reyes that presently exists."

58. *Hearings on S. 1093* (1976), at 327.

59. *Hearings on S. 1093* (1976), at 327.

60. Friedman also submitted a letter from the Inverness Association, which supported the CAC's wilderness recommendations and specified, regarding Drakes Estero: "We urge you to consider the negative consequences (i.e. the allowability of motorized off-road vehicles) were this geologically unstable dune-covered land to be managed as a 'Recreation Area.' The possibility of jeeps and motorcycles having access to the Estero shore and adjoining area is a frightening one." *Hearings on S. 1093* (1976), at 330–31.

61. *Hearings on S. 1093* (1976), at 355–57. Many of the organizations that Mr. Friedman represented also sent in letters of support, all explicitly endorsing the CAC's recommendations. See, for instance, a letter from the president of the Marin Conservation League, Robert F. Raab, *Hearings on S. 1093* (1976), at 369.

62. Jerry Friedman, undated notes in preparation to speak at the PRNS wilderness hearings for H.R. 8003; copy on file with author.

63. Letter from Larry Kolb, vice chair for wilderness issues, Sierra Club, San Francisco Bay Chapter, to the California Resources Agency, dated October 6, 1975; emphasis in original. Copy on file with author.

64. *Hearings on S. 1093* (1976), at 4–6. Burton explained that the areas "should be designated potential wilderness now because they would be ineligible for actual wilderness designation because of a statute on the books of California at the time of the original Point Reyes establishment under the authorship of your former colleague, the late Clem Miller, where the State reserved the subwater mineral rights." He also noted that a portion of the Murphy (Home) Ranch that was in the original wilderness proposal had been taken out entirely (i.e., was not classified even as potential wilderness), "because in its present

situation of operation to call that 'wilderness' would be inconsistent and not make any sense." This last point implies that if the operation of the oyster farm was considered inconsistent with wilderness, it too would likely have been removed from the proposal.

65. *Hearing Held Before the Subcommittee on National Parks and Recreation of the Committee on Interior and Insular Affairs, Markup Session, H.R. 8002, Point Reyes National Seashore,* 94th Congress, 2nd Session (September 14, 1976) (written statement from Richard Curry), at 17–18. Curry testified that "Our report was concerned and did not include the tidelands because the State retained jurisdiction over those mineral rights, but I am sure that we would have no objection to those being designated as 'potential wilderness' additions if the committee chose to act. I think that is consistent with that element in our report" (6).

66. "Wilderness Bill Gets Boost," *Point Reyes Light,* Sept. 16, 1976. The article specifies that "the U.S. Department of the Interior is unwilling to accept as wilderness land over which the federal government has less than total ownership." It also quotes Jerry Friedman as referencing the NPS's 1963 master plan, with its extensive developments and roadways for the estero area, noting that without wilderness designation, a new NPS administration could choose to change the character of the area.

67. The Point Reyes wilderness designation was originally part of the larger omnibus bill, but the large bill got hung up in committee, and so Burton sent the Point Reyes portion up through the House "naked," where it passed. Then the omnibus bill passed as well in a "bonus period" in which the Senate missed its midnight deadline for adjournment by two hours and fifty-two minutes. "This means that President Ford—if he also signs the omnibus bill—will have designated the Point Reyes wilderness twice. It is ironic since for awhile it appeared he wouldn't have a chance to designate it at all"; Dave Mitchell, "Congress Ok's Wilderness Bill," *Point Reyes Light,* Oct. 7, 1976.

68. The two laws represent the same acreage at Point Reyes, and both laws refer to the same map, No. 612–90,000-B. Oddly, that map lists a different acreage size than the two acts: the map refers to a total of 24,200 acres of wilderness and 8,530 acres of potential wilderness. It is not clear why the discrepancy exists, and impossible to tell which acres have been added/left out of each category.

69. *Report No. 94–1680, to Accompany H.R. 8002, Designating Certain Lands in the Point Reyes National Seashore, California, as Wilderness, Designating Point Reyes National Seashore as a Natural Area of the National Park System, and Other Purposes,"* 94th Congress, 2nd Session (September 24, 1976), at 3.

70. National Park Service, *Final Environmental Impact Statement, General Management Plan, Yosemite National Park, California* (National Park Service, 1980), 3.

71. *Hearing Held before the Subcommittee on National Parks, Forests, and Public Lands of the House Committee on Natural Resources, Concerning H.R. 3022, a Bill to Designate the John Krebs Wilderness in the State of California, to Add Certain Lands to the Sequoia-Kings Canyon National Park Wilderness,*

and Other Purposes, 110th Congress, 1st Session (October 30, 2007) (testimony of Karen Taylor-Goodrich, associate director for visitor and resource protection, National Park Service), available online at https://www.doi.gov/ocl /hearings/110/HR3022_103007 (accessed June 13, 2016).

72. *Report 94–1357, to Accompany H.R. 13160, Wilderness Designations within the National Park System*, 94th Congress, 2nd Session, (September 29, 1976), at 3.

73. George Nevin, "Point Reyes Wilderness Established," article published in unknown newspaper, clipping in archival material from Jerry Friedman's collection; on file with author.

74. James W. Feldman, *A Storied Wilderness: Rewilding the Apostle Islands* (Seattle, WA: University of Washington Press, 2011), 12, 225–33.

75. David J. Nemeth and Deborah J. Keirsey, "Elaboration on the Nature of Woody Debris: An Ethical Snag in the Aesthetic Justification for Organized River Cleanup," *APCG Yearbook* 61 (1999): 86–107.

76. Woods, "Federal Wilderness Preservation," 140.

77. "Cherry-stemming" is the practice of drawing the wilderness boundary along the edge of a road, looping around the end of the road, and then back along the other side; technically the road remains outside the legal boundary, even though it is physically present across the landscape.

78. *Hearings on S. 1093* (1976), at 330.

79. National Parks and Recreation Act of 1978, Public Law 95–625, 95th Congress (November 10, 1978), Section 318.

80. Department of the Interior, National Park Service, "Notice of Designation of Potential Wilderness as Wilderness, Point Reyes National Seashore," *Federal Register* 64, no. 222 (Nov. 18, 1999): 63057.

CHAPTER 5. REMAKING THE LANDSCAPE

1. *Hearing on S. 2428, to Establish the Point Reyes National Seashore in the State of California and for Other Purposes*, 86th Congress, 2nd Session (April 16, 1960), at 6.

2. *Hearings on H.R. 2775 and H.R. 3244, to Establish the Point Reyes National Seashore in the State of California, and for Other Purposes*, 87th Congress, 1st Session (March 24, July 6, and August 11, 1961), at 12–13.

3. *Hearings on H.R. 2775 and H.R. 3244* (1961), at 54.

4. R. Scott Anderson, Ana Ejarque, Peter M. Brown, and Douglas J. Hallett, "Holocene and Historical Vegetation Change and Fire History on the North-Central Coast of California, USA," *The Holocene* 23, no. 12 (2013): 1797–1810; also see report by R. Scott Anderson, *Contrasting Vegetation and Fire Histories on the Point Reyes Peninsula During the Pre-Settlement and Settlement Periods: Fifteen Thousand Years of Change* (Flagstaff, AZ: Center for Environmental Sciences and Education, Quaternary Sciences Program, 2005).

5. Ronald A. Foresta, *America's National Parks and Their Keepers* (Washington, DC: Resources for the Future, 1984), 132; also James A. Glass, *The Beginnings of a New National Historic Preservation Program, 1957 to 1969* (Nashville: American Association for State and Local History, 1990).

6. Glass, *Beginnings of a New National Historic Preservation Program*, 10.

7. National Trust for Historic Preservation, *With Heritage So Rich*, 2nd ed. (Washington, DC: Preservation Press, 1983), 207–8; emphasis is mine.

8. Specifically the Act's preamble states that "the historical and cultural foundations of the Nation should be preserved as a living part of our community life and development in order to give a sense of orientation to the American people"; National Historic Preservation Act, Public Law 89–665, 80 Stat. 915 (1966).

9. The name was not officially mandated by the original statute itself; the NPS began referring to the Register by this name in mid-1968, and Congress made it the legal name in the 1980 amendments to the Act; Barry Mackintosh, *The National Historic Preservation Act and the National Park Service: A History* (Washington, DC: History Division, National Park Service, 1986), 21.

10. Barry Mackintosh, *The Historic Sites Survey and National Historic Landmarks Program: A History* (Washington, DC: History Division, National Park Service, 1984), 21. Note that nationally significant properties were already covered by the 1935 Act.

11. Robert Z. Melnick, with Daniel Sponn and Emma Jane Saxe, *Cultural Landscapes: Rural Historic Districts in the National Park System* (Washington, DC: National Park Service, 1984). Also see Melody Webb, "Cultural Landscapes in the National Park Service," *The Public Historian* 9, no. 2 (1987): 77–89, for more detail on the initial interest in the NPS regarding cultural landscapes.

12. Melnick, *Cultural Landscapes*, 10.

13. Linda Flint McClelland, J. Timothy Keller, Genevieve Keller, and Robert Z. Melnick, "Guidelines for Evaluating and Documenting Rural Historic Landscapes," *National Register Bulletin* 30 (Washington, DC: Interagency Resources Division, National Park Service, 1989, rev. 1999).

14. Melnick, *Cultural Landscapes*, 8.

15. Melnick, *Cultural Landscapes*, 1; for definition, see 67. What he called sociocultural landscapes are now termed ethnographic landscapes, but it is not always clear how to distinguish between these and "rural historic" landscapes. The main difference is the perspective from which the landscape is viewed. Ethnographic landscapes are those viewed through the eyes of a specific culture, and mirror their systems of meaning, ideology, belief, and so on, while rural historic districts may be the same places but viewed from an outside perspective. See Donald L. Hardesty, "Ethnographic Landscapes: Transforming Nature into Culture," in *Preserving Cultural Landscapes in America*, eds. Arnold R. Alanen and Robert Z. Melnick (Baltimore: Johns Hopkins University Press, 2000), 169–85.

16. Melnick's definition of a cultural landscape is as follows: "a geographic area, including both natural and cultural resources, including the wildlife or domestic animals therein, that has been influenced by or reflects human activity or was the background for an event or person significant to human history," Melnick, *Cultural Landscapes*, 66.

17. Melnick, *Cultural Landscapes*, 10.

18. Catherine Howett, "Integrity as a Value in Cultural Landscape Preservation," in *Preserving Cultural Landscapes in America*, eds. Arnold R. Alanen and Robert Z. Melnick (Baltimore: Johns Hopkins University Press, 2000), 199.

19. Arnold R. Alanen, "Considering the Ordinary: Vernacular Landscapes in Small Towns and Rural Areas," in *Preserving Cultural Landscapes in America*, eds. Arnold R. Alanen and Robert Z. Melnick (Baltimore: Johns Hopkins University Press, 2000), 142.

20. Bryan McCarthy, transcript of Citizens Advisory Commission on development of Marin County meeting, November 11, 1959, 76; PRNS archives.

21. *Hearings on S. 476, to Establish the Point Reyes National Seashore in the State of California, and for Other Purposes*, 87th Congress, 1st Session (March 28, 30, and 31, 1961), at 111–13; Marin County Supervisor James Marshall made the same point in *Hearing on S. 2428* (1960), at 17.

22. 112 Congr. Rec. S3823 (March 17, 1970) (written statement by Senator Alan Bible), discussing the amendment to repeal Section 4 in the 1962 legislation that established the pastoral zone; emphasis is mine.

23. Initially there was some confusion about how the terms of reservation would work: a letter written to Congress by Margaret McClure pointed out that, as initially written, the reservation of right allowed "you and your sons" to stay until the youngest turned thirty—but her youngest at the time was fifty-two; *Hearings on H.R. 2775 and H.R. 3244* (1961), at 56.

24. This reduction in the purchase price could have major consequences for any properties that had recently been purchased. In the GGNRA lands, for instance, there were several instances of young families who could not afford the reduction in purchase price to obtain a reservation (as they had to pay back their full mortgage), and thus were forced to opt for a less-secure special use permit.

25. The ranches where tenants were initially allowed to remain on leases were: Spaletta (C), Pierce (for Merv McDonald and family), Truttman, R. Giacomini, and Wilkins (for Mary Tiscornia). Tenants had to leave at Heims (N), DeSouza, and Hagmaier.

26. National Parks and Recreation Act of 1978, Public Law 95–625 (November 10, 1978); Section 318 (b) states that "Where appropriate in the discretion of the Secretary, he or she may lease federally owned land (or any interest therein) which has been acquired by the Secretary under this Act, and which was agricultural land prior to its acquisition . . . . Any land to be leased by the Secretary under this section shall be offered first for such a lease to the person who owned such land or was a leaseholder thereon immediately before its acquisition by the United States."

27. Email from PRNS Assistant Superintendent Frank Dean, March 30, 2000. Copy on file with author.

28. Email from PRNS Assistant Superintendent Frank Dean, March 30, 2000. Copy on file with author.

29. Author interview with PRNS Assistant Superintendent Frank Dean, Point Reyes National Seashore, April 4, 2000. Copy on file with author.

30. Letter from Interior Field Solicitor Ralph Mihan to the Department of Justice, dated November 17, 1978. Copy on file with author.

31. Vernita Lea Ediger, "Natural Experiments in Conservation Ranching: The Social and Ecological Consequences of Diverging Land Tenure in Marin County, California," PhD diss. Stanford University, 2005, 18.

32. Most of these programs require applicants to show that they will have "effective control" of the land over the length of the contract, which is often ten years or more.

33. Ediger, "Natural Experiments," iv, 124–25.

34. Ediger, "Natural Experiments," 88.

35. Some details of where removal work was done can be read in the minutes of weekly staff meetings, kept intermittently from 1973–76; file folder code A4031, PRNS archives. The sites would then be cleaned of debris (sometimes hauled away, sometimes buried onsite) and reseeded, so as to appear more "natural." This work was done so thoroughly that a historic archeology study conducted in 2011 at several former ranch sites could only identify approximately five to ten artifacts or landscape features at each; these landscapes were quite thoroughly erased. Paul M. Engle, "Toward an Archeology of Ranching on the Point Reyes Peninsula," Cultural Resources Management masters thesis, Sonoma State University, 2012, 104.

36. This estimate was calculated from D. S. Livingston, *Ranching on the Point Reyes Peninsula: A History of the Dairy and Beef Ranches within Point Reyes National Seashore, 1834–1992* (Point Reyes Station, CA: Point Reyes National Seashore, National Park Service, 1993) and *A Good Life: Dairy Farming in the Olema Valley: A History of the Dairy and Beef Ranches of the Olema Valley and Lagunitas Canyon* (San Francisco: Golden Gate National Recreation Area and Point Reyes National Seashore, National Park Service, 1995), as well as from examination of site photos from land appraisals in PRNS files. This estimate does not include structures such as fences, corrals, gates, or roads.

37. Livingston, *Ranching on the Point Reyes Peninsula*, 179.

38. Livingston, *Ranching on the Point Reyes Peninsula*, 443–90.

39. John McClure's mother, Margaret, had lived at the Lower Ranch; after Pierce was purchased by the investment group, called the Bahia del Norte Land and Cattle Company, the Lower Ranch was utilized occasionally by the group partners, with a schedule for who could use it. Author interview with Webb Otis, San Francisco, June 3, 2009.

40. Dewey Livingston, *Interviews with Merv McDonald, Marin County Rancher* (Point Reyes Station, CA: Marin Resource Conservation District, 2010), 62. In the PRNS staff meeting minutes from November 17, 1975, Superintendent Sansing stated: "I have taken a total of seven people to look at the lower Pierce Point Ranch. They had lots of opinions about what to do with it." The buildings were demolished soon after, and by the December 8 staff meeting the crew was already reseeding the site. Squad meeting minutes 1975–76, file folder code A4031, PRNS archives.

41. Letter to Representative John Burton from C. Malcolm Watkins, senior curator of cultural history at the Smithsonian, March 10, 1976; box H3012, "History, Pierce Point Ranch"; PRNS archive. Watkins wrote, "A lower Pierce Point ranch, nearly as old [as the Upper Ranch], was abandoned when the National Park Service acquired it. It rapidly deteriorated from vandalism and weather until National Seashore Management, in seeming haste, had it demolished last December. I fear that a similar fate may befall the upper ranch, hastened, perhaps, by a narrow interpretation of 'recreational' use as applied to the

Seashore's mission. The immensely popular recreation potential of historic structures seems to be entirely overlooked here."

42. A few on the uplands were retained for use as staff housing, but none on Limantour Spit itself.

43. Joel Reese, "Park Service Razes Historic Farmhouse in Olema," *Point Reyes Light,* Jan. 5, 1995.

44. Livingston, *A Good Life,* 201. Superintendent Sansing also ordered "three rare historic chicken coops and a shed" to be removed at the R. Giacomini Ranch in 1990; Livingston, *A Good Life,* 174.

45. Livingston, *A Good Life,* 258–60.

46. See National Park Service, *General Management Plan, Environmental Analysis: Golden Gate National Recreation Area/Point Reyes National Seashore* (Denver: National Park Service, 1980), 89. Alanen, "Considering the Ordinary," specifically cites national seashores and lakeshores as places where, despite the passage of historic preservation regulations, NPS managers continued to demolish old structures or allow abandoned buildings to deteriorate until they became safety hazards (129).

47. As of April 2000 the park's official list of classified structures (LCS) listed 218 remaining historic structures, excluding NPS-constructed buildings but including gates, corrals, fences, historic roads, and all the buildings at Hamlet. By this estimate, it appears that the approximately 170 structures removed by that point represented roughly half of the built landscape. Since then, cultural landscape inventories at individual ranches have added landscape features to the LCS, which as of the last update in 2006 includes 289 elements.

48. National Park Service, *Land Use Survey and Economic Feasibility Report: Proposed Point Reyes National Seashore* (San Francisco: Region Four Office, National Park Service, 1961), 1.

49. *Hearings on S. 476* (1961) (testimony of Mr. Davis), at 186.

50. *Hearings on S. 476* (1961) (question from Senator Dworshak), at 223.

51. *Hearing on S. 1607, to Establish the Point Reyes National Seashore in the State of California and For Other Purposes,* 89th Congress, 2nd Session (July 27, 1966), at 10; Hartzog repeated this statement almost verbatim in the 1969 House hearings, *Hearing on H.R. 3786 and Related Bills, to Authorize the Appropriation of Additional Funds Necessary for Acquisition of Land at the Point Reyes National Seashore in California,* 91st Congress, 1st Session (May 13, 1969), at 29.

52. National Park Service, *Assessment of Alternatives for the General Management Plan, Golden Gate National Recreation Area and Point Reyes National Seashore* (San Francisco: National Park Service, 1977), at 314.

53. E, F, and N Ranches are currently leased for grazing only. On E Ranch, ranch employees and their families reside in the existing buildings. PRNS tract files, PRNS archive.

54. Presentation on PRNS vegetation management by plant ecologist Barbara Moritsch, October 26, 1999, Point Reyes National Seashore.

55. In addition, Earl Lupton died in September 2001, and his nonfamily business partner was not allowed to continue their seventy-head grazing operation at the ranch. PRNS staff said that the Lupton ranch would stay in agricul-

tural use, but first the land needed to rest, and a ranch plan would to be developed; it was not clear how long this would take. Transcript of meeting of the Citizen's Advisory Commission, GGNRA/PRNS, October 20, 2001; PRNS archives. Eventually the land was leased to the neighboring Stewart Ranch.

56. Ellie Rilla and Lisa Bush, *The Changing Role of Agriculture in Point Reyes National Seashore* (Marin County: University of California Cooperative Extension), 7.

57. Author interview with rancher Sharon Doughty, Point Reyes Station, January 26, 2001.

58. Livingston, *Ranching on the Point Reyes Peninsula*, 200.

59. Dairies, especially those that are organic, are currently seen as leading a resurgence of agricultural production in Marin County. As of this writing there are twenty-nine dairies in the county, of which twenty-five are organic; six of them are located within the Seashore. Nels Johnson, "Organic Milk Steers Marin Farm Production to Record $101 Million," *Marin Independent Journal,* July 21, 2015.

60. New Albion was still operating as a dairy; Laguna and Bear Valley were both beef ranches.

61. James Lundgren's property was condemned in 1975, and while he was granted a fifty-year reservation of occupancy for noncommercial residence on two acres (the old home site), he was not allowed to retain grazing use of the remaining 339 acres of land. He then continued to run livestock on the condemned 339 acres and remained in a legal dispute with the NPS for years. PRNS tract files, PRNS archive.

62. For example, the Truttman brothers retired in 1990 from their Olema Valley ranch, after years of strongly worded warnings and threats of fines from Superintendent Sansing, usually based in misunderstandings about the lease terms. Sansing also arranged to lease the land to another family six years before the Truttmans actually retired, even though the Truttmans had given no sign of doing so. Letter from Superintendent Sansing to Armin and Frank Truttman, February 10, 1984, and memo from Superintendent Sansing to Frank and Armin Truttman, October 30, 1986. PRNS tract files, PRNS archive.

63. Marian Schinske, "Park Service to Oust Tenant to Make Room for PRBO," *Point Reyes Light,* May 7, 1998. From the article: "'I was completely caught by surprise by the whole thing,' [Tiscornia] said. Late last summer, she explained, she got a friendly call from Neubacher, who said he wanted to talk about their lease agreement. 'He said that maybe the PRBO would like to move in to the Rancho *if* I didn't want to stay, and asked me if he could show a few of them around the place. All of a sudden about thirty people showed up—architects, office designers, park people, and bird observatory people. They were telling me, "I'm sorry you have to leave here." That's how I first heard of it.'" Neubacher "claimed he informed her two years ago that her move was imminent."

64. NPS, *General Management Plan*, 59; emphasis is mine.

65. The Clem Miller Environmental Education Center already exists in the park at the old Laguna Ranch site, and the NPS had just established a new Pacific Learning Center for facilitating research in the Seashore at the old

Hagmaier Ranch. There are also a number of other privately run centers in West Marin, including the Audubon Canyon Ranch, located just south of the Wilkins Ranch, the Marine Mammal Center on the Marin Headlands, the Point Reyes Field Seminars Program, the Point Reyes Bird Observatory, Slide Ranch, and the Marconi Center.

66. Sansing specifically stated, "Ms. Tiscornia is an excellent permittee. The range utilization measurements are among the very best in the park. Maintenance and rehabilitation of the historic structures are the very best in the park. Consequently, we have no desire nor intention to evict her or others who live at Rancho Baulines and assist with the operations." Quoted in Marian Schinske, "Park Service to Oust Tenant to Make Room for PRBO," *Point Reyes Light,* May 7, 1998.

67. His note to Tiscornia that accompanied the lease stated, "We hope that the lease provides you with sufficient security so you can continue with your excellent record of repairs and rehabilitation of the facilities." Letter from Superintendent Sansing to Mary Tiscornia, July 19, 1988; PRNS archives.

68. Memo from Superintendent Sansing to NPS Western Regional Director, May 27, 1988; PRNS archives. It also appears that both Edgar Wayburn and Amy Meyer, key members of both the influential group People for the GGNRA and the joint GGNRA/PRNS CAC, opposed this lease and asked that it be discussed by CAC, but Jerry Friedman, chair of the Point Reyes Committee of CAC at that time, declined to hold a hearing on the subject. A memo from Sansing to the Western Regional Director dated September 15, 1988 specified that "Other than those two individuals, we know of no other opposition"; PRNS archives.

69. Tiscornia agricultural use lease, signed December 21, 1993, page 5, item 26: "Six months prior to the expiration of this lease, this lease or a similar lease including changes in conditions negotiated with the leasee *will be issued* for an additional five-year period" (emphasis added). PRNS archives.

70. Gregory Foley, "Bolinas Scolds Park over Rancho Plan," *Point Reyes Light,* November 22, 2000.

71. Stephen Barrett, "Residents Blast Park over Rancho Baulines Eviction," *Point Reyes Light,* October 26, 2000.

72. Steve Matson, "Park Service Shouldn't Be in a Rush to Alter Rancho Baulines' Character," *Point Reyes Light,* December 28, 2000. He also asserted that the Park Service had "erred by removing Rancho Baulines from the 1980 Management Plan update [which had then just begun] without due public notification."

73. Barrett, "Residents Blast Park."

74. Transcript of GGNRA/PRNS CAC meeting, held October 20, 2001, 53; PRNS archives.

75. Patrik Jorgensen, "Park Advisors Back Plan for Rancho Ed Center," *Point Reyes Light,* November 15, 2001.

76. Memo from Richard Curry, watershed analysis consultant, to Stephen Volker and clients, February 16, 2001; copy on file with author. He specified that under Mary Tiscornia's previous management, "the Rancho Baulines lands

are in the best condition of any similar habitats in public or private ownership on the Point Reyes Peninsula."

77. Exhibits at Bear Valley were designed by Daniel Quan and Jane Glickman in 1983. Park staff were included in discussions of what to include in the exhibits. Email to author from John Dell'Osso, chief of interpretation, December 3, 1999, copy on file with author.

78. See, for example, all the references to Drake in the 1960 and 1961 hearings. Note that the Miwok Indians were glossed over as well; for more detail on the Miwok experience with PRNS, see Jennifer Sokolove, Sally K. Fairfax, and Breena Holland, "Managing Place and Identity: Thoughts on the Coast Miwok Experience," *Geographical Review* 92, no. 1 (2002): 23–44.

79. The settlement subtheme included the Mexican rancheros, initial settlement by pioneer Americans, and the Shafter era, but no other mention of ranching. "Unrelated themes" included lighthouses, shipwrecks, the 1906 earthquake, prohibition and bootlegging at Point Reyes, and the development of the National Seashore. May 1968, draft 1 of Historical Resource Management Plan; PRNS archives.

80. May 1968, draft 1 of Historical Resource Management Plan; PRNS archives.

81. Memo from Supervisory Historian Ross Holland to director of Denver Service Center, via manager of Historic Preservation Team, Denver Service Center, May 29, 1973; PRNS archives.

82. Anna Coxe Toogood, *A Civil History of Golden Gate National Recreation Area and Point Reyes National Seashore,* Historic Resource Study, vol. 2 (Denver: Historic Preservation Branch, Pacific Northwest/Western Team, Denver Service Center, National Park Service, 1980), 229.

83. Toogood, *Civil History,* 234–39. The Olema Valley district was actually determined eligible for listing in 1979 but that determination was "lost" by park staff in the interim.

84. I asked the chief of interpretation why the Rogers Ranch did not have a sign, and he responded that he thought it did have one, that perhaps it had fallen down or been vandalized; author interview with John Dell'Osso, March 28, 2000. However, I cannot recall ever seeing a sign there, nor has one been put back in place since I raised the issue.

85. These include the W Ranch at Bear Valley, where old ranch buildings are now used adaptively as the administration headquarters; the former Laguna (T) Ranch, now used as an environmental education center; and the hike-in campground sites.

86. In several older files, such as the squad meeting minutes from 1973–76, PRNS archives, PRNS staff routinely used the name Rancho Baulines. Their recent switch to the older name may be related to the recent controversy regarding evicting the long-term tenant and converting the ranch to an Environmental Education Center.

87. For more on the symbolic power of naming, see Jared Farmer "Renaming the Land," in *On Zion's Mount: Mormons, Indians, and the American Landscape* (Cambridge, MA: Harvard University Press, 2008).

88. Author interview with PRNS Assistant Superintendent Frank Dean, Point Reyes National Seashore, November 30, 1999.

89. For more detailed discussions of NPS architectural styles, see Ethan Carr, *Wilderness by Design: Landscape Architecture and the National Park Service* (Lincoln: University of Nebraska Press, 1998), and Linda Flint McClelland, *Building the National Parks: Historic Landscape Design and Construction* (Baltimore: Johns Hopkins University Press, 1998).

90. In order to move the Randall House, the park staff would have had to do a complete National Register eligibility determination for the building, which the superintendent was apparently reluctant to do. Memo from Tom Mulhern, chief of cultural resource management, Western Regional Office, to Superintendent Sansing, May 24, 1978; PRNS archives.

91. Email messages to the author from John Dell'Osso, PRNS chief of interpretation, December 3, 1999 and March 27, 2000. On file with author.

92. Email message to the author from Gordon White, April 5, 2000. On file with author.

93. Albert Bagshaw, the attorney for the Mendozas, Nunes, and Kehoes, as well as the Point Reyes Milk Producers Association, warned of this possibility early on: "There is no pastoral splendor under a leased operation"; *Hearing on S. 2428* (1960), at 153–56. Similar arguments were made at both the House and Senate hearings in 1961.

94. As of April 2000, the condition of the 218 listed historic structures in the park was 27 percent good, 60 percent fair, and 13 percent poor. PRNS Historic Structures Condition/Responsibility Report, April 2000; PRNS archives.

95. Alanen, "Considering the Ordinary," 137–38.

96. Author's phone interview with Gordon White, chief of integrated resources management, July 2, 2015.

97. During our 2015 interview, White gave two such examples: work on the Lighthouse that has been in the queue since 2002, and repairs to the main barn at the Home Ranch that have been waiting since before White was hired in 1997.

98. Author's phone interview with Gordon White, chief of integrated resources management, July 2, 2015.

99. Paul Sadin, *Managing a Land in Motion: An Administrative History of Point Reyes National Seashore* (Washington, DC: National Park Service, U.S. Department of Interior, 2007), 303; quote is from LeeRoy Brock.

100. Note that a house at N Ranch was used for housing for several years, but was later knocked down in 1976. Squad meeting minutes, file folder A4031, PRNS archives.

101. David Lowenthal, "European Landscape Transformations: The Rural Residue," in *Understanding Ordinary Landscapes* eds. Paul Groth and Todd Bressi (New Haven, CT: Yale University Press, 1997), 180–88.

102. Transcript of CAC meeting, January 23, 1999, 24–25; PRNS archives. The possible uses included housing for NPS staff, a Coast Miwok center, an artists retreat, and meeting space for community groups.

103. Transcript of CAC meeting, January 23, 1999, 81; PRNS archives. One property is currently leased to the Salmon Protection and Watershed Network

(SPAWN), a project of a local nonprofit organization called the Turtle Island Restoration Network.

104. Author's phone interview with Gordon White, chief of integrated resources management, July 2, 2015.

105. Marian Schinske, "Park Service to Raze Most of Hamlet in the Fall," *Point Reyes Light*, February 27, 1997. Jensen further elaborated, "I just get a big kick out of how long those buildings have lasted even though they were neglected by the [Park Service] for ten years. People have come by and have ripped off old doorknobs and wooden panels. It's an old ghost town now. It's a tear-jerker to look at. They might as well bring a [bull]dozer in now. There's nothing left to save."

106. NPS, *Assessment of Alternatives for the General Management Plan*, 5.

107. Richard Francaviglia, "Selling Heritage Landscapes," in *Preserving Cultural Landscapes in America,* eds. Arnold R. Alanen and Robert Z. Melnick (Baltimore: Johns Hopkins University Press, 2000), 58.

108. Historic Structures Report, Pierce Ranch, by Richard Borjes, regional historical architect and Gordon Chappell, regional historian, January 1985, 8; PRNS archives.

109. Box H3012, "History, Pierce Point Ranch," PRNS archives. The file contains only the RFP, not any indication of whether the NPS received any proposals.

110. Budgeted cost from Sadin, *Managing a Land in Motion*, 313.

111. Memo from Regional Historic Preservation Team, Western Regional Office (WRO), to associate regional director, WRO, dated August 9, 1974, regarding an evaluation of the house done on July 12, 1974; PRNS files, code H3015—Historic Structures (Randall House), PRNS archives. They suggested the house be used for residential or office purposes. The barn was removed in December 1975; squad meeting minutes, file folder A4031, PRNS archives.

112. A reply memo from the PRNS superintendent, dated August 22, 1974, disagreed with the preservation team's report, stating that (1) money would be better spent on residences at nearby Hagmaier Ranch, supposedly in better condition, and (2) Randall House, contrary to the report, was not in good condition—the roof leaking and needing replacement, etc. A later memo dated March 22, 1975 from the associate director of NPS professional services, notes a large discrepancy between descriptions of the house's condition in the August 9, 1974 report and in a Board of Survey report dated October 17, 1974. The memo writer suspects the Board of Survey must be "a serious exaggeration of existing conditions even allowing for vandalism" and expresses a belief that the building should be retained, repaired and perhaps used for employee housing. PRNS files, code H3015—Historic Structures (Randall House), PRNS archives.

113. The initial cost estimates for repair varied widely, from $9,840 (estimated by Robert Cox, historical architect, WRO, on January 15, 1976) to $37,800 (chief of maintenance, PRNS, June 18, 1975). PRNS files, code H3015—Historic Structures (Randall House), PRNS archives.

114. Robert Cox revised his initial cost estimate to $63,000 on September 17, 1981; Superintendent Sansing replied two months later on November 17, 1981 with his own revised estimate of $131,690. Eligibility for listing was

formally determined on August 29, 1979 by the California State Historic Preservation Office (SHPO) and National Register staff, despite disagreement from Regional and Washington office NPS historians. PRNS files, code H3015—Historic Structures (Randall House), PRNS archives.

115. Memo from PRNS superintendent to regional director, WRO, November 17, 1981, reporting that there had been twenty-one "fruitless inquiries and discussions" over the past two years with people regarding adaptive use of the Randall House. Because these discussions were only informal, their names and proposals were not recorded. PRNS files, code H3015—Historic Structures (Randall House), PRNS archives.

116. Letter from James E. Zwaal to PRNS superintendent, March 14, 1983. PRNS files, code H3015—Historic Structures (Randall House), PRNS archives.

117. Charles Birnbaum, "A Reality Check for Our Nation's Parks," *Cultural Resources Management* 16, no. 4 (1993): 1–4.

118. Author interview with Bob Page, director, Park Historic Structures and Cultural Landscapes Program, NPS, July 29, 1999.

119. Arnold R. Alanen and Robert Z. Melnick, "Introduction: Why Cultural Landscape Preservation?", in *Preserving Cultural Landscapes in America* eds. Arnold R. Alanen and Robert Z. Melnick (Baltimore: Johns Hopkins University Press, 2000), 17.

120. Alanen and Melnick, "Introduction."

121. Author interview with Bob Page, director, Park Historic Structures and Cultural Landscapes Program, NPS, July 29, 1999.

122. NPS's twenty-eight policy guidelines contain general concepts regarding cultural resources in the parks; available on-line at https://www.nps.gov/policy /MP2006.pdf, Chapter 7 deals specifically with cultural landscapes and general policies toward their management.

123. Author interview with Cari Goetcheus, Park Historic Structures and Cultural Landscapes Program, NPS, March 9, 1999.

124. Dewey Livingston's meeting notes, June 1999, Bob Page speaking: "[Chief of NPS Cultural Resource Management Randy] Biallas considers Point Reyes to be most significant cultural landscape in System." Copy on file with author.

125. The National Register criteria cited by the California SHPO were criteria A, C, and D; PRNS archives.

126. Letter from Cherilyn Widell, State Historic Preservation Officer, to Acting Superintendent LeeRoy Brock, April 3, 1995; PRNS archives.

127. Author interview with Bob Page at PRNS Cultural Landscape Team meeting, Point Reyes National Seashore, October 25, 1999.

128. National Park Service, *Cultural Landscape Inventory, D Ranch, Point Reyes National Seashore,* 1997; National Park Service, *Cultural Landscape Inventories for A, B, C, Home, I, L, M, and Pierce Ranches,* 2004; National Park Service, *Cultural Landscape Inventory, Point Reyes Ranches Historic District,* 2004; National Park Service, *Cultural Landscape Inventories for the Giacomini, Hagmeier, McFadden, Rogers, Truttman, and Wilkins Ranches,* 2011.

129. Author's phone interview with Gordon White, chief of integrated resources management, July 2, 2015. The nominations for the Point Reyes

Ranches Historic District and the Olema Valley/Lagunitas Loop Ranches Historic District remain incomplete at the time of this writing.

130. Author interview with Thom Thompson, Cultural Resources Division, PRNS, February 5, 2001.

131. David Schulyer and Patricia O'Donnell, "The History and Preservation of Urban Parks and Cemeteries," in *Preserving Cultural Landscapes in America* eds. Arnold R. Alanen and Robert Z. Melnick (Baltimore: Johns Hopkins University Press, 2000), 75.

132. Alanen and Melnick, *Preserving Cultural Landscapes*, 17.

133. Emily Wakild, *Revolutionary Parks: Conservation, Social Justice, and Mexico's National Parks, 1910–1940* (Tucson: University of Arizona Press, 2011), 4.

CHAPTER 6. REASSERTION OF THE PARK IDEAL

1. Paul Sadin, *Managing a Land in Motion: An Administrative History of Point Reyes National Seashore* (Washington, DC: National Park Service, 2007), 155, citing his own interview with Sansing.

2. Vernita Lea Ediger, "Natural Experiments in Conservation Ranching: The Social and Ecological Consequences of Diverging Land Tenure in Marin County, California," Ph.D. diss., Stanford University, 2005, 95–96.

3. Sadin, *Managing a Land in Motion*, 181.

4. Richard West Sellars, *Preserving Nature in the National Parks: A History* (New Haven, CT: Yale University Press, 1997), xiii.

5. Sellars, *Preserving Nature*, 201.

6. A. Starker Leopold et al., "Wildlife Management in the National Parks" (1963), reprinted in Lary M. Dilsaver, ed., *America's National Park System: The Critical Documents* (Lanham, MD: Rowman and Littlefield Publishers, 1994), 239–41.

7. Emma Marris, *Rambunctious Garden: Saving Nature in a Post-Wild World* (New York: Bloomsbury, 2011), 12.

8. Sellars, *Preserving Nature*, 287.

9. Ronald A. Foresta, *America's National Parks and Their Keepers* (Washington, DC: Resources for the Future, 1984), 69–70. Note that there was a similar erosion of support from the wealthy elite; particularly after the strong association of the NPS with "Great Society" social programs in the 1960s, these traditional ties weakened, and have not been regained. Foresta, *America's National Parks*, 71–72.

10. For example, the Conservation Foundation recommended in 1985 that the NPS should focus exclusively on an environmental preservation mission, and that historical and cultural sites be assigned to some other agency. The Conservation Foundation, *National Parks for a New Generation: Visions, Realities, Prospects* (Washington, DC: Conservation Foundation, 1985).

11. James Morton Turner, *The Promise of Wilderness: American Environmental Politics Since 1964* (Seattle: University of Washington Press, 2012), 256.

12. Richard White, "Epilogue: Contested Terrain," in *Land in the American West: Private Claims and the Common Good*, eds. William G. Robbins and

James C. Foster (Seattle: University of Washington Press, 2000), 203; emphasis in original.

13. Foresta, *America's National Parks*, 90. In 1995 the agency's hierarchical structure was substantially reorganized (for the first time since 1937), giving even greater autonomy to the parks, and reducing both the size and the oversight of the DC office. See NPS press release, "National Park Service Reorganization Marks Most Significant Organizational Change in Agency's 79-Year History," May 15, 1995 (Washington, DC: Office of Public Affairs).

14. Dwight F. Rettie, *Our National Park System: Caring for America's Greatest Natural and Historic Treasures* (Urbana: University of Illinois Press, 1995), 7: "The Service has usually found new departures difficult to assimilate into its perceived mission."

15. Sellars, *Preserving Nature*, 284.

16. Joseph L. Sax and Robert B. Keiter, "Glacier National Park and Its Neighbors: A Study of Federal Interagency Relations," *Ecology Law Quarterly* 14, no. 2 (1987): 207–63.

17. Sellars, 268.

18. Sellars, *Preserving Nature*, 279–80.

19. Sellars, *Preserving Nature*, 279, quoting the *Vail Agenda*, 95–97.

20. Robert B. Keiter, *To Conserve Unimpaired: The Evolution of the National Park Idea* (Washington DC: Island Press, 2013), 213.

21. Janet A. McDonnell, "Reassessing the National Park Service and the National Park System," *George Wright Forum* 25, no. 2 (2008): 6–14, 10.

22. Eventually, this aspect of the project was done with no NEPA review at all; the *Giacomini Wetland Restoration Project Final EIS/EIR* (2007) states that environmental review of the disposal of approximately 170,000 cubic yards of fill material in the peninsula quarries would be done as a separate project, but no formal review of the possible impacts was ever conducted.

23. Sadin, *Managing a Land in Motion*, 241.

24. Stephen Barrett, "Deer Count Draws Fire from Residents," *Point Reyes Light*, October 26, 2000.

25. Sadin, *Managing a Land in Motion*, 233. They also had to dispose of the carcasses; many were donated to a San Francisco soup kitchen and to local tribal organizations; Sadin, *Managing a Land in Motion*, 242.

26. Judy Schroeter, "Park May Eliminate Exotic Deer; Tule Elk Shoot Delayed," *Point Reyes Light*, November 5, 1992.

27. Sadin, *Managing a Land in Motion*, 242. The Marin County Humane Society played a key role in convincing the NPS to end culling, after receiving calls for years from nearby homeowners "distraught about bloody wounded deer wandering on to their property." M. Martin Smith, "The Deer Departed," *High Country News*, May 28, 2007.

28. Smith, "Deer Departed."

29. National Park Service, *Point Reyes National Seashore Non-native Deer Management Plan: Protecting the Seashore's Ecosystems*, final environmental impact statement (Washington, DC: National Park Service, 2006).

30. John Hart, *An Island In Time: Fifty Years of Point Reyes National Seashore* (Mill Valley, CA: Pickleweed Press, 2012), 123.

31. Interviews with Dorothy McClure, conducted by PRNS Ranger Diana Skiles on June 9, 1977, May 29, 1979, and September 18, October 1, and October 9, 1980; Box H3012, PRNS archive. Also author interview with Web Otis, June 3, 2009.

32. Dale McCullough, *The Tule Elk: Its History, Behavior, and Ecology* (Berkeley: University of California Publications in Zoology, 1969).

33. W. E. Phillips, *The Conservation of the California Tule Elk: A Socio-Economic Study of a Survival Problem* (Edmonton: University of Alberta Press, 1976), 14–17.

34. Dale McCullough, Jon Fischer, and Jonathan Ballou, "From Bottleneck to Metapopulation: Recovery of the Tule Elk in California," in *Metapopulations and Wildlife Conservation,* ed. Dale McCullough (Washington, DC: Island Press, 1996), 375–76. Also McCrea Andrew Cobb, "Spatial Ecology and Population Ecology of Tule Elk *(Cervus canadensis nannodes)* at Point Reyes National Seashore, California," PhD diss., University of California, Berkeley, 2010, 1, notes that early records indicate that tule elk were extirpated from the San Francisco Bay Area by 1872.

35. Phillips, *Conservation of the California Tule Elk,* 17–18.

36. McCullough, Fischer, and Ballou, "From Bottleneck to Metapopulation," 377–78.

37. For a more detailed history of tule elk management across California and at Point Reyes, see Laura A. Watt, "The Continually Managed Wild: Tule Elk at Point Reyes National Seashore," *Journal of International Wildlife Law and Policy* 18, no. 4 (2015): 289–308.

38. Ann Lage, "Oral History Interview with Peter H. Behr," July 1988, California State Archives, State Government Oral History Program, 132–35. Edmiston's organization was called the Committee for the Preservation of Tule Elk.

39. California S.B. 722, 1971 Reg. Sess., Cal. Stat., ch. 1250.

40. In an interview, Dr. Dale McCullough identified himself as having lobbied for Point Reyes as a location for tule elk even back when the Seashore was first being established. Author interview with Dr. Dale McCullough, Kensington, CA, July 13, 2015.

41. Sadin, *Managing a Land in Motion,* 244, citing unpublished memoir by Sansing; emphasis is mine.

42. Tomales Point was not among the three places at Point Reyes initially identified as potential elk sites; memo from Richard Myshak, acting assistant secretary for Fish and Wildlife and Parks, Dept. of Interior, to Albert Bianchi [Mervin McDonald's lawyer], April 7, 1978; PRNS archives. In a recent interview, rancher Merv McDonald recalled that the elk were originally supposed to be on "the other point," near the lighthouse, within a fence, but the ranchers in that area protested, so the elk site was changed to Pierce Ranch/Tomales Point. Author interview with Mervin McDonald, Marshall, CA, June 23, 2015.

43. Philip L. Fradkin, "No Room for Cows on Point Reyes," *Audubon,* July 1978: 98–104, 102.

44. The first ninety-day notice to vacate was issued on February 16, 1973, and described by Philip Fradkin as a "tersely worded three-month notice";

Fradkin, "No Room for Cows," 99. Two weeks later, after meetings between McDonald and NPS staff, Superintendent Sansing sent a letter setting a new deadline of October 1974 to allow time for relocation.

45. This requirement stemmed from the Uniform Relocation Assistance and Real Property Acquisition Policies Act of 1970, Public Law 91–646, 84 Stat. 1894.

46. "Relocation Assistance Furnished Relocatee toward Establishing a Substitute Cattle Ranching Operation Elsewhere," attachment to letter from NPS Western Regional Director Howard Chapman to Mervin McDonald, March 16, 1978; PRNS archives.

47. Letter from NPS Western Regional Director Howard Chapman to Mervin McDonald, March 16, 1978; PRNS archives. The letter describes most ranches as being in the one thousand to twelve hundred dollar per animal unit range, which was twice what McDonald could afford.

48. Memo to Robert Herbst, Assistant Secretary of the Interior, from Albert Bianchi [Mervin McDonald's lawyer], February 17, 1978; PRNS archives.

49. Dewey Livingston, *Interviews with Merv McDonald, Marin County Rancher* (Point Reyes Station, CA: Marin Resource Conservation District, 2010), 72–73; author interview with Mervin McDonald, Marshall, CA, June 23, 2015.

50. Memo to Robert Herbst, Assistant Secretary of the Interior, from Albert Bianchi [Mervin McDonald's lawyer], February 17, 1978; PRNS archives.

51. McDonald v. United States, Civil No. C-78–2155 RHS (N.D. Calif., filed September 13, 1978).

52. B.G. Buttemiller, "Pierce Point's Last Roundup," *Point Reyes Light,* December 6, 1979.

53. Michael Gray, "Evicted Rancher Staying Close," *Point Reyes Light,* May 29, 1980.

54. Tule elk had been living at the San Diego Zoo since 1915. McCullough, Fischer, and Ballou, "From Bottleneck to Metapopulation," 384.

55. Point Reyes National Seashore, *Tule Elk Management Plan and Environmental Assessment* (Point Reyes, CA: National Park Service, July 1998), 8: available at www.nps.gov/pore/learn/management/upload/planning_tule_elk_mp_ea_1998.pdf.

56. Peter Gogan and Reginald Barrett, "Comparative Dynamics of Introduced Tule Elk Populations," *Journal of Wildlife Management* 51, no. 1 (1987): 20–27. The authors noted that the elk herd was supplemented by three additional male elk moved from the Owens Valley in 1981, but that they subsequently disappeared.

57. Sadin, *Managing a Land in Motion,* 245. The original study was Peter Gogan, "Ecology of the Tule Elk Range, Point Reyes National Seashore," PhD diss., University of California, Berkeley, 1986.

58. The 1988 numbers are from "Viewing Tule Elk," online at https://www.nps.gov/pore/planyourvisit/ wildlife_viewing_tuleelk.htm, accessed September 28, 2013. 1994 census information is from Joel Reese, "Park Mulls How to Limit Elk and Non-native Deer," *Point Reyes Light,* August 24, 1995.

59. Point Reyes National Seashore, "Control of Tule Elk Population at Point Reyes National Seashore," draft document, 1992, 4. According to the 1998

Tule Elk Management Plan, this document was "withdrawn from the approval process, [but] the draft assessment and the response by the public helped formulate policy and direct strategies for tule elk at Point Reyes National Seashore." PRNS, *Tule Elk Management Plan and Environmental Assessment* (National Park Service, U.S. Department of Interior, 1998), Appendix B, 85.

60. "Report of the Scientific Advisor Panel on Control of Tule Elk on Point Reyes National Seashore," October 18, 1993, 7; the report also compares (9) its their findings to those of J. W. Bartolome, *Range Analysis of the Tomales Point Tule Elk Range*, preliminary report (Point Reyes National Seashore, 1993).

61. Don Schinske, "Reyes Publica," *Point Reyes Light*, September 2, 1993. The article also quotes McCullough as saying that: "The resilient elk do just fine in inbred situations." He added, "They're the nearest thing to weeds we have."

62. Dale McCullough, Robert A. Garrott, Jay F. Kirkpatrick, Edward D. Plotka, Katherine D. Ralls, and E. Tom Thorne, *Report of the Scientific Advisor Panel on Control of Tule Elk on Point Reyes National Seashore*, October 18, 1993, 5–6; PRNS archives.

63. Judd A. Howell, George C. Brooks, Marcia Semenoff-Irving, and Correigh Greene, "Population Dynamics of Tule Elk at Point Reyes National Seashore, California," *Journal of Wildlife Management* 66, no. 2 (2002): 478–90, 482. The 1996 census numbers were reported by David Rolland, "Limantour May Get Some Elk," *Point Reyes Light*, November 7, 1996.

64. Howell, Brooks, Semenoff-Irving, and Greene, "Population Dynamics," 489. Note that they also found immune-contraception to be effective, but were concerned about the impact of removing individuals from the gene pool, given the low genetic diversity in the herd.

65. Alice Hall, who had lived in the original 1870s-era ranch house, passed away in 1991. D. S. Livingston, *A Good Life: Dairy Farming in the Olema Valley: A History of the Dairy and Beef Ranches of the Olema Valley and Lagunitas Canyon* (San Francisco: Golden Gate National Recreation Area and Point Reyes National Seashore, National Park Service, 1995), 144–46; author interview with Roger and Todd Horick, Chileno Valley (Petaluma), July 29, 2008.

66. Letters from Superintendent Sansing to Vivian Horick, August 30, 1989 and September 10, 1990, PORE 12214, Box 1, Folder 12, PRNS archives.

67. The archive also contains a draft agricultural use lease for Horick from 1991, made up of a copy of Mary Tiscornia's lease but with her information crossed out and handwritten notes on the specific terms of the Horick lease. In pencil, someone had written "(and Successors?)" by Vivian's name. PORE 12214, Box 1, Folder 9, PRNS archives. There is nothing in the file to indicate why PRNS staff switched at that time from using the "more secure" agricultural use lease document (see discussion in chapter 5) to a special use permit, nor why successors' names were not added.

68. Letter from Superintendent Neubacher to Carol, Todd, and Roger Horick, May 26, 1998, PORE 12214, Box 1, Folder 11, PRNS archives. He specified, "we are currently evaluating the situation and conducting an appraisal as to the rental value of the property. The appraisal will be a contract and but will take at least 30 days to complete. We plan to have an appraiser on-site as soon

as possible. After we reach the final appraisal, we should be in position to discuss the future use of the ranch with you."

69. Marian Schinske, "Park Says It Isn't Eyeing Horick Ranch Takeover," *Point Reyes Light,* June 11, 1998.

70. Letter from Max Mickelsen [attorney for the Horick trust] to Dr. John Zimmerman, July 7, 1998; letter from Max Mickelsen to Superintendent Neubacher, July 21, 1998. PORE 9728 S.1, Box 3, Folder 24, PRNS archives.

71. Author interview with Assistant Superintendent Frank Dean, November 30, 1999. At the CAC meeting of October 23, 1999, Assistant Superintendent Dean stated that the three Horick children were not legally permitted to operate the ranch, "and have missed multiple deadlines in the process of getting a special use permit," hence the NPS was terminating use of the ranch as of November 1, 1999. CAC meeting transcript, PRNS archives.

72. Written statement of Todd Horick, dated August 29, 2007. Copy on file with author.

73. Author interview with Roger and Todd Horick, Chileno Valley (Petaluma), July 29, 2008.

74. Author interview with Roger and Todd Horick, Chileno Valley (Petaluma), July 29, 2008.

75. Letter from Max Mickelsen to Ronald Silveira [attorney for Carol Horick Sethman], dated September 24, 1998, citing a phone conversation he had recently had with Superintendent Neubacher. Oddly, a letter several months later from Superintendent Neubacher to Max Mickelsen (dated December 8, 1998) stated, "We must continue to request the family not commit any major funding or rehabilitation on the D Ranch until the estate and permit issues are completed." PORE 9728 S.1, Box 3, Folder 24, PRNS archives.

76. Letter from Superintendent Neubacher to Max Mickelsen, December 8, 1998. This letter also asks for clarity regarding whom to contact about the ranch's management, specifically asking whether Zimmerman still had a role—despite an earlier letter from Silveira to Neubacher on September 29, 1998, which informed him that "Dr. Zimmerman did not want open-ended and vague prospective obligations. Therefore, his duties will terminate" once the last details of the dairy herd sale were completed. PORE 9728 S.1, Box 3, Folder 24, PRNS archives.

77. Letter from Henry Froneberger [attorney for Roger and Todd Horick] to Superintendent Neubacher, January 14, 1999, PORE 9728 S.1, Box 3, Folder 24, PRNS archives. There is a handwritten note on this letter stating that it was faxed to Regional Solicitor Ralph Mihan on February 9, 1999.

78. Letter from Henry Froneberger to Superintendent Neubacher, March 1, 1999; letter from Ronald Silveira, attorney for Carol Horick Sethman, to Superintendent Neubacher, March 3, 1999; letter from Max Mickelsen to Superintendent Neubacher, March 10, 1999; letter from Ronald Silveira to Superintendent Neubacher, March 11, 1999; PRNS archives. The letter from Silveira to Neubacher specifies, "As I indicated to you, counsel representing Roger Horick and Todd Horick and I have established a conceptual approach to determine successor status, but, obviously, that determination will have to await the delineation and determination of terms upon which they, or some of them, may be competing, in a bid process internal to the Trust."

79. Letter from Ronald Silveira to Superintendent Neubacher, April 27, 1999, PORE 9728 S.1, Box 3, Folder 24, PRNS archives; he asks, "I have placed a few phone calls to you to determine status of the appraisal for the permitted property in question and a delineation of the Park's expectation for the permit's renegotiation. Will we be hearing from the Park Service in the foreseeable future?" Also the March 11, 1999 letter from Silveira to Neubacher specifies that he is following up on a phone call they had just completed, confirming that NPS appraisal was still outstanding and that NPS would provide a copy once it was done.

80. Letter from Superintendent Neubacher to Max Mickelsen, May 11, 1999, and letter from Superintendent Neubacher to Max Mickelsen, May 18, 1999; in the latter, he specifies, "We feel the historic complex is in serious disrepair and need to obtain a tenant that is eager, and has the financial capacity, to upgrade the facilities." PORE 9728 S.1, Box 3, Folder 24. PRNS Archives.

81. Author interview with Roger and Todd Horick, Chileno Valley (Petaluma), July 29, 2008. Max Mickelsen, former attorney for the Horick trust, recalled that then-County Supervisor Gary Giacomini talked with Superintendent Neubacher on the Horicks' behalf, asking that he give them more time to let their heifers grow a bit more, and hence be worth more at sale. Author's phone interview with Max Mickelsen, Petaluma, April 4, 2014.

82. Richard White, *Remembering Ahanagran: A History of Stories* (Seattle: University of Washington Press, 1998), 49.

83. In a 2007 statement, Todd wrote, "Regardless of all our work, hours and hours of hard labor and thousands of dollars to bring the property up to Park standards, the lease was never offered to our family, and there was no recourse or appeal of the Superintendent's decision. Despite his continued encouragement during meetings, he never was willing to commit anything on paper, and we were unable to document his promises from the past meetings. We trusted him to live up to his word, and we thought that his job was to uphold the ranching tradition of the original owners." Written statement of Todd Horick, dated August 29, 2007. Copy on file with author.

84. None of the letters from the siblings' lawyers from 1999 asking for the appraisal, nor the September 29, 1998 letter clarifying that Zimmerman's role as ranch manager was ending were filed in the D Ranch Correspondence folders at the PRNS archive; copies can only be found in the separate file of material sent to the CAC in October 1999. This suggests that at some point after the October 1999 CAC meeting, they were intentionally removed from the correspondence file.

85. CAC, transcript of October 23, 1999 meeting; PRNS archives.

86. CAC, transcript of October 20, 2001 meeting, 79–81; PRNS archives.

87. Quote of Livingston is from Gregory Foley, "Future of Ranching in National Seashore Uncertain," *Point Reyes Light,* October 28, 1999, describing discussion from the October 23, 1999 CAC meeting.

88. Marian Schinske, "Park Advisors Given Options for Limiting Tule Elk Herd," *Point Reyes Light,* May 22, 1997.

89. PRNS, *Tule Elk Management Plan,* 40–41.

90. Phillips, *Conservation of the California Tule Elk* (1976) oddly ignored any potential impacts on the local agricultural economy in his analysis.

McCullough, Garrott, Kirkpatrick, Plotka, Ralls, and Thorne, *Report of the Scientific Advisor Panel*, 34.

91. Author interview with Dale McCullough, Kensington, CA, July 13, 2015.

92. Peter Alagona, *After the Grizzly: Endangered Species and the Politics of Place in California* (Berkeley: University of California Press, 2013), 186; emphasis is mine.

93. Marian Schinske, "Tule Elk to Roam Huge Range in Park," *Point Reyes Light*, October 30, 1997. In addition, the article quotes CAC commissioner Merritt Robinson as saying, "We made a promise to the ranchers that we wouldn't damage their economic position. I want the park's tule-elk-management plan to speak to this issue."

94. Elizabeth J. B. Manning, Thomas E. Kucera, Natalie B. Gates, Leslie M. Woods, and Maura Fallon-McKnight, "Testing for *Mycobacterium avium* subsp. *paratuberculosis* infection in Asymptomatic Free-Ranging Tule Elk From An Infected Herd," *Journal of Wildlife Diseases* 39, no. 2 (2003): 323–28. Of the twenty-two animals euthanized, tissue from ten were found to contain the microbe that causes the disease, which surprised the study's authors, "both because of the continued overall health of the source herd and normal clinical status of all study animals."

95. Marian Schinske, "Helicopter Gives Elk Rides from Pierce Point," *Point Reyes Light*, December 3, 1998.

96. Letter from Superintendent Cicely Muldoon to Tess Elliott, editor, *Point Reyes Light*, regarding FOIA request for collar data on tule elk, dated September 17, 2014. Two additional adults apparently passed away between their relocation to Limantour and the herd's release; email to author from PRNS wildlife biologist Dave Press dated November 10, 2015.

97. NPS, *Point Reyes National Seashore 2001 Year in Review*, 11, online at www.nps.gov/pore/learn/ management/upload/yearinreview2001.pdf.

98. Stephen Barrett, "Eighteen Tule Elk Culled from Limantour," *Point Reyes Light*, May 27, 1999, quoting Dr. Sarah Allen. Dr. Allen is also quoted as saying that the NPS had not yet decided on when the quarantined animals would be released, but the release occurred only four days later.

99. PRNS, *Tule Elk Management Plan*, 13.

100. PRNS, *Tule Elk Management Plan*, 51.

101. PRNS, *Tule Elk Management Plan*, 46.

102. PRNS, *Tule Elk Management Plan*, 50.

103. "History of Elk at Drakes Beach," dated July 5, 2011, personal communication from the Point Reyes Seashore Ranchers Association (PRSRA).

104. Personal communication from the Point Reyes Seashore Ranchers Association.

105. Email to author from Dale McCullough, July 15, 2015; email from McCrea Cobb, January 8, 2016. Copy on file with author.

106. Letter from Superintendent Cicely Muldoon to Tess Elliott, editor, *Point Reyes Light*, regarding FOIA request for collar data on tule elk, dated September 17, 2014. Copy on file with author.

107. NPS, *Point Reyes National Seashore 2001 Year in Review*, 11, emphasis added. Had the NPS not intended to maintain separation between the elk herd and the ranches, the 2001 report would not have mentioned its efforts to ensure conflict avoidance.

108. Population numbers from the winter count from Peter Fimrite, "Point Reyes' Tule Elk Dilemma," *San Francisco Chronicle*, May 19, 2016.

109. Cobb, "Spatial Ecology," 150–51.

110. While the ranchers lease the land and do not own the buildings or fences, they are responsible for what is called "cyclical maintenance," or day-to-day maintenance and repair. Constantly needing to repair broken fences or irrigation pipes is not only a monetary cost, but also creates opportunity costs in the time stolen from other tasks that need to be done.

111. Letter from Spaletta family to PRNS Superintendent Cicely Muldoon, dated October 28, 2010. Copy on file with author.

112. According to the California Certified Organic Farmers (CCOF), based on national organic regulations, cows must remain on pasture for a minimum of 120 days per year, and receive 30 percent of their dry matter intake from pasture. At the Spalettas' ranch, their organic certifier from the CCOF has noted conflicts with elk in their pastures for two years in a row. Regulations available online at www.ccof.org/certification/standards/pasture-rule, accessed June 9, 2016.

113. Tim Bernot, "Free Range Elk Observations 9/24/10—3/1/11," NPS, PORE; informal report, copy on file with author. These notes actually run through May 21, 2011, and also document elk from the Limantour herd at the Home Ranch in several different pastures. A few months later PRNS Chief of Interpretation John Dell'Osso erroneously stated that the elk typically moved onto ranch lands only three months out of the year, in the fall; Mark Prado, "Rebounding Elk," *Marin Independent Journal*, September 5, 2011.

114. The family had an existing ranch plan, dating back to 1998, but were not contacted about this new ranch unit plan. They received a copy on August 12, 2011, and it was withdrawn by PRNS staff a year later on August 3, 2012 without being finalized.

115. Letter from PRSRA to Superintendent Cicely Muldoon, dated June 17, 2011. Copy on file with author.

116. Letter from Superintendent Cicely Muldoon to PRSRA, dated July 7, 2011. Copy on file with author.

117. Transcript of July 11, 2011 PRSRA meeting; copy on file with author. The experimental lowered fencing had already been put in place along the Drakes Beach road the week prior, without informing any of the ranchers; while elk were less likely to damage it, the ranchers pointed out that cattle might also be able to jump over the lowered fences, allowing herds to mix or for cattle to get out onto the main road, where they could possibly cause collisions with tourist vehicles. Ranchers would be liable for any damage stemming from such collisions. The lowered section covers just one small segment in miles of fencing, and it is not clear whether the elk preferentially use it.

118. Jeremy Blackman, "Ranchers Foresee Elk Crisis, Fear Park Inaction," *Point Reyes Light*, July 21, 2011. The article quotes John Dell'Osso: "Most of

the anticipation seemed to be that they would travel kind of south and maybe east. Of course, now we know that some split off and moved north . . . . As of now, we know for sure that elk have been spotted going through Home and A, B, and C ranches."

119. From the July 11, 2011 PRSRA meeting transcript: "This whole monitoring program was supposedly set up to watch this and react, much like we had an elk at L, M, and H ranches that was moved back twice. Don's policy at the time was to remove elk if they got into the Pastoral Zone. He was very vocal about that . . . that animal was removed permanently." Copy on file with author.

120. Blackman, "Ranchers Foresee Elk Crisis." Another rancher is quoted as saying, "What we are really worried about now is that [park service officials] are saying they can't move [the elk] off because that would require a new public process . . . . To manage a public failure? At what point does this become just completely ridiculous?"

121. Letter from PRSRA to Superintendent Cicely Muldoon, dated September 27, 2011. Copy on file with author.

122. Letter from PRSRA to Senator Dianne Feinstein, dated November 9, 2011. Copy on file with author.

123. Letter from Marin County Supervisor Steve Kinsey to Senator Dianne Feinstein, dated March 9, 2012. Copy on file with author.

124. Letter from Senator Dianne Feinstein to Secretary of Interior Ken Salazar, dated March 20, 2012. Copy on file with author.

125. Letter from Secretary of Interior Ken Salazar to Senator Dianne Feinstein, dated May 18, 2012. Copy on file with author.

126. Email from PRNS Superintendent Cicely Muldoon, dated July 14, 2015, copy on file with author; also personal communication from the Point Reyes Seashore Ranchers Association.

127. Ann Miller, "Elk Putting National Seashore Ranches at Risk: Part Two," *West Marin Citizen,* November 28, 2013; later in the same article, Press reiterated, "Again, no one anticipated the elk would migrate out into the ranches."

128. McCullough stated that he had "some pretty intense conversations" with former Superintendent Don Neubacher about his vision for free-ranging elk throughout the Seashore, and that Neubacher "certainly got on board." He also clarified that most of the Point Reyes peninsula south of the Limantour Road, in the designated wilderness area, is actually quite poor tule elk habitat, as it is composed mostly of forest and brushlands, whereas tule elk are an open land species; "I knew eventually they would spread out onto the ranches and cause conflict." Author interview with Dale McCullough, Kensington, CA, conducted July 13, 2015.

129. Author interview with Judd Howell, Point Reyes Station, CA, conducted July 2, 2015.

130. Letter from Point Reyes Seashore Ranchers Association to PRNS Superintendent Cicely Muldoon, dated September 19, 2013; copy on file with author. In the interest of full disclosure, I assisted in the editing of this document, on a strictly volunteer basis.

131. Memo from Secretary of Interior Kenneth Salazar to director, National Park Service, regarding Point Reyes National Seashore—Drakes Bay Oyster Company, dated November 29, 2012, page 2. Copy on file with author.

132. Roderick Nash, *Wilderness and the American Mind* 3rd ed. (1967; New Haven, CT: Yale University Press, 1982), 1.

133. Alagona, *After the Grizzly*, 40.

134. Peter Fimrite, "Conservationists Upset as Much of Point Reyes Elk Herd Dies," *San Francisco Chronicle*, April 19, 2015.

135. See direct reference to this possibility in Phillips, *Conservation of the California Tule Elk;* and in Howell, Brooks, Semenoff-Irving, and Greene, "Population Dynamics."

136. Author interview with Dale McCullough, Kensington, CA, July 13, 2015.

137. In fact, over the same two-year time period that the Tomales Point herd declined by 47 percent, the "free ranging" herds located at least part-time in the pastoral zone *increased* by 32 percent; Fimrite, "Conservationists Upset."

138. Email to author from Dale McCullough, July 15, 2015. He also added, "If ranchers are improving the range, that will be even more attractive to the elk than the natural vegetation, especially in the dry season."

139. California Department of Fish and Wildlife, Wildlife Branch, Game Management, *2014 Elk Hunt Statistics*, online at https://www.wildlife.ca.gov /Hunting/Elk#19519270-harvest-statistics, accessed May 18, 2015.

140. See R. Scott Anderson, "Contrasting Vegetation and Fire Histories on the Point Reyes Peninsula during the Pre-Settlement and Settlement Periods: Fifteen Thousand Years of Change" (Flagstaff, AZ: Center for Environmental Sciences and Education, Quaternary Sciences Program, 2005); comparing historic photographs of the southern ranches published in D. S. Livingston, *Ranching on the Point Reyes Peninsula: A History of the Dairy and Beef Ranches within Point Reyes National Seashore, 1834–1992* (Point Reyes Station, CA: Point Reyes National Seashore, National Park Service, 1993) with current vegetation maps documents these landscape changes.

141. Barbara Moritsch, PRNS plant ecologist, presentation on vegetation management, October 26, 1999.

142. Marin Carbon Project, at www.marincarbonproject.org.

CHAPTER 7. THE POLITICS OF PRESERVATION

1. See Matthew Morse Booker, *Down by the Bay: San Francisco's History between the Tides* (Berkeley: University of California Press, 2013), particularly chapter 4, "An Edible Bay."

2. Elinore Barrett, *The California Oyster Industry*, California Department of Fish and Game, Fish Bulletin 123 (1963), available online at http://content .cdlib.org/ark:/13030/kt629004n3/, accessed August 20, 2015.

3. Paul Sadin, *Managing a Land in Motion: An Administrative History of Point Reyes National Seashore* (Washington, DC: National Park Service, U.S. Department of Interior, 2007), 128.

4. *Hearing on S. 2428, to Establish the Point Reyes National Seashore in the State of California and for Other Purposes,* 86th Congress, 2nd Session (April 16, 1960), at 14.

5. National Park Service, *Land Use Survey and Economic Feasibility Report: Proposed Point Reyes National Seashore* (San Francisco: Region Four Office, National Park Service, 1961), 16.

6. CA General Assembly Joint Resolution No. 27, March 1961, included in *Hearings on S. 476, to Establish the Point Reyes National Seashore in the State of California, and for Other Purposes,* 87th Congress, 1st Session (March 28, 30, and 31, 1961) (testimony of Rep. Clem Miller), at 240–41.

7. Sadin, *Managing a Land in Motion,* 89, citing a letter from Assistant Secretary of Interior John A. Carver Jr. to Robert L. Condon, attorney, Martinez, CA, March 26, 1962.

8. Point Reyes National Seashore, *Statement for Management, June 1993 Revision* (National Park Service, 1993), 18.

9. PRNS, *Statement for Management,* 35, 42.

10. NPS, *Drakes Estero: A Sheltered Wilderness Estuary.* This report was revised and reissued a number of times: September 2006, April 1, 2007, May 8, 2007, May 11, 2007, and a version not released publicly dated July 27, 2007. The Department of Interior Inspector General found in 2008 that the report had been written specifically to counter conclusions drawn in Peter Jamison, "Park's Oyster Studies Yield Unexpected Results," *Point Reyes Light,* May 18, 2006. Earl E. Devaney, *Report of Investigation: Point Reyes National Seashore* (Washington, DC: Department of the Interior, Office of Inspector General, 2008), 7.

11. National Research Council, *Shellfish Mariculture in Drakes Estero, Point Reyes National Seashore, California* (Washington, DC: National Academies Press, 2009), 6, 72–73.

12. National Research Council, *Shellfish Mariculture,* at 72–73.

13. Sarah Allen, Jules Evens, and John Kelly, "The Naturalist," *Point Reyes Light,* April 26, 2007.

14. Moreover, Allen described a boat path, water depth, and eelgrass coverage right next to the main seal haul-out location that are inconsistent with DBOC GPS records and NPS aerial photographs.

15. See detail in Marine Mammal Commission, *Mariculture and Harbor Seals in Drakes Estero, California* (Bethesda, MD: Marine Mammal Commission, November 2011), 20–21.

16. PRNS later made these documents available on their website at https://www.nps.gov/pore/learn/management/planning_reading_room_upper_drakes_seal_oyster.htm.

17. Goodman found the reference on June 6, 2010, in an NPS briefing statement dated May 1, 2009. Personal communication.

18. Point Reyes National Seashore, *Draft Environmental Impact Statement, Drakes Bay Oyster Company Special User Permit* (Point Reyes Station, CA: National Park Service, September 2011), 181.

19. Gavin M. Frost, *Public Report on Allegations of Scientific Misconduct at Point Reyes National Seashore, California,* addressed to Will Shafroth, Acting

Assistant Secretary for Fish and Wildlife and Parks (Washington, DC: U.S. Department of Interior, Office of the Solicitor, March 22, 2011), 35. Copy on file with author.

20. Frost, *Public Report on Allegations*, 1.

21. Author interview with Kevin and Nancy Lunny, Olema, CA, July 19, 2010.

22. Letter from Senator Dianne Feinstein to Secretary of Interior Ken Salazar, March 23, 2011. Copy on file with author.

23. Letter from Ralph Mihan, field solicitor, U.S. Department of Interior Office of the Solicitor, San Francisco Field Office, to superintendent, Point Reyes National Seashore, dated February 26, 2004. PRNS Archives.

24. Environment, and Related Agencies Appropriations Act of 2010, Public Law 111–88, Department of the Interior, Section 124 (October 30, 2009).

25. Some ranch permits are marked as having "categorical exclusion" from NEPA review, others are marked as having been included in other approved plans.

26. In describing the fourth option as an alternative "considered but rejected," the NPS stated, "The GMP (NPS 1980) calls for the continuation of an oyster operation within the park. PRNS is currently in the process of updating and revising the existing GMP which will need to address the issue of JOC lease hold interest." Point Reyes National Seashore, *Environmental Assessment/Initial Study Joint Document, Johnson Oyster Company, Marin County* (National Park Service, May 1998), 7.

27. Point Reyes National Seashore, *Finding of No Significant Impact (FONSI), Johnson Oyster Company Replacement and Rehabilitation of Facilities, Point Reyes National Seashore*, August 11, 1998. The original proposal was for the NPS to split the cost of the expansion with Johnson's Oyster Co.; Superintendent Neubacher wrote a letter of support to the Bank of Oakland in November 1996, stating that the NPS "endorses the concept to bring the facility up to modern standards and to improve the general appearance of the site as the oyster company operates within a national park area." Letter from Superintendent Neubacher to the Bank of Oakland, November 22, 1996. Copy on file with author.

28. Marian Schinske, "Johnson Oyster Company Plans Major Improvements," *Point Reyes Light*, May 21, 1998. The proposed new processing facility was designed by Chuck Desler, the same architect who planned the Bear Valley Visitor Center and Clem Miller Environmental Education Center for PRNS.

29. Letter from Superintendent Neubacher to Johnson Oyster Company, June 11, 1999. Copy on file with author.

30. "Reservation of Use and Occupancy for Tract 02–106," signed by NPS Chief of Lands Division, Western Regional Office, on October 16, 1972. Exhibit C, Item 11, states, "Upon expiration of the reserved term, a special use permit may be issued for the continued occupancy of the property for the herein described purposes, provided however, that such permit will run concurrently with and will terminate upon the expiration of State water bottom allotments assigned to the Vendor." The CA Fish and Game Commission renewed the

oyster farm's water bottom leases in 2004, with an expiration date in 2029. Copy on file with author.

31. Author interview with Kevin and Nancy Lunny, Olema, CA, July 19, 2010.

32. Kevin and Nancy Lunny, "Keep on Shucking," *Point Reyes Light,* July 10, 2014.

33. Lunny and Lunny, "Keep on Shucking."

34. Interview with Kevin Lunny, conducted by Jake DeGrazia, February 2, 2013; email to author from Ken Fox dated March 2, 2016. Copies on file with author.

35. Letter from Superintendent Neubacher to Drakes Bay Oyster Company, March 28, 2005. Copy on file with author.

36. In the executive summary, table 2–5 presented a summary of the alternatives, listing eighteen items or issues that each alternative addresses; for twelve of those items (67 percent), the conditions for Alternatives C and/or D simply stated "Same as Alternative B." Point Reyes National Seashore, *Draft Environmental Impact Statement, Drakes Bay Oyster Company Special User Permit* (National Park Service, U.S. Department of Interior, 2011), xxxv–xxxviii.

37. Council of Environmental Quality, *A Citizen's Guide to the NEPA: Having Your Voice Heard* (Washington, DC: Council of Environmental Quality, 2007), 17.

38. Jeremy Blackman, "Oyster Farm Damned by Draft EIS," *Point Reyes Light,* September 29, 2011; emphasis is mine.

39. PRNS, *Draft Environmental Impact Statement,* 294–303.

40. PRNS, *Draft Environmental Impact Statement,* 316–17, 321–22.

41. PRNS, *Draft Environmental Impact Statement,* xv.

42. Corey Goodman, "Simple Numbers with Profound Implications," *Point Reyes Light,* May 2, 2013.

43. Letter from Senator Dianne Feinstein to Secretary of Interior Ken Salazar, March 29, 2012. Copy on file with author.

44. Jeremy Blackman, "Park Misused Sound Data in Oyster EIS," *Point Reyes Light,* March 29, 2012.

45. PRNS, *Draft Environmental Impact Statement,* xvi–xvii.

46. Comment letter from Ellie Rilla, community development advisor, and Lisa Bush, agricultural ombudsman, Cooperative Extension of Marin County, to Superintendent Cicely Muldoon, dated December 7, 2011; emphasis is mine. Copy on file with author.

47. PRNS, *Draft Environmental Impact Statement,* 213.

48. Comment letter from Point Reyes Outdoors, Sea Trek Kayaking Center, and Blue Waters Kayaking, representing 85 percent of the kayaking tours conducted on Drakes Estero, December 13, 2011. Copy on file with author.

49. PRNS, *Draft Environmental Impact Statement,* xviii–xix.

50. National Research Council, *Scientific Review of the Draft Environmental Impact Statement: Drakes Bay Oyster Company Special Use Permit* (Washington, DC: National Academies Press, 2012).

51. Brent S. Stewart, *Evaluation of Time-Lapse Photographic Series of Harbor Seals Hauled Out in Drakes Estero, California, for Detecting and Assessing Disturbance Events,* Hubbs SeaWorld Research Institute, Technical Report

2012–378, submitted to NPS Pacific West Regional Office, May 12, 2012. Stewart, a senior research scientist, did find one instance of seals disturbed by a kayaker, despite the closure of Drakes Estero to kayakers during the seals' pupping season.

52. William A. Lellis, Carrie J. Blakeslee, Laurie K. Allen, Bruce F. Molnia, Susan D. Price, Sky Bristol, and Brent Stewart, *Assessment of Photographs from Wildlife Monitoring Cameras in Drakes Estero, Point Reyes National Seashore, California*, U.S. Geological Survey Open File Report 2012–1249; Point Reyes National Seashore, *Final Environmental Impact Statement, Drakes Bay Oyster Company Special User Permit* (National Park Service, November 2012), 298–99.

53. Michael Ames, "The Oyster Shell Game," *Newsweek*, January 18, 2015.

54. Letter from Secretary of Interior Ken Salazar to the director of the National Park Service, dated November 29, 2012, 4; copy on file with author. The final EIS also included this assertion, stating: "Although the Secretary's authority under Section 124 is 'notwithstanding any other provisions of law,' the Department has determined that it is helpful to generally follow the procedures of NEPA. The EIS provides decision-makers with sufficient information on potential environmental impacts, within the context of law and policy, to make an informed decision on whether or not to issue a new SUP. In addition, the EIS process provides the public with an opportunity to provide input to decision-makers on the topics covered by this document." It then stated, a few pages later, "The NEPA process will be used to inform the decision of whether a new [special use permit] should be issued to DBOC for a period of ten years." PRNS, *Final Environmental Impact Statement*, 2, 5.

55. Letter from Secretary of Interior Ken Salazar to the director of the National Park Service, dated November 29, 2012, 5; copy on file with author.

56. Phyllis Faber, Laura Watt, and Peter Prows, "The Plan to Get Rid of the Ranchers," *Point Reyes Light*, January 9, 2014; email from Phyllis Faber, March 3, 2014, copy on file with author. In the latter Faber added, "When he told me this I was quite overwhelmed."

57. Tim Setnicka, former superintendent of Channel Island National Park, wrote a three-part article about this effort to shut down the Vail family ranch, published in the *Santa Barbara News-Press*, October 8, 15, 22, 2006. Setnicka wrote that he regretted his involvement in what he had come to consider unethical behavior on the part of the NPS, and mailed copies of his articles to the Lunnys in 2008. He also gave a public talk about his experience in Point Reyes Station in October 2014, telling the audience that in his view, the NPS "has no soul." Samantha Kimmey, "Retired Channel Islands Superintendent Delivers a Cautionary Tale," *Point Reyes Light*, October 30, 2014.

58. Federal Register, Volume 62, No. 198, page 53336 (October 14, 1997).

59. Copies of all the letters received to date were given to CAC members at their January 29, 2000 meeting. Copy on file with author.

60. The full text reads: "Dear Superintendent Neubacher: On a recent tour of Point Reyes National Seashore, we were dismayed to see all the cattle operations on what we discovered were National Seashore Lands. Upon further inquiry, we discovered that 17 families and one corporation control 26,000

acres within the seashore that had been purchased by the National Park Service more than 25 years ago. It is our understanding that many of these leases are coming due over the next few years. *We urge you to not renew any leases as they come due.* The needs of the people who love Point Reyes for what it offers in the form of hiking, camping, interpretive centers, wildlife viewing, research, must come ahead of commercial ranching operations which offers nothing to anyone except for those who maintain these operations." Emphasis in original. At least three were received before the NOI was published. Some letter writers modified the text to include their own message, but the basic wording and sentiment were the same; those that were modified are all in a different font than the original.

61. Letter from Bruce Keegan, secretary, Committee for the Preservation of the Tule Elk, to Superintendent Don Neubacher, undated but stamped received October 14, 1999. An earlier comment letter from Sara Vickerman, director, West Coast Office, Defenders of Wildlife, dated April 3, 1998, mentioned that Bruce Keegan had called to inform her office about ranching permits coming due soon, and similarly urged that the leases not be renewed. Copy on file with author.

62. Letter from Helen Wagenvoord, associate director, Pacific Regional Office, NPCA, to Superintendent Don Neubacher, November 30, 1999. Copy on file with author.

63. It should be noted that the counting of the comment letters is sloppy; several in the packet given to the CAC in January 2000 are duplicates, and a number are not official comment letters but requests to be added to the mailing list, etc.

64. Point Reyes National Seashore, "General Management Plan Newsletter," undated. Copy on file with author.

65. Point Reyes National Seashore, "General Management Plan Update, Concepts Newsletter 2003." Copy on file with author.

66. Judy Teichman, "Guest Opinion: A Park Proposal," *Point Reyes Light,* February 12, 2004; in the interest of full disclosure, I contributed to the drafting of concept 6, on a strictly volunteer basis.

67. "Staff Report Summary of Comments from *Concepts Newsletter* Review, Point Reyes National Seashore General Management Plan Update," undated. Copy on file with author.

68. Point Reyes National Seashore, "General Management Plan Update Newsletter," Summer 2008. Copy on file with author.

69. "General Management Plans," Point Reyes National Seashore website, www.nps.gov/pore/parkmgmt/planning_gmp.htm, accessed February 29, 2016.

70. National Park Service, *NPS Management Policies* (Washington, DC: National Park Service, 2006), Section 2.3.1.12, "Periodic Review of General Management Plans."

71. Author's phone interview with Gordon White, PRNS chief of integrated resource management, July 2, 2015.

72. Letter from Gordon Bennett, vice-chair for the Sierra Club Marin Group, to Superintendent Don Neubacher, February 20, 2004. Copy on file with author.

73. Letter from Ralph G. Mihan, field solicitor, U.S. Department of Interior, to Superintendent Don Neubacher, February 26, 2004. Copy on file with author.

74. The Solicitor's letter incorrectly identifies the September 8, 1976, letter from John Kyl, Assistant Secretary of the Interior, to the congressional committee, arguing that the State of California's retention of mineral and fishing rights rendered the tidal areas inconsistent with wilderness, as "the only record in the legislative history that raises this point in the area's wilderness and potential wilderness designation."

75. Amicus brief by Sarah Rolph, October 28, 2013, 4–5; copy on file with author. Her brief contains a computer analysis of the public comments by date sent (6), showing enormous spikes just following the solicitation emails sent by each of the four environmental groups.

76. Jeremy Blackman, "NGO's Gush Letters into Seashore," *Point Reyes Light,* March 8, 2012.

77. PRNS, *Final Environmental Impact Statement,* Appendix F, "Comments and Responses on the Draft Environmental Impact Statement," F-16. For comment letters to count as substantive under NEPA, they must address anticipated environmental effects, not just the action the government is taking. William Murray Tabb, "The Role of Controversy in NEPA: Reconciling Public Veto with Public Participation in Environmental Decisionmaking," *William and Mary Environmental Law and Policy Review* 21, no. 1 (1997): 175–231, 194.

78. Letter from Senator Dianne Feinstein to Secretary of Interior Ken Salazar, March 29, 2012. Copy on file with author.

79. Letter from California Assemblyman William T. Bagley, Congressman John L. Burton, and Congressman Paul N. "Pete" McCloskey Jr., to Interior Secretary Ken Salazar, August 11, 2011. Copy on file with author.

80. Rob Rogers, "Former Legislators Back Point Reyes Oyster Company's Claims," *Marin Independent Journal,* August 13, 2011.

81. Rogers, "Former Legislators."

82. Letter from Secretary of Interior Ken Salazar to the director of the National Park Service, dated November 29, 2012, 6. Copy on file with author.

83. Letter from Ralph G. Mihan, field solicitor, U.S. Department of Interior, to Superintendent Don Neubacher, February 26, 2004; copy on file with author. Mihan incorrectly implied that the Johnson Oyster Company tract was within potential wilderness.

84. DBOC v. Salazar, Complaint for Declaratory and Injunctive Relief (U.S. District Court for the Northern District of California, December 3, 2012). Copy on file with author.

85. DBOC v. Salazar, Notice of Motion and Motion for Preliminary Injunction (December 21, 2012). Copy on file with author.

86. Author's phone interview with Peter Prows, lead attorney from Briscoe Ivester & Bazel LLP, September 21, 2015; also DBOC v. Salazar, Notice of Motion and Motion for Preliminary Injunction (December 21, 2012), 12–13. Copy on file with author.

87. Federal Register, Vol. 77, No. 233 (December 4, 2012), at 71826.

88. Richard Halstead, "Head of Nonprofit Providing Legal Aid to Drakes Bay Oyster Co. Worked for Koch Brothers," *Marin Independent Journal,*

December 8, 2012; Evan Halper, "Koch-Backed Group with Ties to Liberal Causes? Critics Call It a Charade," *Los Angeles Times*, February 7, 2015.

89. Helen Grieco, "Stealing the Heart of a National Park: Oyster Company's Alliance with Big Oil," *Huffington Post*, April 24, 2013: www.huffingtonpost .com/helen-grieco/stealing-the-heart-of-a-n_b_3112347.html.

90. Helen Grieco, "Drakes Bay Oyster Company's PR Face Lift Won't Fix Its Koch Brothers Blemishes," *Huffington Post*, June 5, 2013, www.huffingtonpost .com/helen-grieco/drakes-bay-oyster-company_b_3387269.html.

91. Letter from Mary Beth Hutchins, communications director, Cause of Action, to Michael Getler, ombudsman, Public Broadcasting Service, May 17, 2013. Copy on file with author.

92. Letter from Dr. Corey Goodman, Nancy Lunny, and Kevin Lunny, to Michael Getler, ombudsman, Public Broadcasting Service, May 26, 2013. Copy on file with author.

93. Amicus brief of Alice Waters, the Hayes Street Grill, Tomales Bay Oyster Company, Marin County Agricultural Commissioner Stacy Carlsen, California Farm Bureau Federation, Marin County Farm Bureau, Sonoma County Farm Bureau, Food Democracy Now, Marin Organic, and Alliance for Local Sustainable Agriculture, dated March 13, 2013. Copy on file with author.

94. U.S. Court of Appeals for the Ninth Circuit, Opinion No. 13–15227, (September 3, 2013), 2, 23. Copy on file with author.

95. U.S. Court of Appeals for the Ninth Circuit, Opinion No. 13–15227, (September 3, 2013), 37. Copy on file with author.

96. U.S. Court of Appeals for the Ninth Circuit, Opinion No. 13–15227, (September 3, 2013), 47–48. Copy on file with author.

97. Amy Trainer, "Don't Be Fooled by Drakes Bay Oyster Company's Supreme Court Petition," *East Bay Express*, April 14, 2014. Amy Trainer, "Close to Home: Welcome West Coast's Only Marine Wilderness," *Santa Rosa Press Democrat*, December 29, 2014.

98. Michael Ames, "The Oyster Shell Game," *Newsweek*, January 18, 2015.

99. Michael Ames, "The Oyster Shell Game," *Newsweek*, January 18, 2015.

100. Letter from Frederick Smith, executive director, Environmental Action Committee of West Marin, published in the *Point Reyes Light*, January 21, 2009.

101. National Research Council, *Scientific Review of the Draft Environmental Impact Statement*, 79.

102. William Cronon, "The Trouble with Wilderness: A Response," *Environmental History* 1, no. 1 (1996): 47–55, 55.

103. Letter from Michael Pollan to Senator Dianne Feinstein, October 5, 2012. Copy on file with author.

104. Guy Kovner, "Cleanup Transforms Drakes Estero," *Santa Rosa Press Democrat*, March 23, 2015.

CONCLUSION

1. For more on the explosion of the sustainable food movement in the Bay Area, see Sally K. Fairfax, Louise Nelson Dyble, Greig Tor Guthey, Lauren ·

Gwin, Monica Moore, and Jennifer Sokolove, *California Cuisine and Just Food* (Cambridge, MA: MIT Press, 2012).

2. Ellie Rilla and Lisa Bush, *The Changing Role of Agriculture in Point Reyes National Seashore* (Marin County: University of California Cooperative Extension, 2009).

3. *Marin Countywide Plan*, adopted November 6, 2007.

4. Nora Mitchell, Mechtild Rössler, and Pierre-Marie Tricaud, *World Heritage Cultural Landscapes: A Handbook for Conservation and Management* (Paris: UNESCO, 2009), 35–36.

5. www.nps.gov/history/cultural_landscapes/, accessed June 9, 2016.

6. See, for instance, Nathan Sayre, *Working Wilderness: The Malpai Borderlands Group and the Future of the Western Range* (Tucson, AZ: Rio Nuevo Publishers, 2005).

7. Adam Broderick, "Department of Interior Says Greater Sage Grouse Listing Not Warranted," *Crested Butte News*, September 24, 2015.

8. Susan Charnley, Thomas E. Sheridan, and Gary P. Nabhan, eds., *Stitching the West Back Together: Conservation of Working Landscapes* (Chicago: University of Chicago Press, 2014).

9. David Lowenthal, "Is Wilderness 'Paradise Enow'? Images of Nature in America," *Chicago University Forum* 7, no. 2 (1964): 40.

10. Christopher Conte, "Creating Wild Places from Domesticated Landscapes: The Internationalization of the American Wilderness Concept," in *American Wilderness: A New History,* ed. Michael Lewis (New York: Oxford University Press, 2007), 225.

11. Mark Dowie, *Conservation Refuges: The Hundred-Year Conflict between Global Conservation and Native Peoples* (Cambridge, MA: MIT Press, 2009).

12. Gary P. Nabham, "Cultural Parallax in Viewing North American Habitats," in *Reinventing Nature? Responses to Postmodern Deconstruction,* eds. Michael Soulé and Gary Lease (Washington DC: Island Press, 1995), 87–101.

13. Dolly Jørgensen, "Rethinking Rewilding," *Geoforum* 65 (2015): 482–88, 487.

14. Peter Alagona, *After the Grizzly: Endangered Species and the Politics of Place in California* (Berkeley: University of California Press, 2013), 195–96.

15. See Urban Coyote Research, at http://urbancoyoteresearch.com, for examples; also see Gavin Van Horn and Dave Aftandilian, eds., *City Creatures: Animal Encounters in the Chicago Wilderness* (Chicago: University of Chicago Press, 2015).

16. Aldo Leopold, "Wilderness as a Form of Land Use" (1925), in *The River of the Mother of God, and Other Essays by Aldo Leopold,* eds. Susan L. Flader and J. Baird Callicott (Madison: University of Wisconsin Press, 1992), 135–36.

17. Rolf Diamant, Jeffrey Roberts, Jacquelyn Tuxill, Nora Mitchell, and Daniel Laven, *Stewardship Begins with People: An Atlas of Places, People, and Handmade Products,* Conservation and Stewardship Publication No. 14 (Woodstock, VT: Conservation Study Institute, National Park Service, 2007), 45.

18. Countryside Conservancy, www.cvcountryside.org/farm-farming-home.htm, accessed February 29, 2016. Also see "The Countryside Initiative," https://

countrysideconservancy.worldsecuresystems.com/countryside-initiative-program for more information on the program.

19. Cuyahoga Valley National Park, *Countryside Initiative Request for Proposals* (National Park Service, 2015), 11.

20. CVNP, *Countryside Initiative Request for Proposals*, 18; emphasis is mine.

21. Letter from Senator Dianne Feinstein to PRNS and PRSRA, dated January 6, 2009. Copy on file with author.

22. Rilla and Bush, *Changing Role of Agriculture*, 15–19.

23. Rob Rogers, "Former Legislators Back Point Reyes Oyster Company's Claims," *Marin Independent Journal*, August 13, 2011.

24. Frederick Law Olmsted, "The Yosemite Valley and the Mariposa Big Trees: A Preliminary Report" (1865), reprinted in *Landscape Architecture* 43 (1952): 12–25, 14.

25. Alfred Runte, *Yosemite: The Embattled Wilderness* (Lincoln: University of Nebraska Press, 1990), 25; the case is *Hutchings v. Low* (1872). Note that neither claim was technically legal, as the area wasn't yet formally surveyed by the General Land Office at the time that they were filed—yet the 1868 State District Court still decided the case in their favor, only to be overturned by the State Supreme Court and then upheld by the U.S. Supreme Court. In an ironic footnote, in 1874 the California legislature voted to compensate Hutchings, Lamon, and several other claimants for their improvements in the park, and allocated sixty thousand dollars to do so. For greater detail, see Jen A. Huntley, *The Making of Yosemite: James Mason Hutchings and the Origin of America's Most Popular National Park* (Lawrence: University Press of Kansas, 2011).

26. Geographer Kenneth Olwig sees the 1868 removal of an orchard and farm in the Yosemite Valley as setting precedent for ouster of farms in Shenandoah, forming a precedent for the removal of a rural cultural landscape in favor of an (artificial) "natural" landscape. Kenneth R. Olwig, "Reinventing Common Nature: Yosemite and Mount Rushmore: A Meandering Tale of a Double Nature," in *Uncommon Ground: Toward Reinventing Nature*, ed. William Cronon (New York: Norton, 1995), 395.

27. Conte, "Creating Wild Places," 233.

28. For instance, see quote from EAC Executive Director Amy Trainer saying "A deal's a deal. Drakes Estero is supposed to become wilderness," in Robert Gammon, "Dianne Feinstein's War," *East Bay Express*, June 13, 2012.

29. Amy Leinbach Marquis, "A Raw Deal: Marine Wilderness Is at Stake in the Ecological Heart of Point Reyes National Seashore," *National Parks* 86, no. 2 (2012): 1–2.

30. Center for Biological Diversity press release, April 16, 2015, available at www.biologicaldiversity.org/news/press_releases/2015/tule-elk-04-16-2015.html, accessed June 9, 2016.

31. Memo from Secretary of Interior Kenneth Salazar to director, National Park Service, regarding Point Reyes National Seashore / Drakes Bay Oyster Company, dated November 29, 2012, 2. Copy on file with author.

32. Letter to NPS regional director, Pacific West Region, from NPS Director Jonathan Jarvis, dated January 31, 2013, "Delegation of Authority for Point

Reyes National Seashore Agricultural Leases and Directions to Implement the Secretary's Memorandum of November 29, 2012": "This delegation authorizes the issuance of lease/permits for the purpose of grazing cattle and operating beef and dairy ranches, along with associated residential uses by lessees and their immediate families and their employees' immediate families, within the pastoral zone of Point Reyes National Seashore and the northern District of Golden Gate National Recreation area administered by Point Reyes National Seashore." Copy on file with author.

33. Samantha Kimmey, "Seashore Will Write New Dairy and Ranch Plan," *Point Reyes Light,* December 6, 2013.

34. Tiscornia Agricultural Use Lease, Lease No. 8530-4-L001, Point Reyes National Seashore, signed December 21, 1993, item 26, pp. 5, emphasis added; PRNS archives.

35. Grossi Historic "M" Ranch Beef Cattle Agricultural Use Lease/Permit, Lease No. AGR 8530–1000–3020, Point Reyes National Seashore, dated October 7, 2003, item 5.3, pp. 6; PRNS archives. The same language appears in the 2003 McClure and Kehoe leases, and in the 2010 Rogers Ranch lease.

36. For instance, special use permit for Jo-Ann Stewart, Permit #AGRI-8530–2600–06–6003, signed January 2007, item 26; emphasis added. PRNS Archives.

37. See www.coydavidson.com/leasing-tips/negotiating-the-renewal-option/, accessed June 9, 2016.

38. Megan Foster, "Resident Inclusion in Protected Land Use Decisions: The Role of a Citizens Advisory Commission at Point Reyes National Seashore," masters thesis, University of California, Davis, December 2015.

39. Leigh Raymond, "Localism in Environmental Policy: New Insights from an Old Case," *Policy Sciences* 35, no. 2 (2002): 179–201.

40. William Tweed, *Uncertain Path: A Search for the Future of National Parks* (Berkeley: University of California Press, 2010), 157.

41. Tweed, *Uncertain Path,* 206.

42. See Emma Marris, *Rambunctious Garden: Saving Nature in a Post-Wild World* (New York: Bloomsbury, 2011), particularly chapter 2.

43. Ben A. Minteer and Stephen J. Pyne, "Writing on Stone, Writing in the Wind," in *After Preservation: Saving American Nature in the Age of Humans,* eds. Ben A. Minteer and Stephen J. Pyne (Chicago: University of Chicago Press, 2015), 6.

EPILOGUE

1. Center for Biological Diversity press release, "Conservation Groups Sue over Missing Point Reyes Management Plan," February 10, 2016.

2. Jeremy P. Jacobs, "National Parks: Ranchers in Bulls-Eye of Legal Brawl at Calif. Seashore," *Greenwire,* March 14, 2016, www.eenews.net/stories /1060033938.

3. Paul Rogers, "Point Reyes: Lawsuit Challenges Historic Ranching Operations at Iconic Park," *San Jose Mercury News,* February 11, 2016.

4. Phyllis Faber, Laura A. Watt, and Peter Prows, "The Plan to Get Rid of the Ranchers," *Point Reyes Light,* January 9, 2014.

5. Tony Davis, "'Firebrand Ways': A Visit with One of the Founders of the Center for Biological Diversity," *High Country News,* December 28, 2009; in the interview, CBD Executive Director Kieran Suckling describes the group's legal approach specifically as an exertion of power over federal land managers: "They feel like their careers are being mocked and destroyed—and they are. So they become much more willing to play by our rules and at least get something done. Psychological warfare is a very underappreciated aspect of environmental campaigning." He also asserts that "the core talent of a successful environmental activist is not science and law. It's campaigning instinct."

6. Gordon Bennett, Bridger Mitchell, and Amy Meyer, "More Sustainable Ranching a Worthy Goal for Point Reyes," *Marin Independent Journal,* January 30, 2016, www.marinij.com/opinion/20160130/marin-voice-more-sustainable-ranching-a-worthy-goal-for-point-reyes.

7. Gordon Bennett, "The Renaissance of Seashore Ranching: The Salazar Playbook," *Point Reyes Light,* March 17, 2016. He also suggests the ranching plan should include "the development of wildlife-friendly ranching practices, an appropriate sharing of grassland forage between wildlife and cattle, and a commitment to allowing the public to access their lands."

8. "Some environmentalists go so far as to compare Lunny and the Point Reyes ranchers to Cliven Bundy, who faces 16 felony charges for his role in a 2014 armed confrontation with federal land managers in Nevada, and two of Bundy's sons and others involved in the recent armed standoff at a federal wildlife refuge in southeastern Oregon"; Jacobs, "National Parks." Similarly, an advertisement for a June 2016 panel featuring a Resources Renewal Institute representative and Ken Brower at the David Brower Center, titled "Point Reyes, Ranching, and the Fate of Our Public Lands" described the Seashore as being beset by "an accelerating campaign by Western cattlemen and Eastern oil billionaires to privatize public lands."

9. Nancy Langston, *Where Land and Water Meet: A Western Landscape Transformed* (Seattle: University of Washington Press, 2003). U.S. Fish and Wildlife Service, *Malheur National Wildlife Refuge: Comprehensive Conservation Plan,* May 2013, http://catalog.data.gov/dataset/malheur-national-wildlife-refuge-comprehensive-conservation-plan.

10. Nancy Langston, "Beyond the Oregon Protests: The Search for Common Ground," *Yale Environment 360,* January 28, 2016, http://e360.yale.edu/feature/beyond_the_oregon_protests_the_search_for_common_ground/2952/.

11. Gloria Flora, "Former Forest Supervisor: Why I Resigned over a Public Lands Dispute," *Time,* January 13, 2016. She goes on to write, "We Americans love our landscapes. Those who spent a lifetime on the land, like ranchers, connect with it even more deeply. That attachment, our sense of place, is part of our identities."

12. Franz Vera, "The Shifting Baseline Syndrome in Restoration Ecology," in *Restoration and History: The Search for a Usable Environmental Past,* ed. Marcus Hall (New York: Routledge Publishers, 2010).

13. Rolf Diamant, "Letter from Woodstock: Find Your System," *George Wright Forum* 32, no. 3 (2015): 214–20, 219.

14. David Lowenthal, "Pioneering Stewardship: New Challenges for CRM," *CRM Journal* 1, no. 1 (2003): 7–13, 12.

15. For more specifically on the Malpai Group, see Nathan F. Sayre, *Working Wilderness: The Malpai Borderlands Group and the Future of the Western Range* (Tucson, AZ: Rio Nuevo Publishers, 2005).

Selected Bibliography

A note of explanation as to the numbering systems for archival material at Point Reyes National Seashore (PRNS):

The initial research for this project was conducted for my doctoral dissertation, in 1998–2001; at this time, the archives at PRNS had not yet been formally organized, and while most files had code numbers associated with them, they were stored in various file drawers and boxes at the Seashore's headquarters. I kept photocopies of quite a few key documents. Since then, the archives have been organized and catalogued, so materials researched more recently for the book have the detailed file, box, and folder numbers of the present archive (such as "PORE 9728 S.1, Box 3, Folder 24").

GOVERNMENT DOCUMENTS, REPORTS, AND PUBLICATIONS

Barrett, Elinore. *The California Oyster Industry.* Sacramento: California Department of Fish and Game, 1963.

Bartolome, James W. *Range Analysis of the Tomales Point Tule Elk Range.* Report for the Point Reyes National Seashore, 1993.

Brown, Sharon A. *Administrative History: Jefferson National Expansion Memorial, 1935–1980.* Washington, DC: National Park Service, U.S. Department of Interior, 1984.

Buck, Paul Herman. *The Evolution of the National Park System of the United States.* Washington, DC: U.S. Government Printing Office, for the National Park Service, U.S. Department of the Interior, 1946.

Calling for the Preparation of a Report on the Proposed Point Reyes National Seashore Recreational Area, Marin County, California. 85th Congress, 2nd Session, Committee on Interior and Insular Affairs. House Report 2463. August 5, 1958.

Cockrell, Ron. *A Green Shrouded Miracle: The Administrative History of Cuyahoga Valley National Recreation Area, Ohio.* Omaha, NE: Midwest Regional Office, National Park Service, U.S. Department of the Interior, 1992.

Congressional Record. Vol. 104, Issue 14, pps 18215–9386. Washington, DC: Government Printing Office. September 6–13, 1961.

———. Vol. 108, Issue 11, pps 14239–5598. Washington, DC: Government Printing Office. July 20-August 3, 1962.

Council on Environmental Quality. *A Citizen's Guide to the NEPA: Having Your Voice Heard.* Washington, DC, 2007.

Cuyahoga Valley National Park. *Countryside Initiative Request for Proposals.* National Park Service, U.S. Department of Interior, 2015.

Devaney, Earl E. *Report of Investigation: Point Reyes National Seashore.* Washington, DC: Department of the Interior, Office of Inspector General, 2008.

Diamant, Rolf, Jeffrey Roberts, Jacquelyn Tuxill, Nora Mitchell, and Daniel Laven. *Stewardship Begins with People: An Atlas of Places, People, and Handmade Products.* Woodstock, VT: The Conservation Study Institute, National Park Service, U.S. Department of Interior, 2007.

Frost, Gavin M. *Public Report on Allegations of Scientific Misconduct at Point Reyes National Seashore, California.* Washington, DC: Department of the Interior, Office of the Solicitor, 2011.

General Accounting Office. *The Federal Drive to Acquire Public Lands Should Be Reassessed.* Washington, DC: General Accounting Office, 1979.

Hearing Held before the Subcommittee on National Parks and Recreation of the Committee on Interior and Insular Affairs, Markup Session, H.R. 8002, Point Reyes National Seashore. 94th Congress, 2nd Session, Committee on Interior and Insular Affairs, Subcommittee on National Parks and Recreation. September 14, 1976.

Hearing Held before the Subcommittee on National Parks, Forests, and Public Lands of the House Committee on Natural Resources, Concerning H.R. 3022, a Bill to Designate the John Krebs Wilderness in the State of California, to Add Certain Lands to the Sequoia-Kings Canyon National Park Wilderness, and Other Purposes, 110th Congress, 1st Session. October 30, 2007.

Hearing on H.R. 3786 and Related Bills, to Authorize the Appropriation of Additional Funds Necessary for Acquisition of Land at the Point Reyes National Seashore in California. 91st Congress, 1st Session, Committee on Interior and Insular Affairs, Subcommittee on National Parks and Recreation. May 13, 1969.

Hearing on S. 1530 and H.R. 3786 to Authorize the Appropriation of Additional Funds Necessary for the Acquisition of Lands at the Point Reyes National Seashore in California. 91st Congress, 2nd Session, Committee on Interior and Insular Affairs, Subcommittee on Parks and Recreation. February 26, 1970.

Hearing on S. 1607, to Establish the Point Reyes National Seashore in the State of California and for Other Purposes. 89th Congress, 2nd Session, Committee on Interior and Insular Affairs, Subcommittee on Parks and Recreation. July 27, 1966.

Hearing on S. 2428, to Establish the Point Reyes National Seashore in the State of California and for Other Purposes. 86th Congress, 2nd Session, Committee on Interior and Insular Affairs, Subcommittee on Public Lands. April 16, 1960.

Hearings on H.R. 2775 and H.R. 3244, to Establish the Point Reyes National Seashore in the State of California, and for Other Purposes. 87th Congress, 1st Session, Committee on Interior and Insular Affairs, Subcommittee on National Parks and Recreation. March 24, July 6, and August 11, 1961.

Hearings on H.R. 13562 and H.R. 13563, to Designate Certain Lands in the National Park System as Wilderness. 93rd Congress, 2nd Session, Committee on Interior and Insular Affairs, Subcommittee on National Parks and Recreation. March 22, 25, and 26, 1974.

Hearings on S. 476, to Establish the Point Reyes National Seashore in the State of California, and for Other Purposes. 87th Congress, 1st Session, Committee on Interior and Insular Affairs, Subcommittee on Public Lands. March 28, 30, and 31, 1961 1961.

Hearings on S. 1093, to Designate Certain Lands in the Point Reyes National Seashore, California, as Wilderness, and on S. 2472, to Designate Certain Lands in the Point Reyes National Seashore, California, as Wilderness; to Designate Point Reyes National Seashore as a Natural Area of the National Park System, and for Other Purposes. 94th Congress, 2nd Session, Committee on Interior and Insular Affairs, Subcommittee on Parks and Recreation. March 2, 1976.

Karamanski, Theodore J. *A Nationalized Lakeshore: The Creation and Administration of Sleeping Bear Dunes National Lakeshore.* Washington, DC: National Park Service, U.S. Department of Interior, 2000.

Lage, Ann. *Oral History Interview with Peter H. Behr.* Sacramento: California State Archives, State Government Oral History Program, 1988.

Lellis, William A., Carrie J. Blakeslee, Laurie K. Allen, Bruce F. Molnia, Susan D. Price, Sky Bristol, and Brent S. Stewart. *Assessment of Photographs from Wildlife Monitoring Cameras in Drakes Estero, Point Reyes National Seashore, California.* U.S. Geological Survey, U.S. Department of Interior, 2012.

Leopold, A.S., S.A. Cain, C.M. Cottam, I.N. Gabrielson, and T.L. Kimball. "Wildlife Management in the National Parks." In *America's National Park System: The Critical Documents*, edited by Lary M. Dilsaver, 237–52. Lanham, MD: Rowman and Littlefield, 1994. Original report dated March 4, 1963.

Livingston, D.S. (Dewey). *Ranching on the Point Reyes Peninsula: A History of the Dairy and Beef Ranches within Point Reyes National Seashore, 1834–1992.* Point Reyes Station, CA: Point Reyes National Seashore, National Park Service, 1993.

———. *A Good Life: Dairy Farming in the Olema Valley: A History of the Dairy and Beef Ranches of the Olema Valley and Lagunitas Canyon.* San Francisco: Golden Gate National Recreation Area and Point Reyes National Seashore, National Park Service, 1995.

———. *Interviews with Merv Mcdonald, Marin County Rancher.* Point Reyes Station, CA: Marin County Conservation District, 2010.

Louter, David. *Contested Terrain: North Cascades National Park Service Complex, an Administrative History*. Seattle: National Park Service, U.S. Department of Interior, 1998.

Mackintosh, Barry. *The Historic Sites Survey and National Historic Landmarks Program: A History*. Washington, DC: History Division, National Park Service, 1984.

———. *The National Parks: Shaping the System*. Washington, DC: Division of Publications, National Park Service, 1985.

———. *The National Historic Preservation Act and the National Park Service: A History*. Washington, DC: History Division, National Park Service, 1986.

Marine Mammal Commission. *Mariculture and Harbor Seals in Drakes Estero, California*. Bethesda, MD: Marine Mammal Commission, 2011.

McClelland, Linda Flint, J. Timothy Keller, and Robert Z. Melnick. *Guidelines for Evaluating and Documenting Rural Historic Landscapes*. Washington, DC: U.S. Department of Interior, National Park Service, Interagency Resources Division, 1987.

McCullough, Dale, Robert A. Garrott, Jay F. Kirkpatrick, Edward D. Plotka, Katherine D. Ralls, and E. Tom Thorne. *Report of the Scientific Advisor Panel on Control of Tule Elk on Point Reyes National Seashore*. National Park Service, 1993.

Melnick, Robert Z., with Daniel Sponn, and Emma Jane Saxe. *Cultural Landscapes: Rural Historic Districts in the National Park System*. Washington, DC: U.S. Department of Interior, 1984.

Mitchell, Nora, Mechtild Rössler, and Pierre-Marie Tricaud. *World Heritage Cultural Landscapes: A Handbook for Conservation and Management*. Paris: UNESCO, 2009.

National Park Service. *Recreational Use of Land in the United States*. Report prepared for the Land Planning Committee of the National Resources Board. Washington, DC: U.S. Government Printing Office, 1938.

———. *Our Vanishing Shoreline*. Washington, DC: National Park Service, U.S. Department of Interior, 1955.

———. *Seashore Recreation Area Survey of the Atlantic and Gulf Coasts*. Washington, DC: National Park Service, U.S. Department of Interior, 1955.

———. *Our Fourth Shore: Great Lakes Shoreline, Recreation Area Survey*. Washington, DC: National Park Service, U.S. Department of Interior, 1959.

———. *Pacific Coast Recreation Area Survey*. Washington, DC: National Park Service, U.S. Department of Interior, 1959.

———. *Land Use Survey and Economic Feasibility Report: Proposed Point Reyes National Seashore*. San Francisco: Region Four Office, National Park Service, 1961.

———. *Assessment of Alternatives for the General Management Plan, Golden Gate National Recreation Area and Point Reyes National Seashore*. San Francisco: National Park Service, 1977.

———. *Final Environmental Impact Statement, General Management Plan, Yosemite National Park, California*. National Park Service, U.S. Department of Interior, 1980.

————. *General Management Plan, Environmental Analysis: Golden Gate National Recreation Area/Point Reyes National Seashore*. Denver: Denver Service Center, National Park Service, 1980.

————. "Notice of Designation of Potential Wilderness as Wilderness, Point Reyes National Seashore." *Federal Register* 64, no. 222 (November 18, 1999): 63057.

————. *Drakes Estero: A Sheltered Wilderness Estuary*. Point Reyes National Seashore: National Park Service, U.S. Department of Interior, 2006.

————. *NPS Management Policies*. Washington, DC: National Park Service, U.S. Department of Interior, 2006.

————. *Point Reyes National Seashore Non-Native Deer Management Plan: Protecting the Seashore's Ecosystems*. Washington, DC: National Park Service, U.S. Department of Interior, 2006.

National Research Council. *Shellfish Mariculture in Drakes Estero, Point Reyes National Seashore, California*. Washington, DC: National Academies Press, 2009.

————. *Scientific Review of the Draft Environmental Impact Statement: Drakes Bay Oyster Company Special Use Permit*. Washington, DC: National Academies Press, 2012.

Page, Robert R., Cathy A. Gilbert, and Susan A. Dolan. *A Guide to Cultural Landscape Reports: Contents, Process and Techniques*. Washington, DC: Park Cultural Landscapes Program, National Center for Cultural Resources Stewardship, National Park Service, 1998.

Pitcaithley, Dwight T. *Let the River Be: A History of the Ozark's Buffalo River*. Santa Fe: Southwest Cultural Resources Center, Southwest Regional Office, National Park Service, U.S. Department of the Interior, 1989.

Point Reyes National Seashore. *Statement of Management Objectives*. National Park Service, U.S. Department of Interior, 1970.

————. *Wilderness Study*. National Park Service, U.S. Department of Interior, 1971.

————. *Wilderness Recommendation*. National Park Service, U.S. Department of Interior, 1972.

————. *Final Environmental Statement for Proposed Wilderness*. Western Region, National Park Service, 1974.

————. *Environmental Assessment/Initial Study Joint Document, Johnson Oyster Company, Marin County*. National Park Service, U.S. Department of Interior, 1988.

————. *Statement for Management, June 1993 Revision*. National Park Service, U.S. Department of Interior, 1993.

————. *Tule Elk Management Plan and Environmental Assessment*. National Park Service, U.S. Department of Interior, 1998.

————. *Draft Environmental Impact Statement, Drakes Bay Oyster Company Special User Permit*. National Park Service, U.S. Department of Interior, 2011.

————. *Final Environmental Impact Statement, Drakes Bay Oyster Company Special User Permit*. National Park Service, U.S. Department of Interior, 2012.

Report No. 94–1680, to Accompany H.R. 8002, Designating Certain Lands in the Point Reyes National Seashore, California, as Wilderness, Designating Point Reyes National Seashore as a Natural Area of the National Park System, and Other Purposes. 94th Congress, 2nd Session, Committee on Interior and Insular Affairs. September 24, 1976.

Report to Accompany H.R. 5892. 86th Congress, 1st Session, Committee on Interior and Insular Affairs. 1959.

Roise, Charlene K., Edward W. Gordon, and Bruce C. Fernald. *Minute Man National Historic Park: An Administrative History.* NPS North Atlantic Region: National Park Service, U.S. Department of Interior, 1989.

Sadin, Paul. *Managing a Land in Motion: An Administrative History of Point Reyes National Seashore.* Washington, DC: National Park Service, U.S. Department of Interior, 2007.

Santa Monica Mountains National Recreation Area. *Final General Management Plan and Environmental Impact Statement.* National Park Service, U.S. Department of Interior, 2002.

Stewart, Brent S. *Evaluation of Time-Lapse Photographic Series of Harbor Seals Hauled out in Drakes Estero, California, for Detecting and Assessing Disturbance Events.* Hubbs SeaWorld Research Institute, 2012.

Toogood, Anna Coxe. *George Washington Carver National Monument, Diamond, Missouri: Historic Resource Study and Administrative History.* Denver: Denver Service Center, Historic Preservation Team, National Park Service, 1973.

———. *A Civil History of Golden Gate National Recreation Area and Point Reyes National Seashore.* Denver: Historic Preservation Branch, Pacific Northwest/Western Team, Denver Service Center, National Park Service, 1980.

U.S. Fish and Wildlife Service. *Malheur National Wildlife Refuge: Comprehensive Conservation Plan.* U.S. Department of the Interior, 2013.

Westmacott, Richard. *Managing Culturally Significant Agricultural Landscapes in the National Park System.* Washington, DC: Park Cultural Landscapes Program, National Center for Cultural Resources Stewardship, National Park Service, 1998.

Wirth, Conrad L. *Study of a National Seashore Recreation Area, Point Reyes Peninsula, California.* Washington, DC: National Park Service, U.S. Department of Interior, 1935.

SCHOLARLY ARTICLES AND BOOKS

Alagona, Peter. *After the Grizzly: Endangered Species and the Politics of Place in California.* Berkeley: University of California Press, 2013.

Alanen, Arnold R. "Considering the Ordinary: Vernacular Landscapes in Small Towns and Rural Areas." In *Preserving Cultural Landscapes in America,* edited by Arnold R. Alanen and Robert Z. Melnick, 112–42. Baltimore: Johns Hopkins University Press, 2000.

Alanen, Arnold R., and Robert Z. Melnick. "Introduction: Why Cultural Landscape Preservation?" In *Preserving Cultural Landscapes in America,* edited by Arnold R. Alanen and Robert Z. Melnick, 1–21. Baltimore: Johns Hopkins University Press, 2000.

———, eds. *Preserving Cultural Landscapes in America.* Baltimore: Johns Hopkins University Press, 2000.

Albright, Horace M., and Robert Cahn. *The Birth of the National Park Service: The Founding Years, 1913–33.* Salt Lake City: Howe Brothers, 1985.

Anderson, R. Scott. "Contrasting Vegetation and Fire Histories on the Point Reyes Peninsula during the Pre-Settlement and Settlement Periods: Fifteen Thousand Years of Change." Flagstaff, AZ: Center for Environmental Sciences and Education, Quaternary Sciences Program, 2005.

Anderson, R. Scott, Ana Ejarque, Peter M. Brown, and Douglas J. Hallett. "Holocene and Historical Vegetation Change and Fire History on the North-Central Coast of California, USA." *The Holocene* 23, no. 12 (2013): 1797–810.

Barringer, Mark Daniel. *Selling Yellowstone: Capitalism and the Construction of Nature.* Lawrence: University Press of Kansas, 2002.

Bender, Barbara. "Introduction: Landscape—Meaning and Action." In *Landscape: Politics and Perspective,* edited by Barbara Bender, 1–17. Oxford: Berg, 1993.

———, ed. *Landscape: Politics and Perspective.* Oxford: Berg, 1993.

Birnbaum, Charles. "A Reality Check for Our Nation's Parks." *Cultural Resources Management* 16, no. 4 (1993): 1–4.

Booker, Matthew Morse. *Down by the Bay: San Francisco's History between the Tides.* Berkeley: University of California Press, 2013.

Bowles, Samuel. *Across the Continent.* Springfield, MA: Samuel Bowles and Company, 1865.

Brooks, Shelley. "Inhabiting the Wild: Land Management and Environmental Politics in Big Sur." *Western Historical Quarterly* 44, no. 3 (2013): 295–317.

Bruner, Edward M. "Epilogue: Creative Persona and the Problem of Authenticity." In *Creativity/Anthropology,* edited by Smadar Lavie, Kirin Narayan, and Renato Rosaldo, 321–34. Ithaca, NY: Cornell University Press, 1993.

Bunnell, Lafayette Houghton. *Discovery of the Yosemite, and the Indian War of 1851, Which Led to that Event.* First published in 1880, reprinted by General Books, 2009.

Burling, Francis P. *The Birth of Cape Cod National Seashore.* Plymouth, MA: Leyden Press, 1979.

Burnham, Philip. *Indian Country, God's Country: Native Americans and the National Parks.* Washington, DC: Island Press, 2000.

Callicott, J. Baird, and Michael P. Nelson. "Introduction." In *The Great New Wilderness Debate: An Expansive Collection of Writings Defining Wilderness from John Muir to Gary Snyder,* edited by J. Baird Callicott and Michael P. Nelson, 1–20. Athens: University of Georgia Press, 1998.

———, eds. *The Great New Wilderness Debate: An Expansive Collection of Writings Defining Wilderness from John Muir to Gary Snyder.* Athens: University of Georgia Press, 1998.

Cameron, Jenks. *The National Park Service: Its History, Activities and Organization.* Baltimore: Institute for Government Research, Johns Hopkins Press, 1922.

Campbell, Carlos C. *Birth of a National Park in the Great Smoky Mountains: An Unprecedented Crusade Which Created, as a Gift of the People, the Nation's Most Popular Park.* Knoxville: University of Tennessee Press, 1960.

Cannavò, Peter F. *The Working Landscape: Founding, Preserving, and the Politics of Place.* Cambridge, MA: MIT Press, 2007.

Carr, Ethan. *Wilderness by Design: Landscape Architecture and the National Park Service.* Lincoln: University of Nebraska Press, 1998.

———. *Mission 66: Modernism and the National Park Dilemma.* Amherst: University of Massachusetts Press, 2007.

Castree, Noel. "The Nature of Produced Nature: Materiality and Knowledge Construction in Marxism." *Antipode* 27, no. 1 (1995): 12–48.

Catton, Theodore. *Inhabited Wilderness: Indians, Eskimos, and National Parks in Alaska.* Albuquerque: University of New Mexico Press, 1997.

Chamberlin, E. R. *Preserving the Past.* London: J. M. Dent and Sons, 1979.

Charnley, Susan, Thomas E. Sheridan, and Gary P. Nabhan, eds. *Stitching the West Back Together: Conservation of Working Landscapes.* Chicago: University of Chicago Press, 2014.

Chase, Alston. *Playing God in Yellowstone: The Destruction of America's First National Park.* San Diego, CA: Harcourt Brace Jovanovich, 1987.

Cobb, McCrea Andrew. "Spatial Ecology and Population Ecology of Tule Elk (*Cervus canadensis nannodes*) at Point Reyes National Seashore, California." PhD dissertation, Department of Environmental Science, Policy, and Management, University of California, Berkeley, 2010.

The Conservation Foundation. *National Parks for a New Generation: Visions, Realities, Prospects.* Washington, DC: The Conservation Foundation, 1985.

Conte, Christopher. "Creating Wild Places from Domesticated Landscapes: The Internationalization of the American Wilderness Concept." In *American Wilderness: A New History,* edited by Michael Lewis, 223–41. New York: Oxford University Press, 2007.

Corey, Steven H., and Lisa Krissoff Bohem, eds. *The American Urban Reader: History and Theory.* New York: Routledge, 2011.

Cosgrove, Denis. *Social Formation and Symbolic Landscape.* London: Croon Helm, 1984.

Cronon, William. *Nature's Metropolis: Chicago and the Great West.* New York: W. W. Norton and Co., 1991.

———. "The Trouble with Wilderness; or, Getting Back to the Wrong Nature." In *Uncommon Ground: Toward Reinventing Nature,* edited by William Cronon, 69–90. New York: W. W. Norton, 1995.

———. "The Trouble with Wilderness: A Response." *Environmental History* 1, no. 1 (1996): 47–55.

———, ed. *Uncommon Ground: Toward Reinventing Nature.* New York: W. W. Norton, 1995.

Dana, Samuel Trask, and Sally K. Fairfax. *Forest and Range Policy: Its Development in the United States.* 2nd ed. New York: McGraw-Hill, 1980. First published in 1956.

DeLyser, Dydia. "Good, by God, We're Going to Bodie! Landscape and Social Memory in a California Ghost Town." PhD diss., Department of Geography, Syracuse University, May 1998.

DeRooy, Carola, and Dewey Livingston. *Point Reyes Peninsula: Olema, Point Reyes Station, and Inverness.* Charleston, SC: Arcadia Publishing, 2008.

Diamant, Rolf. "From Management to Stewardship: The Making and Remaking of the U.S. National Park System." *George Wright Forum* 17, no. 2 (2000): 31–45.

———. "Letter from Woodstock: Find Your System." *George Wright Forum* 32, no. 3 (2015): 214–20.

Dilsaver, Lary M., ed. *America's National Park System: The Critical Documents.* Lanham, MD: Rowman and Littlefield, 1994.

Doe, Douglas W. "The New Deal Origins of the Cape Cod National Seashore." *Historical Journal of Massachusetts* 26, no. 2 (1997): 136–56.

Dowie, Mark. *Conservation Refuges: The Hundred-Year Conflict between Global Conservation and Native Peoples.* Cambridge, MA: MIT Press, 2009.

Dunn, Durwood *Cades Cove: The Life and Death of a Southern Appalachian Community, 1818–1937.* Knoxville: University of Tennessee Press, 1988.

Ediger, Vernita Lea. "Natural Experiments in Conservation Ranching: The Social and Ecological Consequences of Diverging Land Tenure in Marin County, California." PhD diss., Department of Anthropological Studies, Stanford University, March 2005.

Engbeck Jr., Jospeh H. *The Enduring Giants.* Berkeley: University Extension, University of California, 1973.

Engle, Paul M. "Toward an Archeology of Ranching on the Point Reyes Peninsula." Masters thesis, Cultural Resources Management Program, Department of Anthropology, Sonoma State University, 2012.

Everhart, William C. *The National Park Service.* 2nd ed. Boulder, CO: Westview Press, 1983.

Fairfax, Sally K., Louise Nelson Dyble, Greig Tor Guthey, Lauren Gwin, Monica Moore, and Jennifer Sokolove. *California Cuisine and Just Food.* Cambridge, MA: MIT Press, 2012.

Fairfax, Sally K., Lauren Gwin, Mary Ann King, Leigh Raymond, and Laura A. Watt. *Buying Nature: The Limits of Land Acquisition as a Conservation Strategy, 1780–2003.* Cambridge, MA: MIT Press, 2005.

Farmer, Jared. *On Zion's Mount: Mormons, Indians, and the American Landscape.* Cambridge, MA: Harvard University Press, 2008.

———. *Trees in Paradise: A California History.* New York: W. W. Norton and Co., 2013.

Feldman, James W. *A Storied Wilderness: Rewilding the Apostle Islands.* Seattle: University of Washington Press, 2011.

Foresta, Ronald A. *America's National Parks and Their Keepers.* Washington, DC: Resources for the Future, 1984.

Foster, Charles H. W. *The Cape Cod National Seashore: A Landmark Alliance.* Hanover, NH: University Press of New England, 1985.

Foster, Megan. "Resident Inclusion in Protected Land Use Decisions: The Role of a Citizens Advisory Commission at Point Reyes National Seashore." Masters thesis, Community Development, University of California, Davis, December 2015.

Fradkin, Philip L. "No Room for Cows on Point Reyes." *Audubon*, July 1978, 98–104.

Francaviglia, Richard. "Selling Heritage Landscapes." In *Preserving Cultural Landscapes in America*, edited by Arnold R. Alanen and Robert Z. Melnick, 44–69. Baltimore: Johns Hopkins University Press, 2000.

Frank, Jerry J. *Making Rocky Mountain National Park: The Environmental History of an American Treasure*. Lawrence: University of Kansas Press, 2013.

Freyfogle, Eric T. *The Land We Share: Private Property and the Common Good*. Washington, DC: Island Press, 2003.

Gilliam, Harold. *Island in Time: The Point Reyes Peninsula*. San Francisco: Sierra Club Books, 1962.

Glass, James A. *The Beginnings of a New National Historic Preservation Program, 1957 to 1969*. Nashville: American Association for State and Local History, 1990.

Gogan, Peter. "Ecology of the Tule Elk Range, Point Reyes National Seashore." PhD diss., University of California, Berkeley, 1986.

Gogan, Peter, and Reginald Barrett. "Comparative Dynamics of Introduced Tule Elk Populations." *Journal of Wildlife Management* 51, no. 1 (1987): 20–27.

Greiff, Constance M. *Independence: The Creation of National Park*. Philadelphia: University of Pennsylvania Press, 1987.

Groth, Paul. "Frameworks for Cultural Landscape Study." In *Understanding Ordinary Landscapes*, edited by Paul Groth and Todd Bressi, 1–21. New Haven, CT: Yale University Press, 1998.

Hardesty, Donald L. "Ethnographic Landscapes: Transforming Nature into Culture." In *Preserving Cultural Landscapes in America*, edited by Arnold R. Alanen and Robert Z. Melnick, 169–85. Baltimore: Johns Hopkins University Press, 2000.

Hart, John. *Farming on the Edge: Saving Family Farms in Marin County, California*. Berkeley: University of California Press, 1991.

———. *An Island in Time: Fifty Years of Point Reyes National Seashore*. Mill Valley, CA: Pickleweed Press, 2012.

Harvey, Mark. *A Symbol of Wilderness: Echo Park and the American Conservation Movement*. Seattle: University of Washington Press, 1994.

———. *Wilderness Forever: Howard Zahniser and the Path to the Wilderness Act*. Seattle: University of Washington Press, 2005.

Hausdoerffer, John. *Caitlin's Lament: Indians, Manifest Destiny, and the Ethics of Nature*. Lawrence: University of Kansas Press, 2009.

Hays, Samuel P. *Conservation and the Gospel of Efficiency: The Progressive Conservation Movement, 1890–1920*. Pittsburgh: University of Pittsburgh Press, 1999. First published in 1959 by Harvard University Press.

Hemmat, Steven A. "Parks, People, and Private Property: The National Park Service and Eminent Domain." *Environmental Law* 16, no. 4 (1986): 935–61.

Hills, Patricia. "Picturing Progress in the Era of Westward Expansion." In *The West as America: Reinterpreting Images of the Frontier 1820–1920*, edited by William H. Truettner, 97–147. Washington, DC: Smithsonian Institute, 1991.

Hosmer Jr., Charles B. *Presence of the Past: A History of the Preservation Movement in the United States before Williamsburg*. New York: G.P. Putnam's Sons, 1965.

———. *Preservation Comes of Age: From Williamsburg to the National Trust, 1926–1949*. Charlottesville: University Press of Virginia, 1981.

Howell, Judd A., George C. Brooks, Marcia Semenoff-Irving, and Correigh Greene. "Population Dynamics of Tule Elk at Point Reyes National Seashore, California." *Journal of Wildlife Management* 66, no. 2 (2002): 478–90.

Howett, Catherine. "Integrity as a Value in Cultural Landscape Preservation." In *Preserving Cultural Landscapes in America*, edited by Arnold R. Alanen and Robert Z. Melnick, 186–207. Baltimore: Johns Hopkins University Press, 2000.

Huntley, Jen A. *The Making of Yosemite: James Mason Hutchings and the Origin of America's Most Popular National Park*. Lawrence: University of Kansas Press, 2011.

Huntsinger, Lynn, James W. Bartolome, and Carla M. D'Antonio. "Grazing Management on California's Mediterranean Grasslands." In *Ecology and Management of California Grasslands*, edited by J. Corbin, M. Stromberg and C. D'Antonio. Berkeley: University of California Press, 2015.

Irwin, William. *The New Niagara: Tourism, Technology and the Landscape at Niagara Falls*. University Park: Pennsylvania State University Press, 1996.

Ise, John. *Our National Park Policy: A Critical History*. Washington, DC: Resources for the Future, 1961.

Jackson, John Brinckerhoff. *Discovering the Vernacular Landscape*. New Haven, CT: Yale University Press, 1984.

Jackson, Peter. *Maps of Meaning: An Introduction to Cultural Geography*. London: Unwin Hyman, 1989.

Jacobs, John. *A Rage for Justice: The Passion and Politics of Phillip Burton*. Berkeley: University of California Press, 1995.

Jolley, Harley E. *The Blue Ridge Parkway*. Knoxville: University of Tennessee Press, 1969.

Jørgensen, Dolly. "Rethinking Rewilding." *Geoforum* 65 (2015): 482–88.

Kaufman, Herbert. "Emerging Conflicts in the Doctrine of Public Administration." *American Political Science Review* 50, no. 4 (1956): 1057–73.

Keiter, Robert B. *To Conserve Unimpaired: The Evolution of the National Park Idea*. Washington, DC: Island Press, 2013.

Lage, Ann, and William J. Duddleston. "Saving Point Reyes National Seashore, 1969–1970: An Oral History of Citizen Action in Conservation." Regional Oral History Office, Bancroft Library, University of California, Berkeley, 1993.

Lambert, Darwin. *The Undying Past of Shenandoah National Park*. Boulder, CO: Roberts Rinehart, 1989.

Langston, Nancy. *Where Land and Water Meet: A Western Landscape Transformed.* Seattle: University of Washington Press, 2003.

Lefebvre, Henri. *Critique of Everyday Life.* Translated by John Moore. London: Verso, 1991.

———. *The Production of Space.* Translated by Donald Nicholson-Smith. Oxford: Basil Blackwell, 1991.

Leopold, Aldo. *The River of the Mother of God and Other Essays.* Edited by Susan L. Flader and J. Baird Callicott. Madison: University of Wisconsin Press, 1991.

———. "Wilderness as a Form of Land Use." In *The River of the Mother of God, and Other Essays by Aldo Leopold,* edited by Susan L. Flader and J. Baird Callicott, 134–42. Madison: University of Wisconsin Press, 1991. Essay first published in 1925.

———. "The Farmer as a Conservationist." In *For the Health of the Land: Previously Unpublished Essays and Other Writings,* edited by J. Baird Callicott and Eric T. Freyfogle, 161–175. Washington, DC: Island Press, 1999. Essay first published in 1939.

Lewis, Peirce. "The Future of the Past: Our Clouded Vision of Historic Preservation." *Pioneer America* 7, no. 2 (July 1975): 1–20.

———. "Axioms for Reading the Landscape: Some Guides to the American Scene." In *The Interpretation of Ordinary Landscapes: Geographic Essays,* edited by Donald W. Meinig, 11–32. New York: Oxford University Press, 1979.

Lightfoot, Kent, and Otis Parrish. *California Indians and Their Environments: An Introduction.* Berkeley: University of California Press, 2009.

Liles, Jim. "Boxley Valley: Buffalo National River: A National Park Service Historic District in Private Hands." *OzarksWatch* 4, no. 1 (1990): 14–18.

Lowenthal, David. "Is Wilderness 'Paradise Enow'? Images of Nature in America." *Chicago University Forum* 7, no. 2 (1964): 40.

———. "The American Way of History." *Columbia University Forum* 9 (1966): 27–32.

———. "Introduction." In *Our Past before Us: Why Do We Save It?* edited by David Lowenthal and Marcus Binney, 9–16. London: Temple Smith, 1981.

———. *The Past Is a Foreign Country.* Cambridge: Cambridge University Press, 1985.

———. "European Landscape Transformations: The Rural Residue." In *Understanding Ordinary Landscapes,* edited by Paul Groth and Todd Bressi, 180–88. New Haven, CT: Yale University Press, 1997.

———. "Pioneering Stewardship: New Challenges for CRM." *CRM Journal* 1, no. 1 (2003): 7–13.

Lynch, Kevin. *What Time Is This Place?* Cambridge, MA: M.I.T. Press, 1972.

Macpherson, C. B. *Property: Mainstream and Critical Positions.* Toronto: University of Toronto Press, 1978.

Manning, Elizabeth J. B., Thomas E. Kucera, Natalie B. Gates, Leslie M. Woods, and Maura Fallon-McKnight. "Testing for *Mycobacterium Avium Subsp. Paratuberculosis* Infection in Asymptomatic Free-Ranging Tule Elk from an Infected Herd." *Journal of Wildlife Diseases* 39, no. 2 (2003): 323–28.

Marquis, Amy Leinbach. "A Raw Deal: Marine Wilderness Is at Stake in the Ecological Heart of Point Reyes National Seashore." *National Parks* 62, no. 2 (2012): 1–2.

Marris, Emma. *Rambunctious Garden: Saving Nature in a Post-Wild World*. New York: Bloomsbury, 2011.

Marsh, Kevin. *Drawing Lines in the Forest: Creating Wilderness Areas in the Pacific Northwest*. Seattle: University of Washington Press, 2007.

Mayerson, Harvey. *Nature's Army: When Soldiers Fought for Yosemite*. Lawrence: University of Kansas Press, 2001.

McClelland, Linda Flint. *Building the National Parks: Historic Landscape Design and Construction*. Baltimore: Johns Hopkins University Press, 1998.

McCullough, Dale. *The Tule Elk: Its History, Behavior, and Ecology*. Berkeley: University of California Publications in Zoology, 1969.

McCullough, Dale, Jon Fischer, and Jonathan Ballou. "From Bottleneck to Metapopulation: Recovery of the Tule Elk in California." In *Metapopulations and Wildlife Conservation*, edited by Dale McCullough, 374–404. Washington, DC: Island Press, 1996.

McDonnell, Janet A. "Reassessing the National Park Service and the National Park System." *George Wright Forum* 25, no. 2 (2008): 6–14.

McGreevy, Patrick. *Imagining Niagara: The Meaning and Making of Niagara Falls*. Amherst: University of Massachusetts Press, 1994.

McKinsey, Elizabeth. "An American Icon." In *Niagara: Two Centuries of Changing Attitudes, 1697–1901*, edited by Jeremy Elwell Adamson, 83–101. Washington, DC: Corcoran Gallery of Art, 1985.

———. *Niagara Falls: Icon of the American Sublime*. Cambridge: Cambridge University Press, 1985.

Meine, Curt. *Aldo Leopold: His Life and Work*. Madison: University of Wisconsin Press, 1988.

Merchant, Carolyn. "Shades of Darkness: Race and Environmental History." *Environmental History* 8, no. 3 (2003): 380–94.

Minteer, Ben A., and Stephen J. Pyne. "Writing on Stone, Writing in the Wind." In *After Preservation: Saving American Nature in the Age of Humans*, edited by Ben A. Minteer and Stephen J. Pyne, 1–8. Chicago: University of Chicago Press, 2015.

Mitchell, Don. *The Lie of the Land: Migrant Workers and the California Landscape*. Minneapolis: University of Minnesota Press, 1996.

Mitchell, Timothy. *Colonizing Egypt*. Cambridge: Cambridge University Press, 1988.

Muir, John. *The Yosemite*. Madison: University of Wisconsin Press, 1986. First published in 1912.

———. *Our National Parks*. San Francisco: Sierra Club Books, 1991.

Nabhan, Gary P. "Cultural Parallax in Viewing North American Habitat." In *Reinventing Nature? Responses to Postmodern Deconstruction*, edited by Michael Soulé and Gary Lease, 87–101. Washington, DC: Island Press, 1995.

Nash, Roderick. *Wilderness and the American Mind*. 3rd ed. New Haven, CT: Yale University Press, 1982. First published in 1967.

National Trust for Historic Preservation. *With Heritage So Rich.* 2nd ed. Washington, DC: Preservation Press, 1983. First published in 1966.

Newhall, Nancy Wayne. *A Contribution to the Heritage of Every American: The Conservation Activities of John D. Rockefeller.* New York: Knopf, 1957.

Oelschlaeger, Max. *The Idea of Wilderness: From Prehistory to the Age of Ecology.* New Haven, CT: Yale University Press, 1991.

Olmsted, Frederick Law. "The Yosemite Valley and the Mariposa Big Trees: A Preliminary Report." *Landscape Architecture* 43, no. 1 (October 1952): 12–25. Essay first published in 1865.

Olmsted, Frederick Law, Jr. "The Distinction between National Parks and National Forests." *Landscape Architecture* 6, no. 3 (1916): 114–15.

Olwig, Kenneth R. "Sexual Cosmology: Nation and Landscape at the Conceptual Interstices of Nature and Culture; or What Does Landscape Really Mean?" In *Landscape: Politics and Perspectives,* edited by Barbara Bender, 307–43. Oxford: Berg Press, 1993.

———. "Reinventing Common Nature: Yosemite and Mount Rushmore: A Meandering Tale of a Double Nature." In *Uncommon Ground: Toward Reinventing Nature,* edited by William Cronon, at 379–408. New York: W. W. Norton, 1995.

———. "Recovering the Substantive Nature of Landscape." *Annals of the Association of American Geographers* 86, no. 4 (1996): 630–53.

———. "Landscape: The Lowenthal Legacy." *Annals of the Association of American Geographers* 93, no. 4 (2003): 871–77.

Pahre, Robert. "Introduction: Patterns of National Park Interpretation in the West." *Journal of the West* 50, no. 3 (2011): 7–14.

Phillips, W. E. *The Conservation of the California Tule Elk: A Socio-Economic Study of a Survival Problem.* Edmonton: University of Alberta Press, 1976.

Raymond, Leigh. "Localism in Environmental Policy: New Insights from an Old Case." *Policy Sciences* 35, no. 2 (2002): 179–201.

Raymond, Leigh, and Sally K. Fairfax. "Fragmentation of Public Domain Law and Policy: An Alternative to the 'Shift-to-Retention' Thesis." *Natural Resources Journal* 39, no. 4 (Fall 1999): 649–753.

Reich, Justin. "Re-Creating the Wilderness: Shaping Narratives and Landscapes in Shenandoah National Park." *Environmental History* 6, no. 1 (2001): 95–117.

Rettie, Dwight F. *Our National Park System: Caring for America's Greatest Natural and Historic Treasures.* Urbana: University of Illinois Press, 1995.

Righter, Robert W. *Crucible for Conservation: The Creation of Grand Teton National Park.* Jackson Hole, WY: Grand Teton Natural History Association, 1982.

Rilla, Ellie, and Lisa Bush. "The Changing Role of Agriculture in Point Reyes National Seashore." Marin County: University of California Cooperative Extension, 2009.

Rome, Adam. *The Bulldozer in the Countryside: Suburban Sprawl and the Rise of American Environmentalism.* Cambridge: Cambridge University Press, 2001.

Rothman, Hal K. *Preserving Different Pasts: The American National Monuments.* Champaign-Urbana: University of Illinois Press, 1989.

———. *Devil's Bargain: Tourism in the Twentieth-Century American West.* Lawrence: University of Kansas Press, 1998.

———. *The New Urban Park: Golden Gate National Recreation Area and Civic Environmentalism.* Lawrence: University of Kansas Press, 2004.

Rosenzweig, Roy, and Elizabeth Blackmar. *The Park and the People: A History of Central Park.* Ithaca, NY: Cornell University Press, 1992.

Runte, Alfred. *National Parks: The American Experience.* 2nd, revised ed. Lincoln: University of Nebraska Press, 1987. First published in 1979.

———. *Yosemite: The Embattled Wilderness.* Lincoln: University of Nebraska Press, 1990.

Sax, Joseph L. "Helpless Giants: The National Parks and the Regulation of Private Lands." *Michigan Law Review* 75, no. 2 (December 1976): 239–74

———. "Buying Scenery: Land Acquisitions for the National Park Service." *Duke Law Journal,* no. 4 (1980): 709–40.

———. *Mountains without Handrails: Reflections of the National Parks.* Ann Arbor: University of Michigan Press, 1980.

———. "Do Communities Have Rights? The National Parks as a Laboratory of New Ideas." *University of Pittsburgh Law Review* 45, no. 3 (1984): 499–511.

———. "Heritage Preservation as a Public Duty: The Abbé Grégoire and the Origins of an Idea." *Michigan Law Review* 88, no. 5 (1990): 1142–69.

Sax, Joseph L., and Robert B. Keiter. "Glacier National Park and Its Neighbors: A Study of Federal Interagency Relations." *Ecology Law Quarterly* 14, no. 2 (1987): 207–63.

Sayre, Nathan. *Working Wilderness: The Malpai Borderlands Group and the Future of the Western Range.* Tucson, AZ: Rio Nuevo Press, 2005.

Schrepfer, Susan R. *The Fight to Save the Redwoods: A History of Environmental Reform, 1917–1978.* Madison: University of Wisconsin Press, 1983.

Schuyler, David, and Patricia M. O'Donnell. "The History and Preservation of Urban Parks and Cemeteries." In *Preserving Cultural Landscapes in America,* edited by Arnold R. Alanen and Robert Z. Melnick, 70–93. Baltimore: Johns Hopkins University Press, 2000.

Scott, Doug. *The Enduring Wilderness: Protecting Our Natural Heritage through the Wilderness Act.* Golden, CO: Fulcrum Publishing, 2004.

Scott, James C. *Seeing Like a State: How Certain Schemes to Improve the Human Condition Have Failed.* New Haven, CT: Yale University Press, 1998.

Sears, John F. *Sacred Places: American Tourist Attractions in the Nineteenth Century.* Oxford: Oxford University Press, 1989.

Sellars, Richard West. *Preserving Nature in the National Parks: A History.* New Haven, CT: Yale University Press, 1997.

Sherman, E. A. "The Forest Service and the Preservation of Natural Beauty." *Landscape Architecture* 6, no. 3 (1916): 115–19.

Smith, James. "The Gateways: Parks for Whom." In *National Parks for the Future,* edited by the Conservation Foundation. Washington, DC: Conservation Foundation, 1972.

Smith, Kenneth L. *Buffalo River Handbook*. Little Rock: University of Arkansas Press, 2004.

Sokolove, Jennifer, Sally K. Fairfax, and Breena Holland. "Managing Place and Identity: Thoughts on the Coast Miwok Experience." *Geographical Review* 92, no. 1 (2002): 23–44.

Spence, Mark David. *Dispossessing the Wilderness: Indian Removal and the Making of the National Parks*. New York: Oxford University Press, 1999.

Stegner, Wallace. "The Best Idea We Ever Had." *Wilderness* 46, no. 160 (1983): 4–13.

Sutter, Paul S. *Driven Wild: How the Fight against Automobiles Launched the Modern Environmental Movement*. Seattle: University of Washington Press, 2002.

Swain, Donald C. "The Passage of the National Park Service Act of 1916." *Wisconsin Magazine of History* 50, no. 1 (Autumn 1966): 4–17.

Tabb, William Murray. "The Role of Controversy in NEPA: Reconciling Public Veto with Public Participation in Environmental Decisionmaking." *William and Mary Environmental Law and Policy Review* 21, no. 1 (1997): 175–231.

Thomas, Charlotte E. "The Cape Cod National Seashore: A Case Study of Federal Administrative Control over Traditionally Local Land Use Decisions." *Boston College Environmental Affairs Law Review* 12, no. 2 (1985): 225–72.

Tindall, Gillian. *The Fields Beneath: The History of One London Village*. London: Granada/Paladin, 1980.

Truettner, William H. "Ideology and Image: Justifying Western Expansion." In *The West as America: Reinterpreting Images of the Frontier, 1820–1920*, edited by William H. Truettner, 27–53. Washington, DC: Smithsonian Institute, 1991.

———, ed. *The West as America: Reinterpreting Images of the Frontier, 1820–1920*. Washington, DC: Smithsonian Institute, 1991.

Tuan, Yi-Fu. "Perceptual and Cultural Geography: A Commentary." *Annals of the Association of American Geographers* 93, no. 4 (2003): 878–81.

Turner, James Morton. *The Promise of Wilderness: American Environmental Politics since 1964*. Seattle: University of Washington Press, 2012.

Tweed, William. *Uncertain Path: A Search for the Future of National Parks*. Berkeley: University of California Press, 2010.

Van Horn, Gavin, and Dave Aftandilian, eds. *City Creatures: Animal Encounters in the Chicago Wilderness*. Chicago: University of Chicago Press, 2015.

Vera, Franz. "The Shifting Baseline Syndrome in Restoration Ecology." In *Restoration and History: The Search for a Usable Environmental Past*, edited by Marcus Hall, 98–110. New York: Routledge, 2010.

Wakild, Emily. *Revolutionary Parks: Conservation, Social Justice, and Mexico's National Parks, 1910–1940*. Tucson: University of Arizona Press, 2011.

Watt, Laura A. "Managing Cultural Landscapes: Reconciling Local Preservation and Institutional Ideology in the National Park Service." Ph.D. diss., Department of Environmental Science, Policy, and Management, University of California, Berkeley, December 2001.

———. "The Trouble with Preservation, or, Getting Back to the Wrong Term for Wilderness Protection: A Case Study at Point Reyes National Seashore." *Yearbook of the Association of Pacific Coast Geographers* 64 (2002): 55–72.

———. "Reimagining Joshua Tree: Applying Environmental History to National Park Interpretation." *Journal of the West* 50, no. 3 (2011): 15–20.

———. "The Continually Managed Wild: Tule Elk at Point Reyes National Seashore." *Journal of International Wildlife Law and Policy* 18, no. 4 (2015): 289–308.

Watt, Laura A., Leigh Raymond, and Meryl L. Eschen. "On Preserving Ecological and Cultural Landscapes." *Environmental History* 9, no. 4 (2004): 620–47.

Webb, Melody. "Cultural Landscapes in the National Park Service." *The Public Historian* 9, no. 2 (1987): 77–89.

Wells, Christopher W. *Car Country: An Environmental History.* Seattle: University of Washington Press, 2012.

White, Richard. *"It's Your Misfortune and None of My Own": A New History of the American West.* Norman: University of Oklahoma Press, 1991.

———. "'Are You an Environmentalist, or Do You Work for a Living?' Work and Nature." In *Uncommon Ground: Toward Reinventing Nature,* edited by William Cronon, 171–85. New York: W. W. Norton, 1995.

———. *Remembering Ahanagran: A History of Stories.* Seattle: University of Washington Press, 1998.

———. "Epilogue: Contested Terrain." In *Land in the American West: Private Claims and the Common Good,* edited by William G. Robbins and James C. Foster, 190–206. Seattle: University of Washington Press, 2000.

———. "From Wilderness to Hybrid Landscapes: The Cultural Turn in Environmental History." In *Companion to American Environmental History,* edited by Douglas Cazaux Sackman, 183–90. Hoboken, NJ: Wiley-Blackwell, 2010.

Williams, Raymond. *The Country and the City.* New York: Oxford University Press, 1973.

Wilson, James Q. "The Rise of the Bureaucratic State." *The Public Interest* 41 (1975): 77–103.

Winks, Robin W. "Dispelling the Myth." *National Parks* 70, nos. 7–8 (1996): 52–53.

Wirth, Conrad L. *Parks, Politics, and the People.* Norman: University of Oklahoma Press, 1980.

Wondrak, Alice "Teton Dreams: Horace M. Albright, John D. Rockefeller, Jr., and the Making of the Landscape in Grant Teton National Park." Unpublished manuscript. Boulder: University of Colorado, 1997.

Woods, Mark. "Federal Wilderness Preservation in the United States: The Preservation of Wilderness?" In *The Great New Wilderness Debate: An Expansive Collection of Writings Defining Wilderness from John Muir to Gary Snyder,* edited by J. Baird Callicott and Michael P. Nelson, 131–53. Athens: University of Georgia Press, 1998.

Index

Abbotts Lagoon (Point Reyes), 66*fig.*, 73, 111–12, 121, 210
Acadia National Park (Maine), 45–46, 57, 245nn53–54
access, public, 7, 23, 25, 28; access corridors, 119; access roads, 112, 195; and acquisition, 7, 70, 73–75, 82, 93; beach access, 7, 70, 73–74, 114, 120; disability access, 146; and oyster farm, 195; and public vs. private ownership, 44, 47–49, 53–54, 57, 64; and wilderness, 112, 114, 119–120, 222–23
acquisition, 33, 67–97, 255n79; and alternatives, 68, 70–71, 77, 89–95; and Cape Cod National Seashore, 68–71, 77, 80–81, 84, 89–90, 96; consequences of, 95–97; and cultural landscape, 94–95, 128–29; and Cuyahoga Valley National Park, 91–92, 219; deadlines for, 92; and development, 70, 74–77, 76*fig.*, 79–80, 82–90, 93, 95–97, 250n14, 251n22, 252n39, 253n52, 254n70, 254n76; and funding problems, 81–89; and "hole in the donut" lands, 80, 95, 254n63; and land exchange plan, 81, 84–86; and land values, 81–83, 87, 96, 129, 219; opposition to, 77–82, 90, 92–94; and PRNS, 67–69, 71–91, 76*fig.*, 95; and public vs. private ownership, 46, 56, 58; and urban areas, 64, 68–71; and wilderness, 102, 104, 109, 115

Adams, Ansel, 97
Adams, David, 74, 254n70
adaptive reuse, 134, 143, 146–151, 171, 203, 273n85
Administrative Procedures Act, 207
Advisory Board on National Parks, Historic Sites, Buildings and Monuments, 63, 91
aesthetics, 16–18, 20, 32, 37, 237n18; and acquisition, 90; and cultural landscape, 125, 127, 145; and public vs. private ownership, 44, 51, 53, 55–56, 61; and riparian restoration programs, 119; spatial aesthetics, 16; upper-class aesthetic, 18; and wilderness, 44, 105, 119, 223
African Americans, 6, 14
agencies, federal, 15; and acquisition, 70, 97, 259n142; and cultural landscape, 217; and public comment, 9, 23, 49, 108–9, 112; and public vs. private ownership, 47–49, 52; and wilderness, 99–100, 104, 106–8, 112–13, 116. *See also* NPS; *names of other federal agencies*
agriculture, 1–2, 4–5, 7–8, 17, 41; and acquisition, 72–74, 77–78, 80, 83–85, 88–91, 95–97; agricultural easements, 47, 83–84, 90–91, 96–97; agricultural land trusts, 96; and Central Park, 5; commercial agriculture, 44; and cultural landscape, 124–25, 128–131, 134–35,

210; and park ideal, 158–59, 167, 183; and public vs. private ownership, 46, 49, 51–55, 57, 59, 61, 243n24, 244n35, 246n59; tourist shops, 127; and tule elk, 167; and wilderness, 102, 104–6, 108, 189

Townsend's big-eared bats (*Plectotus townsendi*), 150

tradition. *See* history/heritage

trails, 13, 36, 53, 96, 111, 144, 224; fire trails, 113; walking paths, 148

Trainer, Amy, 209

T Ranch. *See* Laguna Ranch (Point Reyes)

transportation, 28–29, 31–32, 60, 72, 74. *See also* automobiles/motor vehicles; railroads

Treaty of Guadalupe Hidalgo (1848), 48

Trust for Public Land, 229

Truttman Ranch (Point Reyes), 102, 133–35, 142, 268n25, 271n62

Tuan, Yi-Fu, 20

tule elk (*Cervus canadensis nannodes*), 7–9, 148, 156*fig.*, 158, 162, 164–68, 172–183, 203, 223–24, 279n40, 279n42, 280n56, 280n59, 281n61, 285n107; and contraceptives/steriliza- tion, 168, 173–74, 281n64; and copper deficiency, 166; culling of, 167–68, 173, 182; and environmental groups, 181–82, 223, 229–230; free-ranging herds of, 158, 164–65, 172–183, 176*map*, 229–230, 284n93, 285nn112– 113, 286nn119–120, 286n128, 287n137; quarantined/monitored, 174–75, 286n119; and radio collars, 174–75, 284n96

Tunney, John, 108–9, 114

Turner, Jay, 106

Turney, Sayles, 251n22

Turtle Island Restoration Network, 274n103

Tweed, William, 226–27

U, W, Y, and Z Ranches (Point Reyes), 82, 107, 143, 273n85. *See also* Bear Valley Ranch (Point Reyes)

Udall, Stewart, 74

Uncertain Path: A Search for the Future of National Parks (Tweed), 226–27

UNESCO, 216

Union Pacific Railroad, 240n72. *See also* railroads

University of California, Berkeley, 2, 73, 168, 179; Agricultural Extension, 196;

Cooperative Extension, 2, 220–21; and Marin Carbon project, 184

University of Oregon, 230

urban areas, 41, 64; and acquisition, 67–71, 73–75, 89, 91–94, 258n127; "backyard urban wilderness," 107; and riparian restoration programs, 119; and urbanization, 62; urban renewal projects, 125; and urban sprawl, 74; and urban wildlife, 218; and wilderness, 101, 107, 218. *See also* names of cities and urban areas

USGS (U.S. Geological Survey), 179, 198; Biological Resources Division, 179

Vail Conference on National Parks (1991), 160

vandalism, 92, 269n41, 273n83, 275n105, 275n112

Vedanta Society, 78, 80, 93, 95

Vera, Franz, 232

vernacular landscapes, 18, 33–35, 144–45

Victorian ranches, 41, 145, 149, 168, 172

Virginia, 58–62; Arlington National Cemetery, 248n81; Blue Ridge Parkway, 60–61, 247nn70–73; and Byrd, 62, 248n83; Public Park Condemnation Act (1928), 246n62; Shenandoah National Park, 58–60, 69, 89, 246n61, 246n65, 247n67; State Highway Commission, 247n73

Virginia City (Nev.), 21

Virgin Islands National Park, 247n70

visitors. *See* tourism/tourists

visual aspects of landscape, 119–120; and oyster farm, 193

voting rights, 48–49

Voyageurs National Park (Minn.), 93

Wakild, Emily, 154

Walker, Peter, 230–31

Walt Disney Company, 47

war, 15, 38, 62. *See also* names of wars

War Department, U.S., 62

Washington State, 35, 93, 128; Ebey's Landing, 35, 128, 218; North Cascades Complex, 93; Rainier, Mount, 32

water: wastewater treatment systems, 193; water nutrient levels, 196; water pollution, 170; water pumps, 165–66, 169; water quality, 135, 162, 189, 200–201, 211; water supply, 42, 182; water towers, 133

Waters, Alice, 208